# Digital System Design using FSMs

# Digital System Design using FSMs

## A Practical Learning Approach

**Peter D. Minns**
*Formerly at Northumbria University*
*Newcastle upon Tyne, UK*

WILEY

This edition first published 2021


*Registered Offices*
John Wiley & Sons, Inc., 111 River Street, Hoboken, NJ 07030, USA
John Wiley & Sons Ltd, The Atrium, Southern Gate, Chichester, West Sussex, PO19 8SQ, UK

*Editorial Office*
The Atrium, Southern Gate, Chichester, West Sussex, PO19 8SQ, UK

For details of our global editorial offices, customer services, and more information about Wiley products visit us at www.wiley.com.

Wiley also publishes its books in a variety of electronic formats and by print-on-demand. Some content that appears in standard print versions of this book may not be available in other formats.



*Library of Congress Cataloging-in-Publication Data*

Name: Minns, Peter D., author.
Title: Digital system design using FSMs : a practical learning approach / Peter D. Minns.
Description: Hoboken, NJ : Wiley, 2021. | Includes bibliographical references and index.
Identifiers: LCCN 2021015256 (print) | LCCN 2021015257 (ebook) | ISBN 9781119782704 (hardback) | ISBN 9781119782711 (adobe pdf) | ISBN 9781119782728 (epub)
Subjects: LCSH: Sequential machine theory. | Digital electronics.
Classification: LCC QA267.5.S4 M56 2021 (print) | LCC QA267.5.S4 (ebook) | DDC 621.381501/51135–dc23
LC record available at https://lccn.loc.gov/2021015256
LC ebook record available at https://lccn.loc.gov/2021015257

Cover Design: Wiley
Cover Image: (inset) Image by Peter Minns, (background) © Govindanmarudhai/Getty Images

Set in 10.5/13pt Times by Straive, Pondicherry, India

10 9 8 7 6 5 4 3 2 1

# Contents

# Preface

This book is, in large part, a development of *FSM-Based Digital Design using Verilog HDL* (Minns and Elliott 2008), a book I wrote with Ian Elliott. It is rather unusual in that it forms a linear programmed learning text in all chapters to help readers learn on their own.

The intention in this current version is to make use of programmed learning methods in which the chapters are made up of frames that must be read in a sequential manner. It is hoped that the book will help readers in their study of the material. There is also new content in Chapter 6, Appendix A5, and Appendix A6, as well as consideration of unused states in finite state machines (FSMs).

**It is assumed that the reader has a good understanding of Verilog HDL; however, the interested reader will find that Chapters 6, 7, and 8 of Minns and Elliott (2008) provide a very good account of Verilog HDL. Wiley make it possible to purchase these chapters on request for a small fee.**

Note that in this version of the book the reader is given help to assist them as they progress through this book.

Indeed, Chapters 3, 4, 5, and 7 as well as some of the appendices include examples of FSMs with Verilog HDL for illustrated examples. Use is also made of the Digital logic simulation program **Logisim** to help the reader become familiar with using FSMs in the development of their work. This **Logisim Simulator** is freely available throughout the world to run on Windows, OS X, and Linux Operating Systems (see Appendix A2 for details).

The chapters are organized as follows.

**Chapter 1** covers the introductory ideas of what FSMs are and how to represent them using a state diagram.

**Chapter 2** covers the use of external devices and how to control them with an FSM.

**Chapter 3** looks at how to synthesize FSMs using *T* type flip-flops, then *D* type flip-flops.

**Chapter 4** introduces asynchronous FSM design.

**Chapter 5** looks at the use of the one hot method of synchronous FSM design applied to clocked FSM designs.

**Chapter 6**, a new chapter, looks at applying an FSM to event-driven systems, and considers one hot ideas and the one hot method.

**Chapter 7** deals with Petri nets and how they can be used to synthesize electronic circuits using both sequential and parallel state machine design. This allows FSM-based systems to support both sequential and parallel structures.

There are six appendices covering the necessary aspects of FSM systems that the reader needs to understand in the support of the FSM work. These are written in a more formal manner (i.e. not using frame method or programmed learning).

**Appendix A1** looks at the logic gates and Boolean algebra used in this book. This should help those readers who may not have done much work on Boolean algebra for some time.

**Appendix A2** is a tutorial on how to use the simulation programs and the Verilog Hardware Descriptive Language (HDL) with the SynaptiCAD system as well as a short introduction on the use of the gate logic simulator **Logisim**.

**Appendix A3** covers the use of counters and shift registers as used in a number of the tutorials in this book. The reader should find them very useful.

**Appendix A4** covers the use of behavioural Verilog HDL with some examples to help the reader become familiar with its use in the design of FSM-based systems.

**Appendix A5** looks at the way an FSM can be designed using digital hardware that can be programmed to produce programmable FSM systems. Tutorial examples are introduced to illustrate how this works.

**Appendix A6** looks at how a rotation direction indicator can be implemented using an event-driven FSM.

This book provides enough information for the reader to learn how to design their own FSMs and simulate them using the hardware descriptive language Verilog HDL.

I hope that the content of this book is both interesting and useful, and that it helps readers to learn more about digital system design using FSMs. I have used these techniques for many years, in lectures and when working with companies, and have found them really helpful.

**Peter Minns BSc(h) PhD CENG MIET (retired)**

Access to Wiley Web for my Verilog HDL and Logisim files:
www.wiley.com\go\minns\digitalsystemdesign

# Acknowledgements

I would like to thank all those who helped in the proofreading of this book, in particular those who have looked at specific chapters and especially Kathleen Minns for her help in the proofreading of the entire manuscript.

Also thanks to Ian Elliott for introducing the Logisim digital simulation program.

Special thanks go to my commissioning editor Sandra Grayson and managing editor Juliet Booker for their kind support and guidance in the preparation of this book as well as the Wiley publishing team.

Thanks are also extended to: Dr C. Burch – Logisim, IET Digital Library, Oxford Publishing Limited, and Sage Publishing, USA for their various permissions, and these are referred to in the relevant sections of the book.

Any errors are, of course, entirely the responsibility of the author.

# About the Companion Website

This book is accompanied by a companion website

www.wiley.com/go/minns/digitalsystemdesign

This website includes:

- An Index for Logisim Circuits
- An Index for Verilog Listings
- Guide to Supplementary Resources
- Verilog HDL
- Logisim Files

# Guide to Supplementary Resources

The reader will be aware that this book is supported by working Logisim circuits and Verilog listings and these can be accessed via the Wiley website locator.

However in order to be able to access these Logisim and Verilog circuits/listings the reader needs to download two programs namely Logisim and SynaptiCAD Verilog HDL from the internet and install them on their computer (either PC or MAC) as explained in Appendix 2 of this book.

Each of the Logisim and Verilog HDL folders on the Wiley website has an individual index that shows the appropriate chapter/appendix reference locator to which they refer.

Once the appropriate software has been installed the reader will need to click onto either the .circ or .v files located in 'Logisim Circuits for Web Folder' or 'Verilog Listings for Web Folder' which they wish to access.

In both folders there is additional material and ideas that do not appear in this book but maybe of interest to the reader.

# 1

# Introduction to Finite State Machines

This chapter (like all other chapters) is written in the form of a linear frame, programmed learning text. This is to help you learn the basic skills required to design clocked finite state machines (FMSs) so that you can develop your own designs based on traditional $T$ flip-flops and $D$ flip-flops. Later, other techniques will be introduced, such as 'one hot' and 'asynchronous finite state machines', but these will be developed along the same lines as the work covered in this chapter.

The text is organized into 'frames'. Each frame follows on consecutively from the previous one, but at times you may be redirected to other frames, depending upon your response to the questions you are asked. Do not cheat, but follow the frames as indicated.

## *1.1 SOME NOTES ON STYLE*

**Bold** denotes **questions for you to answer** to check your understanding of the material, highlights **important points**, or indicates an **aside when further ideas are presented**.

Please read this chapter first, and attempt all the questions before moving on to the later chapters. Note that the book can be read as a textbook. The programmed aspect of the book makes it more suitable for individuals to read and learn in their own time.

*Digital System Design using FSMs: A Practical Learning Approach*, First Edition. Peter D. Minns.

Companion website: www.wiley.com/go/minns/digitalsystemdesign

### *Frame 1.1 What is a Finite State Machine?*

A finite state machine (FSM) is a digital sequential circuit that can follow a number of predefined states under the control of one or more inputs. Each state is a stable entity that the machine can occupy. It can move from this state to another state under the control of an outside world input.

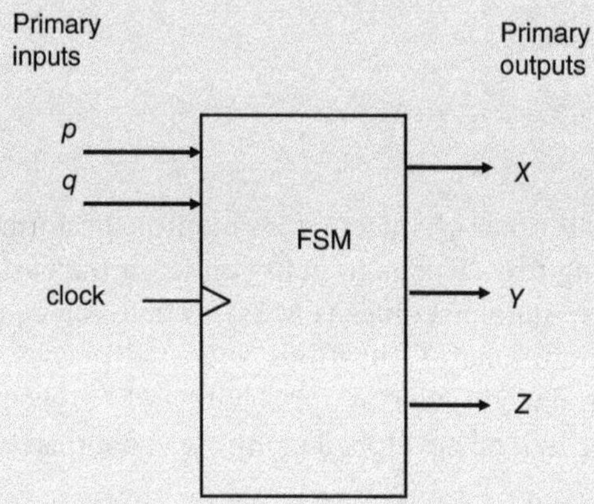


**Figure 1.1** Block diagram of an FSM-based application.

In Figure 1.1, we see an FSM with three outside world inputs (*p*, *q*, and the clock) and three outside world outputs (*X*, *Y*, and *Z*). Note some FSMs have a clock input; those that don't belong to a type of FSM called 'asynchronous FSM'. However, this chapter deals with the more usual synchronous FSM, which **do** have a clock input. Only Chapter 4 and Chapter 6 will look at asynchronous FSM.

Synchronous FSM can move between states only if a clock pulse occurs.

*Task:* **Draw a block diagram for an FSM with five inputs (*x*, *y*, *z*, *t*, and a clock) and with two outputs (*P* and *Q*).**

When you have done this, turn to **Frame 1.2**.

### *Frame 1.2*

The FSM with five inputs (*x*, *y*, *z*, *t*, and a clock) and two outputs (*P* and *Q*) is shown in Figure 1.2.

If you did not get this answer, go back and re-read Frame 1.1. Don't worry about using a mixture of both upper- and lower-case letters here; the only thing that matters is that the same letters are used.

Each state of the FSM needs to be identifiable. This is achieved by using a number of internal flip-flops within the FSM block. An FSM with four states would require two flip-flops since two flip-flops can store $2^2 = 4$ state numbers.

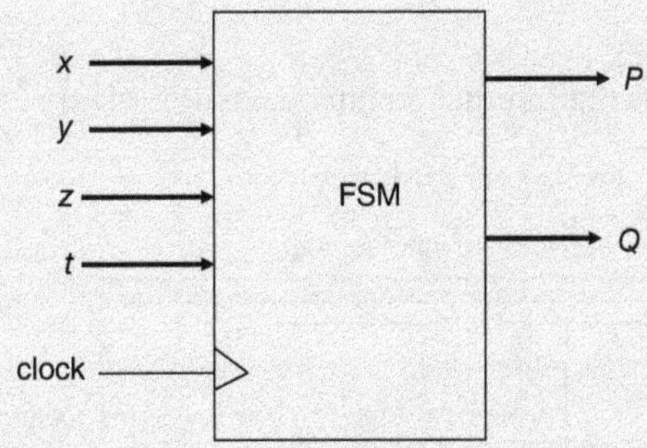


**Figure 1.2** Block diagram with five inputs and two outputs.

Each state has a unique state number, and states are usually assigned numbers, such as s0 (state 0), s1, s2, and s3 (for a four-state example).

As you can see, the rule here is $2^{\text{number of flip-flops}}$.

So an FSM with 13 states would require $2^4$ flip-flops (i.e. 15 states of which 13 are used in the FSM and states 14 and 15 remain unused.

**How many flip-flops would be required for an FSM using 34 states?**

**What would the state numbers be for this FSM?**

When you have answered these questions, turn to **Frame 1.3**.

### *Frame 1.3*

The answer to the previous question is:

$$2^6 = 64, \text{which would accommodate 34 states.}$$

In general: $2^4 = 16$ states, $2^5 = 32$ states, $2^6 = 64$ states, $2^7 = 128$ states, and so on.

**What would the state number be for this FSM?**

s0, s1, s2, s3, s4, s5, s6, s7, s8, s9, s10, s11, s12, s13, s14, s15, s16, s17, s18, s19, s20, s21, s22, s23, s24, s25, s26, s27, s28, s29, s30, s31, s32, s33.

Answer:

The unused states would be s34 through to s63.

**Note that the states run through from s0 to s*n*−1, for *n* states.**

As well as containing flip-flops to uniquely define the individual states of the FSM, there is also combinational logic, which defines the outside world outputs. In addition, the outside world inputs connect to combinational logic, which supplies the flip-flops' inputs.

Please turn to **Frame 1.4**.

### *Frame 1.4*

Figure 1.3 illustrates the internal architecture for a Mealy FSM.

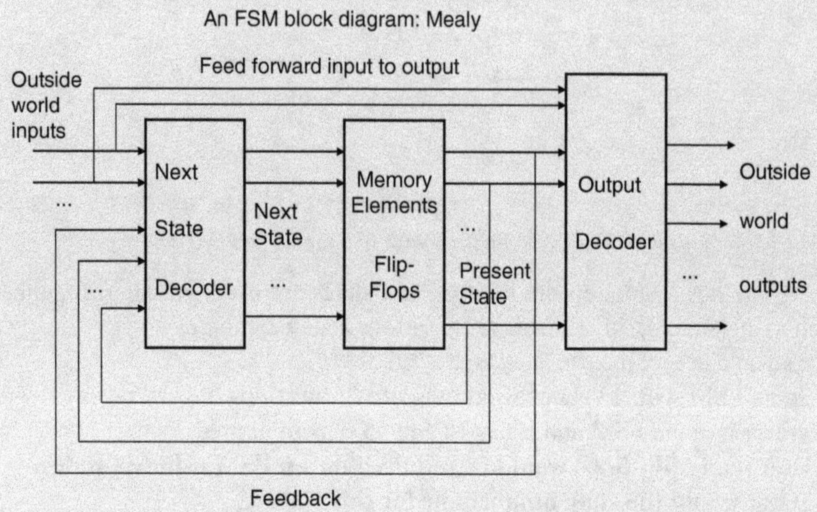


**Figure 1.3** Block diagram of a Mealy state machine structure.

**Note the feed forward paths between the outside world inputs and the input to the output decoder.**

The figure shows that the FSM has a number of inputs that connect to the **next state decoder** (combinational) logic. The *Q* outputs of the memory element flip-flops connect to the **output decoder** logic, which in turn connects to the outside world outputs via the output decoder.

The flip-flop outputs are used as **next state** inputs to the **next state decoder**, and it is these that determine the next state that the FSM will move to. Once the FSM has moved to this **next state**, its flip-flops acquire a new **present state** as dictated by the **next state decoder**.

Note that some of the **outside world inputs** connect **directly** to the **output decoder** logic. This is the main feature of the **Mealy** type of FSM. This affects the outputs of the FSM.

Please turn to **Frame 1.5**.

### *Frame 1.5*

Another architectural form for an FSM is the **Moore** FSM, as shown in Figure 1.4.

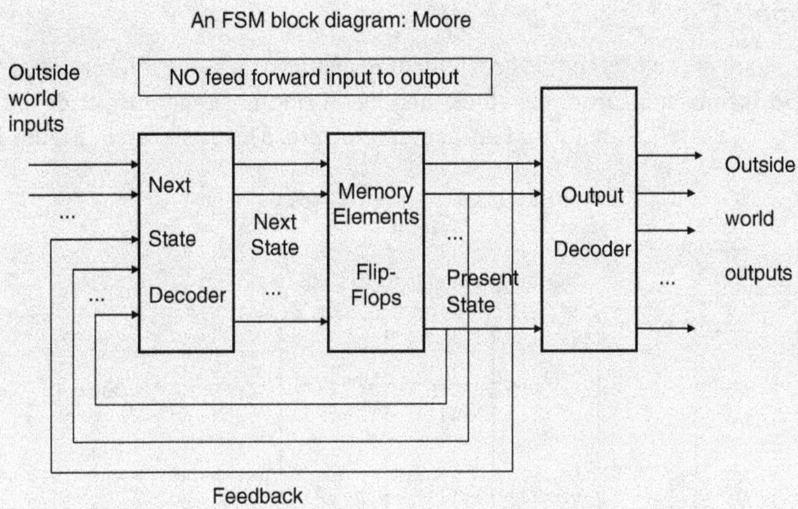


**Figure 1.4** Block diagram of a Moore state machine structure.

This FSM differs from the **Mealy** FSM in that it does not have the feed forward paths.

This type of FSM is very common. Note that the **outside world outputs** are a function of the flip-flop **outputs** only (unlike the **Mealy FSM** architecture where the **outside world outputs** are a function of **flip-flop outputs** and **some outside world inputs**).

We will be using **both Moore and Mealy FSM** in our designs.

Please turn to **Frame 1.6**.

### *Frame 1.6*

Complete the following:

- **A Moore FSM differs to that of a Mealy FSM in that it has…**
- **This means that the Moore FSM outputs depend on…**
- **Whilst the Mealy FSM outputs can depend upon…**

If you cannot complete the above sentences, go back and read Frame 1.4 and Frame 1.5.

When you have completed these questions, please go to **Frame 1.7**.

### *Frame 1.7*

If we look at the Moore FSM architecture again and remove all of the **outside world inputs** apart from the **clock**, and we also remove the **output decoding logic**, we are left with a very familiar architecture. This is shown in Figure 1.5.

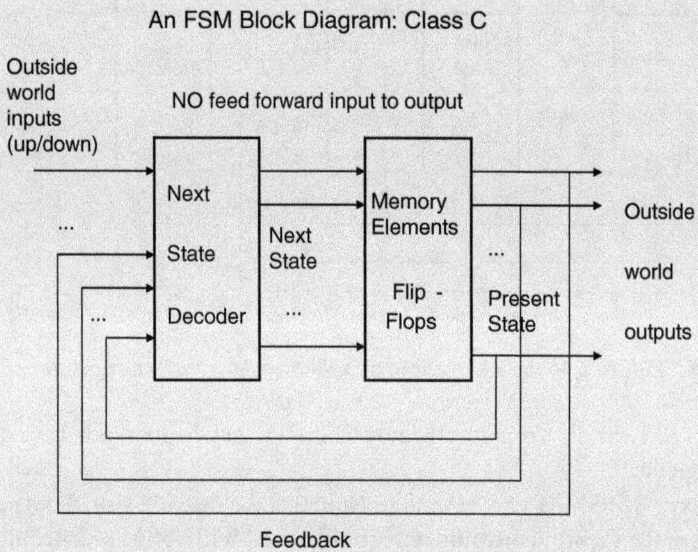


**Figure 1.5** Block diagram of a Class C state machine structure.

This architecture is in fact the synchronous counter the reader may have already seen in previous studies. Note that an **up/down** counter would have the additional outside world input 'up/down', which would be used to control the direction of counting.

The flip-flop outputs in this architecture are used to connect directly to the outside world.

Please move on to **Frame 1.8**.

### *Frame 1.8*

Historically, two types of state diagrams have evolved, one for the design of the **Mealy** FSM the other for the design of the **Moore** FSM. The two are known as 'Mealy state diagrams' and 'Moore state diagrams'.

These days we use a more general type of state diagram, which can be used to design both the **Mealy** and **Moore** type of FSM. This is the type of state diagram we use throughout this book. **As you will learn, it allows you to build a lot of ideas into the FSM diagram.**

Figure 1.6 shows each state of the FSM and the transitions to and from that state to other states.

The states are usually drawn as **circles** (but some people like to use a **square box**).

The **transitions** between states are shown as an **arrowed line** connected between the states.

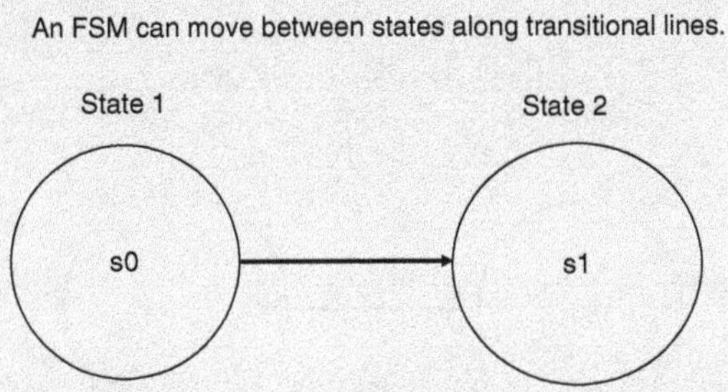


**Figure 1.6** Transition between states.

In addition to the transitional line between states, there is an **input** signal name.

**The right**-angled lines _| represent the clock input (in this case a rising edge 0 to 1) **(Figure 1.7).**

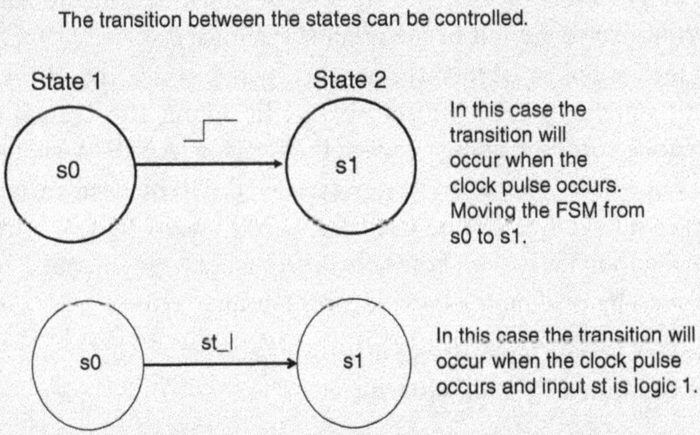


**Figure 1.7** Transition with and without outside world inputs.

In Figure 1.7, the transition between states s0 and s1 will occur at the clock pulse in the upper state diagram, while in the lower state diagram it will **only** occur if the **outside world input set to 1** 'st = 1' **and** a '0 to 1' transition occurs on the clock input.

**What changes would be needed to Figure 1.7 to make the transition between s0 and s1 occur when input st = 0?**

Turn to **Frame 1.9** after you have attempted this question.

## *Frame 1.9*

The answer is shown in Figure 1.8.

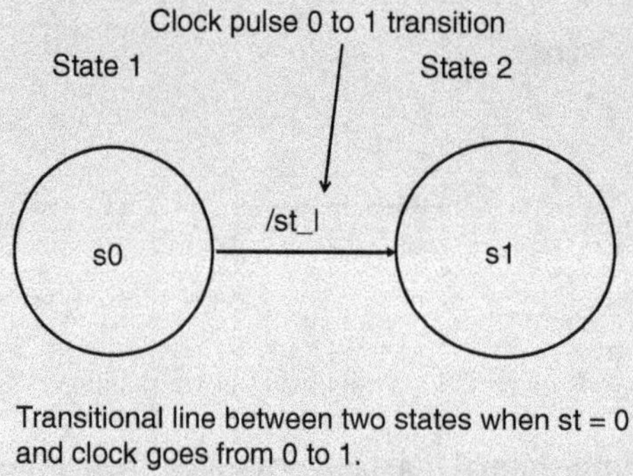


**Figure 1.8** Outside world input between states.

Since in this case the outside world input ‘st’ must be equal to zero (denoted by the inverting bar to the left of the input st (as in /st).

That is / means **not** so /st means **not st**, i.e. when st = 0, then /st = 1.

**Note that outside world inputs always lie along the transitional lines. Also, the reader could be using ‘·’ as well as ‘*’ for ‘AND’. Also, ‘+’ for ‘OR’ in Boolean equations; however, in most cases ‘.’ will be used rather than ‘*’. In some cases no symbol will be used for ‘AND’, as in ‘AB’ to mean ‘A·B’.**

The state diagram must also show how the ‘outside world outputs’ are affected. This is achieved by placing the outside world outputs either:

- **inside the state circle (or square); or**
- **alongside the state circle (or square).**

Figure 1.9 shows the outside world outputs *P* and *Q* inside the state circles. In this particular case, *P* is logic 1 in state s0, and changes to logic 0 when the FSM moves to state s1. Output *Q* does not change in the above transaction, remaining at logic 0 in both states.

Draw a block diagram showing inputs and outputs for the state diagram.

Then turn to **Frame 1.10**.

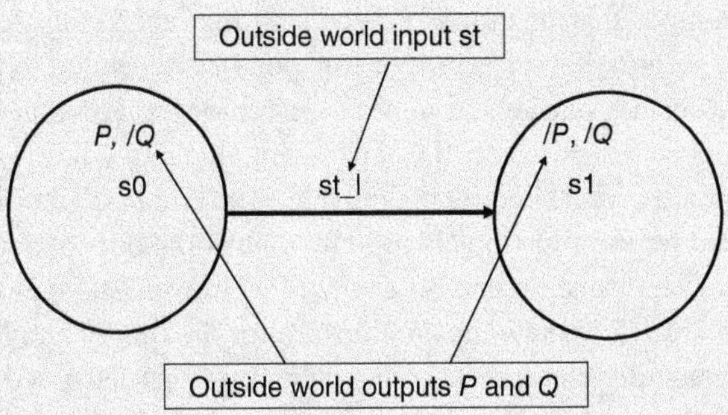


**Figure 1.9** Placement of outside world outputs.

### *Frame 1.10*

The block diagram will look like that shown in Figure 1.10.

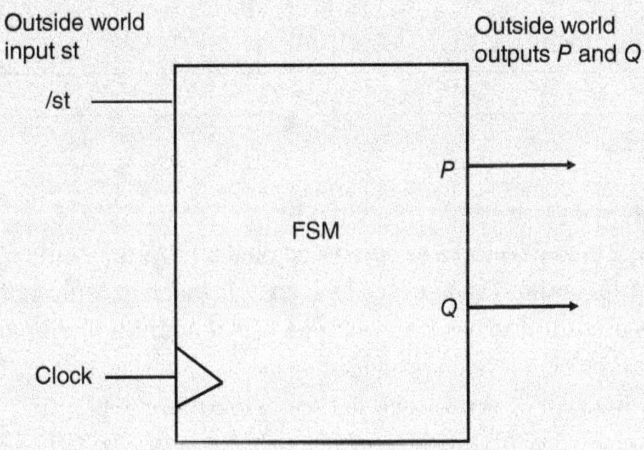


**Figure 1.10** The block diagram for the state diagram shown in Figure 1.9.

Sometimes we show a negating circle to imply that the input is actually inverted (see later).

It is easily obtained from the state diagram since inputs lie along transitional lines and outputs lie inside (or alongside) the state circle. The input st would normally have a **negating circle** to show it is an **active low** input. This is common practice.

You may remember that in Frame 1.2 we said that each state had to have a unique state number and that a number of flip-flops were needed to perform this task. These flip-flops are part of the internal design of the FSM and are used to produce an internal count sequence; they are essentially acting like a synchronous counter, but one that is controlled by the outside world inputs. The internal count sequence produced by the flip-flops is used to control the outside world decoder so that outputs can be turned on and off as the FSM moves between states.

In Frames 1.4 and 1.5 we saw the architecture for the Mealy and Moore FSM. In both cases, the memory elements shown are the flip-flops discussed in the previous paragraph. We look at how the internal flip-flops are coded in a later chapter.

At this stage it is perhaps worth looking at a simple FSM design in detail. We can then bring together all the ideas discussed so far, as well as introducing a few new ones. Try answering the following questions before moving on:

1. **A Mealy FSM differs from a Moore FSM in? (See Frames 1.4 and 1.5.)**
2. **The circles in a state diagram are used to? (See Frames 1.8 and 1.9.)**
3. **Outside world inputs are shown in a state diagram where? (See Frames 1.8 and 1.9.)**
4. **Outside world outputs are shown where? (See Frame 1.9.)**
5. **The internal flip-flops in an FSM are used to do what? (See Frame 1.10.)**

Please turn to **Frame 1.11**.

### *Frame 1.11*

Figure 1.11 shows an example of a single-pule circuit FSM.

The idea here is to develop a circuit based on the FSM that will produce a single output pulse at its output *P* whenever its input *s* is taken to logic 1. The FSM is to be clock driven so it also has an input clock. An additional output *L* is used to indicate that a *P* pulse has been produced so the user can see effect of 'fast' pulses.

The block diagram of this circuit is shown in Figure 1.11.

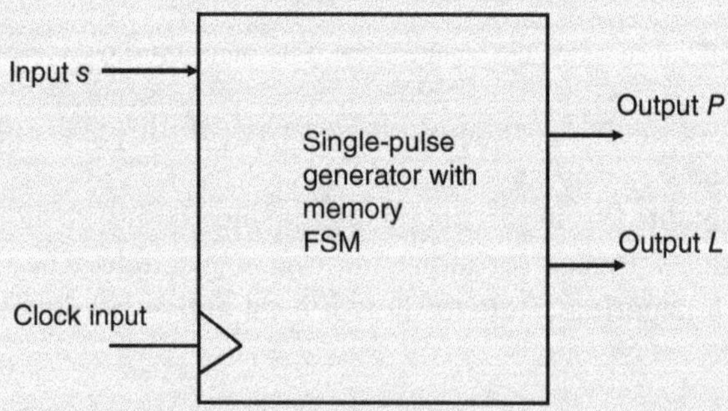


**Figure 1.11** Block diagram of single pulse with memory FSM.

Figure 1.12 shows a suitable state diagram.

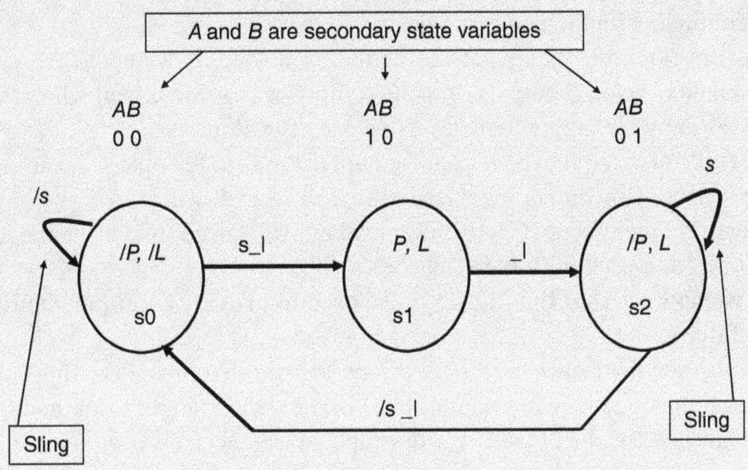


**Figure 1.12** State diagram for single pulse with memory FSM.

In this state diagram the **sling** (loop */s* going to and from s0) indicates that while input *s* is logic 0 (*/s*) the FSM will remain in state s0 regardless of how many clock pulses are applied to the FSM. Only when input *s* goes to logic 1 will the FSM move from state s0 to s1, and then only when a clock pulse arrives. Once in state s1, the FSM will set its output *P* to logic 1, and on the next clock pulse the FSM will move from state s1 to s2.

The reason why the FSM will stay in state s1 for only one clock pulse is because in state s1 the transition from this state-to-state s2 occurs **on a clock pulse only**. Once the FSM arrives in state s2, it will remain there whilst input *s* = 1. As soon as input *s* goes to logic 0 (*/s*) the FSM will move back to state s0 **on the next clock pulse**.

Since the FSM remains in state s1 for only a single clock pulse, and since $P = 1$ only in state s1, the FSM will produce a single output pulse.

**Note in the FSM state diagram that each state has a unique state identity: s0, s1, and s2.**

Also note that each state has been allocated a unique combination of flip-flop states, for example:

- State s0 uses the flip-flop combination $A = 0$ $B = 0$, e.g. both flip-flops reset.
- State s1 uses the flip-flop combination $A = 1$ $B = 0$, e.g. flip-flop *A* is set.
- State s2 uses the flip-flop combination $A = 0$ $B = 1$, e.g. flip-flop *A* is reset, flip-flop *B* is set.

Now move on to **Frame 1.12**.

### *Frame 1.12*

Let's continue with the one-pulse design.

The flip-flop outputs are seen to define each state. If we could see nothing more than the *A* and *B* outputs of the two flip-flops, we could tell what state the FSM was in by the output logic levels on each flip-flop.

We could also tell in which state the output *P* was to be logic 1, i.e. in state s1 where the flip- flop output logic levels are *A* = 1 and *B* = 0.

Therefore, the output *P* = *A*/*B*. (Remember, *AB* is used to indicate the logical AND operation used in Boolean algebra.)

**So we now see that the flip-flops are used to provide a unique identity for each state.**

We also see that, since each state can be defined in terms of the flip-flop output states, the outside world outputs can also be defined in terms of the flip-flop output states since the outside worlds output states themselves are a function of these states.

*L* is logic 1 in states s1 and s2 and is defined in terms of the flip-flop outputs *A*/*B* + /*AB*.

Therefore, *L* = *A*/*B* + /*AB* = *A*/*B* + /*AB*. No Boolean reduction is possible in this case.

The allocation of unique values of flip-flop outputs **to each state** is rather an arbitrary process. In theory, we can use any values so long as **each state has a unique combination**. This means that we cannot have more than one state with the flip-flop values of, say, *A*/*B* (i.e. both states cannot have the same value).

In practice it is common to assign flip-flop values so that the transition between each state involves **only one flip-flop changing state**. This is known as 'following a **unit distance pattern**': only one flip-flop changes state.

The above example does not use a unit distance pattern since there are two flip-flop changes between states s1 and s2. However, the reader will be going on to make use of the unit distance code idea.

The reader could also make the single-pulse state diagram (Figure 1.12) follow a **unit distance pattern** by adding an extra state. This **extra state** could be inserted between states s2 and s0, having the same output for *P* as state s0. In the state diagram the new state would also have the value of *L*, the same as that in state s2, since the reader does not want *L* to change until *s* goes to 0.

**Try re-drawing the state diagram with this additional state and assign a unit distance pattern to the flip-flops.**

When you have done this, go to **Frame 1.13**.

### *Frame 1.13*

A completed state diagram with unit distance patterns for flip-flops is shown in Figure 1.13.

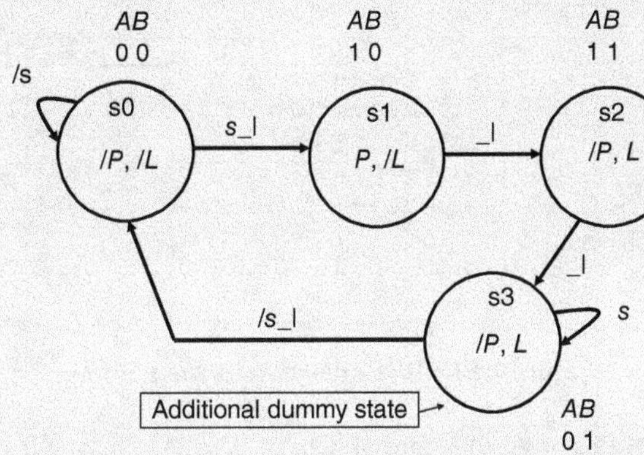


**Figure 1.13** State diagram for single-pulse generator with memory and dummy state.

Note that the added state has the unique name of s3 and the unique flip-flop assignment of $A = 0$ and $B = 1$.

**Also note that s2 uses the $A$ and $B$ values of $A = 1$ and $B = 1$. This provides the required unit distance coding.** It also has the output $P = 0$, as it would in state s0 (the state it is going to go to when $s = 0$).

In this design the addition of the extra state has not added any more flip-flops to the design since two flip-flops can have a maximum of $2^2 = 4$ states (remember Frames 1.2 and 1.3).

The addition of this extra state is usually called a **dummy state**.

Look carefully at the state diagram in Frame 1.13 and satisfy yourself that the state diagram is doing the same thing as the one in Frame 1.11. If you cannot see this, consider reading Frames 1.11–1.13 again.

Now let us add an additional input called $r$ to our state diagram.

Input $r$ is to be added so that if $r = 1$ the FSM will continue to pulse output $P$ (on and off) until $r$ is made 0. At this point the FSM will return to state s0 but only if input $s = 0$.

**Draw the block diagram for the FSM.**

**Draw the state diagram for this modified FSM.**

Take your time and think about what you are doing.

Turn to **Frame 1.14** when you have completed this task.

*Frame 1.14*

The block diagram in Figure 1.14 is changing the behaviour of the state diagram.

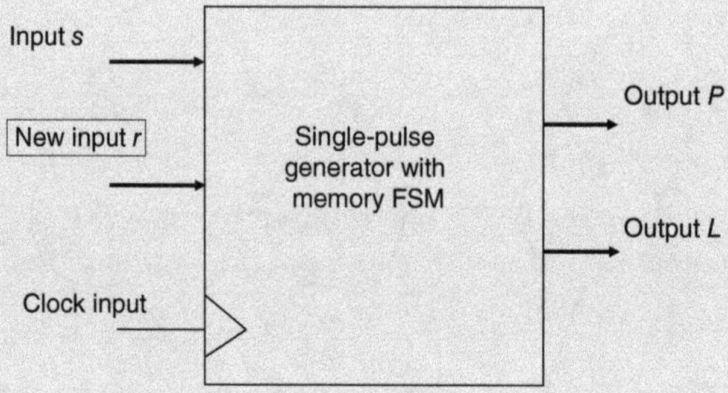


**Figure 1.14** Block diagram for the FSM.

The state diagram is shown in Figure 1.15.

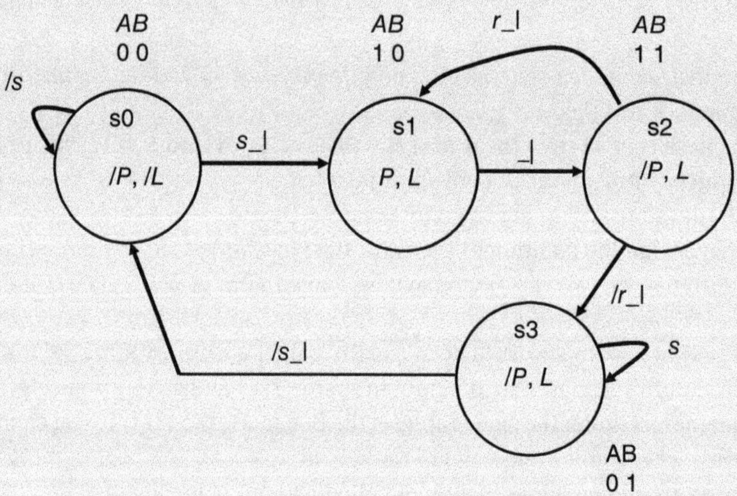


**Figure 1.15** State diagram of single-pulse generator with a multipulse feature.

The new state diagram is essentially the same as that in Frame 1.13 except that now if $r = 1$ the state diagram sequence (with $s = 1$) becomes s0 → s1 → s2 → s1 → s2, and so on.

However, if $s = 1$ and $r = 0$ the sequence will be s0 → s1 → s2 → s3, which then stops until $s = 0$ then goes to s0.

From this you should understand that if we make $r = 0$ before we make $s = 1$ the state diagram will follow a single pulse on output $P$.

**The Boolean equation for $L$ can be active high or active low.**

As active high, we look for states when the value of $L$ is 1.

$$L = s1 + s2 + s3 = A/B + AB + /AB = A\cdot(/B + B) + B(A + /A) = A + B.$$

The Boolean equation for $P$ was $P = s1 = A/B$ (see Frame 1.12).

The Boolean equation for $L = s1 + s2 + s3 = A + B$ since only in states s1, s2 and s3 is the output $L = 1$.

Note that an alternative equation for $L$ could be the **inverse** equation for $L$, otherwise known as 'active low'. Here we look for the states that make $L = 0$.

$$/L = /s0 = /(/A/B) \text{ which is a NAND gate.}$$

In practice there is a tendency to show this active low output as:

$$L\,(\text{active low}) = /(/A/B).$$

**Note that to obtain the $L = 0$ the reader needs to invert s0 (/s0). This idea will be used later.**

This is less complex so it may be used.

The latter equation is in terms of NOT $L$. This means that when in state s1, s2, or s3, $L$ will be logic 1. Only when the FSM is NOT in any of these states will $L = 0$.

So in summary:

- Input $r$ is used with a two-way branch from state s2.
- If $r = 0$ there is no change in the operation of the FSM (single pulse from $P$).
- However, if $r = 1$ the FSM will keep looping between s1 and s2 so as to keep turning the output $P$ on and off, **in effect using input $r$ to change the operation of the FSM.**

This shows how easily it is to change the behaviour of the FSM (in this case).

**Note that this change was done in the state diagram then transferred to the actual FSM.**

**Please turn to** Frame 1.15.

## *Frame 1.15*

In the previous frames we have considered the flip-flop output patterns. These are often referred to as the **secondary state variables (SSVs)** (Figure 1.16).

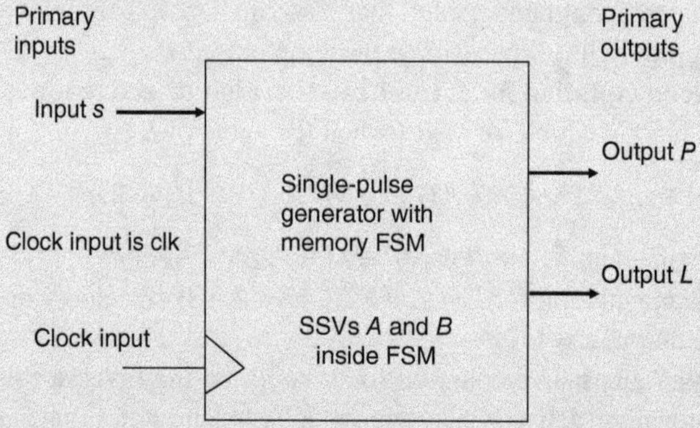


**Figure 1.16** Block diagram showing secondary state variables in the FSM.

They are called 'secondary state variables' because they are (from the FSM architecture viewpoint) internal to the FSM (i.e. secondary not primary). If we consider the outside world inputs and outputs as being primary then it seems sensible to call the flip-flop outputs SSVs (and state variables because they define the states of the state machine).

## *Moore and Mealy state diagram*

The outputs in our FSM are seen to be dependent upon the SSVs or flip-flops internal to the FSM. If you look back to Frame 1.5 you will see that Moore FSM outputs are dependent upon the flip-flop outputs only. The output decoding logic in our *P* pulse example is:

$P = \text{s1} = A/B$ (see Frames 1.12 and 1.13); and
$L = /\text{s0} = /(/A/B)$, an active low.

That is it consists of an AND gate and a NAND gate. This means that a single *P* pulse is a Moore FSM.

**How could we make our single-pulse design into a Mealy FSM?**

One way would be to make the output *P* depend on the FSM being in state s1 (*A/B*), but we could say that the output was to be the width of a single logic 0 of the clock pulse.

**How would we modify our state diagram to do this?**

Try doing this, and then turn to **Frame 1.16** to find out if you got it right.

### *Frame 1.16*

The modified state diagram is shown in Figure 1.17 (the *r* signal here has been dropped so we are back to a simple one-pulse FSM). Also, /clk is clk inverted.

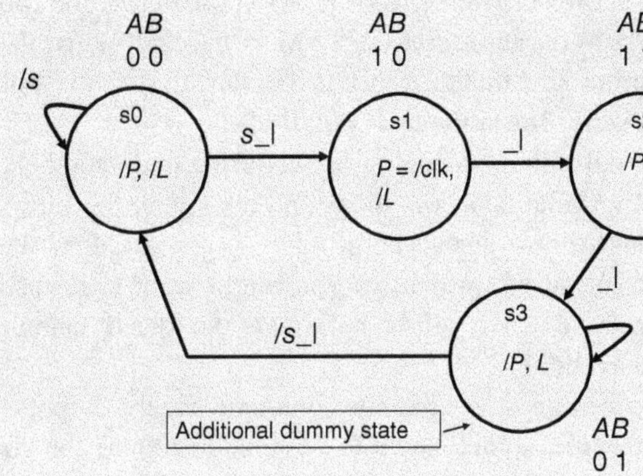


**Figure 1.17** State diagram with Mealy output *P*.

Notice that now the output *P* is only equal to logic 1 when:

- FSM is in state s1 where flip-flop outputs are $A = 1$ and B = 0.
- The clock signal is logic 0, as indicated by $P = /\text{clk}$.

Therefore, when the FSM enters state s1, the *P* output will be equal to logic 0 since /clk = 0 when the clock 'clk' is logic 1.

**The clock will be logic 1 when the FSM enters state 1 (0 to 1 transition) the clock clk will then go to logic 0 (whilst still in state s1), and *P* will go to logic 1, then, when the clock clk goes back to logic 1 the FSM will move to state s2 and the flip-flop outputs will no longer be *A*/*B* so the *P* output will not go high again. Therefore, the *P* output will only be logic 1 for a single clock clk pulse in state s1 when clk = 0 at the end of the state s1.**

Figure 1.18 illustrates this more clearly.

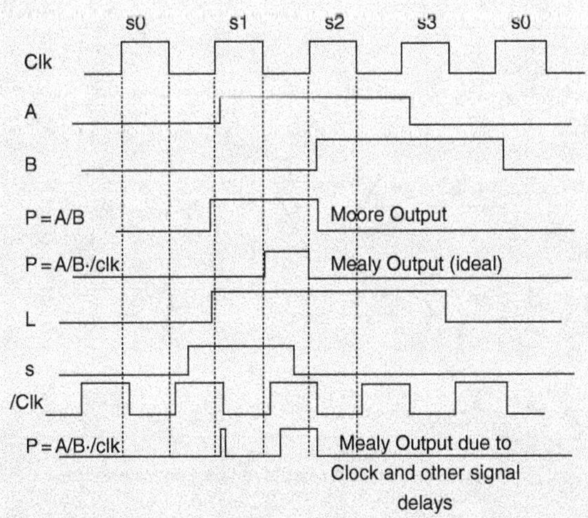


**Figure 1.18** Timing diagram showing Moore and Mealy outputs.

The waveforms show both versions of *P*. As you can see, the Moore version raises *P* for the whole duration that the FSM is in state s1, whilst the Mealy version raises *P* for the time that the clock is low during state s1. **Note in Figure 1.18 *P* = *A*/*B*·/Clk is not the same as *P* = *A*/*B*·clk.**

Note the narrow pulse in the *P* Mealy signal during the second clk pulse. This is a glitch caused by signal delays in the *A* and clk signal lines. (See Chapter 4 for more on this, but for now just accept it.)

There is quite a lot going on here so you might want to re-read this frame again to fully understand it. We will be looking at the idea of using more Mealy outputs later on in this book.

It is possible to design state diagrams without Mealy outputs, using only Moore outputs. However, sometimes it is possible to reduce the size of a state diagram (less number of states) by using Mealy outputs. Examples are shown later.

Now for something a bit different.

**Try producing a state diagram for an FSM that will produce a 101 pattern in response to an *s* input going high. Signal *s* must be returned low before another 101 pattern can be produced. In this example you are trying to use each state to produce the 101 pattern.**

When you have attempted this task, turn to **Frame 1.17**.

## *Frame 1.17*

The solution to this problem is to use the basic arrangement of the single-pulse state diagram and insert more states to generate the required 101 pattern (Figure 1.19).

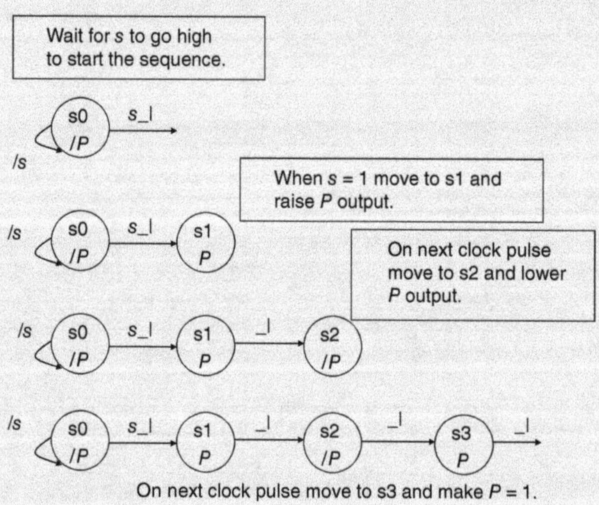


**Figure 1.19** Development of a 101 pattern generator sequence.

It will develop state by state so you can see how it is done.

Note that we must leave state s3 on a clock pulse so that $P = 1$ for the duration of a single clock pulse only.

The final state required is to monitor for the input $s = 0$ condition.

This state should return the FSM back to state s0.

**Complete the FSM state diagram then turn to Frame 1.18.**

***Frame 1.18***

The completed state diagram is shown in Figure 1.20.

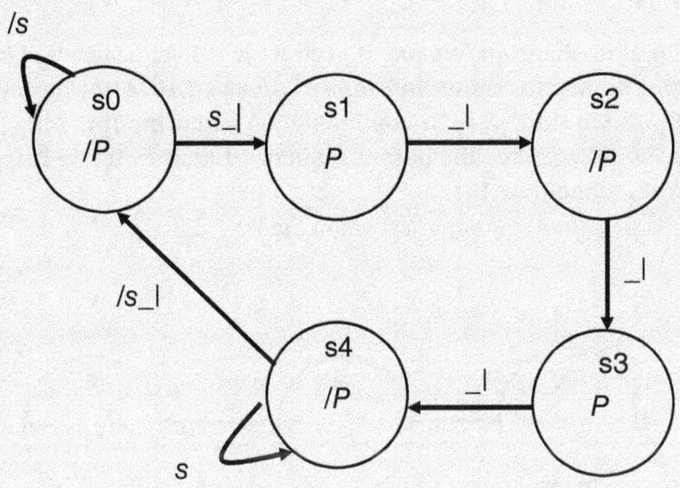


**Figure 1.20** Complete state diagram for the 101 pattern generator.

The Boolean equation for $P$ in this diagram is $P = s1 + s3$. However, we could make the $P$ output a Mealy output that is only equal to 1 when in states s1 and s3, and only if the clock pulse is equal to 1.

Then, $P = s1 \cdot clk + s3 \cdot clk$ since $P$ must be high in both states s1 and s3, but only when the clock input is high.

**Try writing an account of how this FSM works in your own words. If you get stuck, just re-read Frames 1.16 and 1.17 again to refresh your memory.**

**Now try modifying the state diagram to make it produce a 101 sequence of clock pulses (in the same manner as shown in Frames 1.17 and 1.18).**

**Also, arrange for the *P* output pulse in state s3 to be conditional on a new input called *x*. If $x = 0$, the FSM should produce the output sequence 100 at *P*. If $x = 1$, the output sequence at *P* should be 101.**

The reader may have noticed that the state diagram does not need to use slings. This is because slings are not really necessary with modern state diagrams. In fact, they are really only included for cosmetic reasons, to improve the readability of the design. From now on, only use slings where they improve the readability of the state diagram.

**When you have done this, draw your state diagram and then turn to Frame 1.19.**

### *Frame 1.19*

**Modify the state diagram to make it produce a 101 sequence of clock pulses (in the same manner as shown in Frames 1.15 and 1.16. Also, arrange for the *P* output pulse in state s3 to be conditional on a new input called *x*. If $x = 0$, the FSM should produce the output sequence 100 at *P*. If $x = 1$, the output sequence at *P* should be 101.**

The modified state diagram is shown in Figure 1.21.

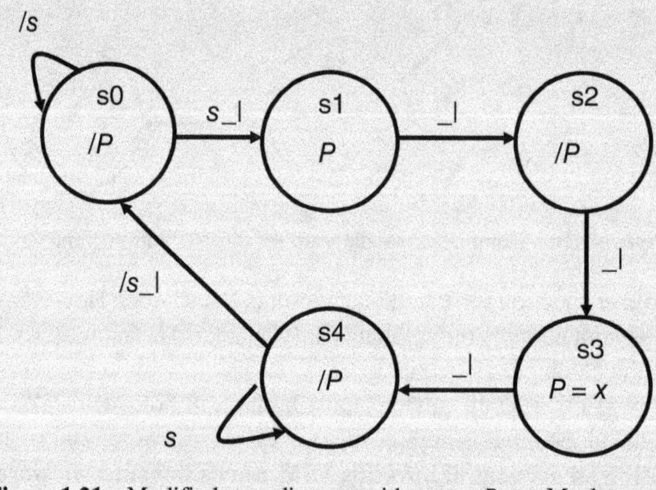


**Figure 1.21** Modified state diagram with output *P* as a Mealy output.

In Figure 1.21, the clock is used as a qualifier in states s1 and s3 so that the output *P* is only logic 1 in these two states. However, state s3 has an additional qualifier *x*, so in s3 $P = 1$ only when in s3 and then only if input signal *x* is true in s3.

Then in state s3, the output $P$ will only obtain a clock pulse if the $x$ input happens to be logic 1.

You can see that if $x = 0$ then, when the input $s$ is raised to logic 1, the FSM will produce the sequence 100 at output $P$. Therefore, $P = s1 + s3 \cdot x$. If $x = 1$ then, when $s$ is raised to logic 1, the FSM will produce a 101 sequence at the output $P$.

This FSM is an example of a Mealy FSM since the output $P$ is a function of both the state and the inputs clock and $x$, i.e. both clock and $x$ are fed forward to the output decoding logic.

**The reader could easily modify the FSM so that the 100 sequence at *P* was produced if *x* = 1, and the 101 sequence produced if *x* = 0. Therefore, now:**

- **Produce the Boolean equation for *P* in state s3 that would satisfy this requirement.**
- **Then assign a unit distance code to the state diagram; see Frames 1.12 and 1.13.**
- **Finally, when you have done that, try producing a timing diagram of the modified FSM.**

When you've finished, turn to **Frame 1.20**.

### *Frame 1.20*

**Produce the Boolean equation for *P* in state s3 that would satisfy this requirement.**

The Boolean equation for $P$ which will produce a 101 sequence when $x = 0$ is:

$$P = s3 \cdot /x.$$

Note you do not need to indicate the clk, as it is assumed. Also note that in this case qualifying with NOT $x$ ($/x$), rather than with $x$, as in Figure 1.22.

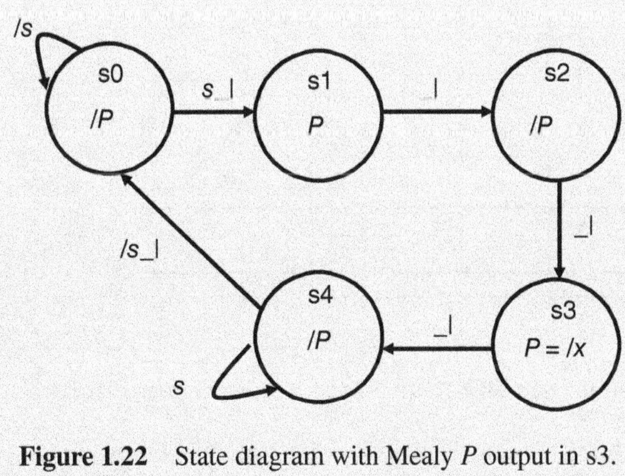


**Figure 1.22** State diagram with Mealy $P$ output in s3.

It is very likely that you came up with a different set of values for the secondary state assignments to those obtained. This is fine since there is no real preferred set of assignments, apart from trying to obtain a unit distance coding (*ABC* values not shown at this stage).

**Try re-drawing the state diagram with the dummy state and modified coding.**

Note: care should be taken where you place the dummy state. If you added a dummy state between states s1 and s2, for example, it would alter the *P* output sequence so that, instead of producing, say, 101, the sequence 1001 would be produced.

A safe place to add a dummy state would be between states s3 and s4, or between states s4 and s0 since they are outside the 'critical *P*' sequence generating in this part of the state diagram.

Turn to **Frame 1.21** for the timing waveform diagram solution.

## *Frame 1.21 The Timing Waveform Diagram Solution*

The solution is, of course, based on the secondary state assignments used, so your solution could well be different if you have used a different SSV pattern.

In this solution (Figure 1.23), the author has deliberately arranged for the *x* input to change to logic 0 inside of the clock pulse equal to 1 in state s3 just to illustrate the effect that this would have on the output *P*. You can see that the output pulse on *P* is not a full clock high period.

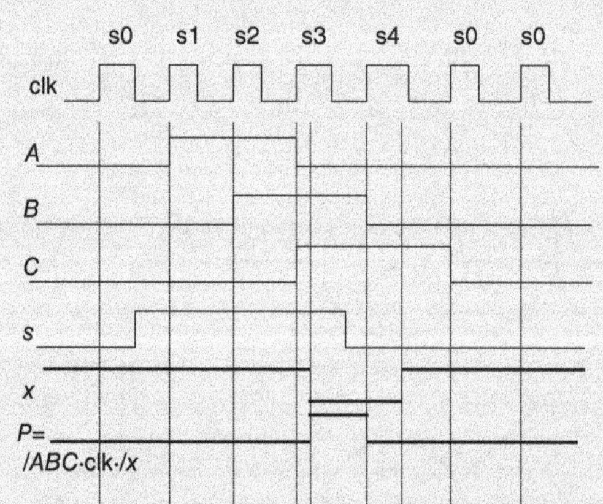


Note that *P* does not become logic 1 until *x* = 0 in state s3, and that *P* goes to logic 0 when FSM leaves state s3, even though *x* is still logic 0.

**Figure 1.23** Timing diagram showing the effect of input *x* on output *P*.

**This is a very realistic event since the outside world input *x* (and, indeed, any outside world input) can occur at any time.**

Turn to **Frame 1.22**.

***Frame 1.22***

At this point in the course we have covered the basics of what an FSM is and how a state diagram can be developed for a particular FSM design.

The reader has also seen how the outputs of the FSM depend upon the SSVs (these are covered in Chapter 3).

The SSVs can be arbitrarily assigned, but that following a unit distance code is good practice.

The reader has looked at a number of simple designs and seen how a Mealy or Moore FSM can be realized in the way in which the output equations are formed.

We have not yet seen how the state diagram can be realized as a circuit made up of logic gates and flip-flops, but this part of the development process is very much a mechanized activity which is covered in detail in Chapter 3.

The next section looks at a number of FSM designs in an attempt to give you some feel for the design of state diagrams for FSMs. The pace will be a little quicker as I will assume that you have understood the previous work.

You may like to take a well-earned break at this point!

For more details, see Minns (1995).

# 2

# Using FSMs to Control External Devices

## 2.1 INTRODUCTION

In this chapter the use of finite state machines (FSMs) to control external devices like hardware timers, counters, analogue-to-digital converters (ADC), and memory devices will be looked at and how they can be controlled by the FSM.

**Frame 2.1**

One of the most common requirements in an FSM is the need to wait in a state for some predefined period. You might, for example, wish to turn on an outside world output for a certain period of time, and then turn it off again. This could be done by just allocating a number of consecutive states, with the required output held high, but this would be very wasteful of states (and the corresponding flip-flops needed to implement the FSM) for all but very short delays.

The best way of dealing with this kind of requirement is to use an external timer unit, which can be controlled by the FSM.

A typical timer unit might look something like that shown in Figure 2.1.

*Digital System Design using FSMs: A Practical Learning Approach*, First Edition. Peter D. Minns.

Companion website: www.wiley.com/go/minns/digitalsystemdesign

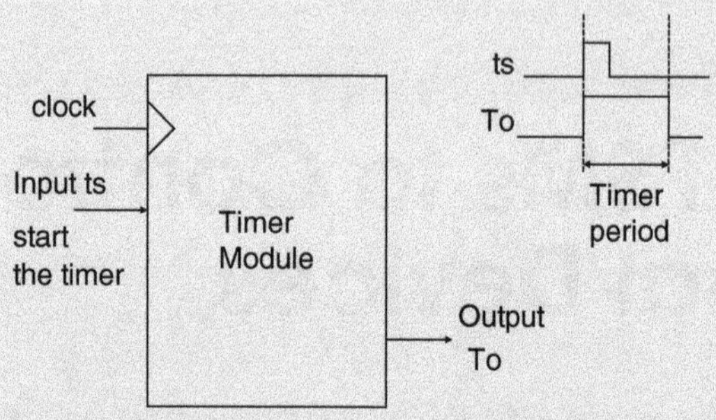


**Figure 2.1** A timing module.

As you can see, the timer unit has two inputs, the clock input clock clk and the start timer input ts, and a single output To.

From the state diagram Figure 2.2, to is an output, and Ts is an input.

In Figure 2.1 the timing diagram for the timer unit output To will go high when the timer start input ts makes a 0 to 1 transition. The output To will remain high until the time period has elapsed, after which it will go low.

So we can use the timer to provide an output from the FSM (Figure 2.2) which is used to start the timer in the state before the one you wish to use the timer in. Then, on the next clock pulse, the FSM moves into the time out state (Figure 2.2 state s*n* + 2). In this state, the ts signal is returned low and the timer output signal to is monitored by the FSM, which is looking for the to signal going low (signalling the end of the timer period). At this point, the FSM will leave the time out state 's*n* + 2'.

**Aside: timers can be made from either analogue circuits (look at the 555 timer chip) or digital circuits like a digital counter (can be up/down counter).**

A small segment of the state diagram shows this sequence in Figure 2.2.

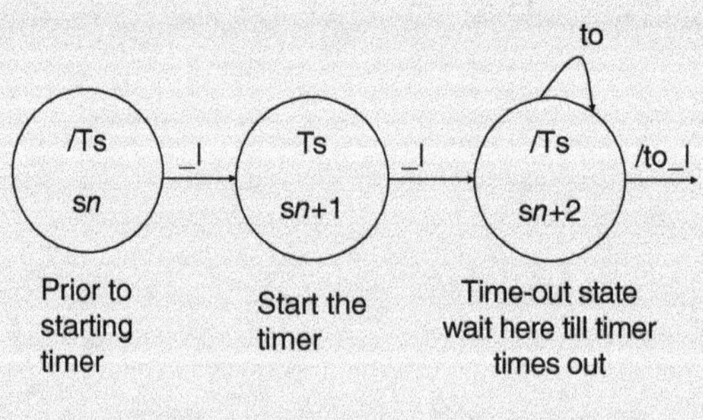


**Figure 2.2** Wait state sequence to control the timing module.

Note that the FSM will wait in state s*n* + 2 until the signal to goes low. Therefore the timer unit will dictate the time spent in state s*n* + 2.

Again note that in Figure 2.2 to is an input to the state diagram, while Ts is an output.

Please turn to **Frame 2.2**.

### *Frame 2.2*

An example of how the timer unit could be used is shown in Figure 2.3.

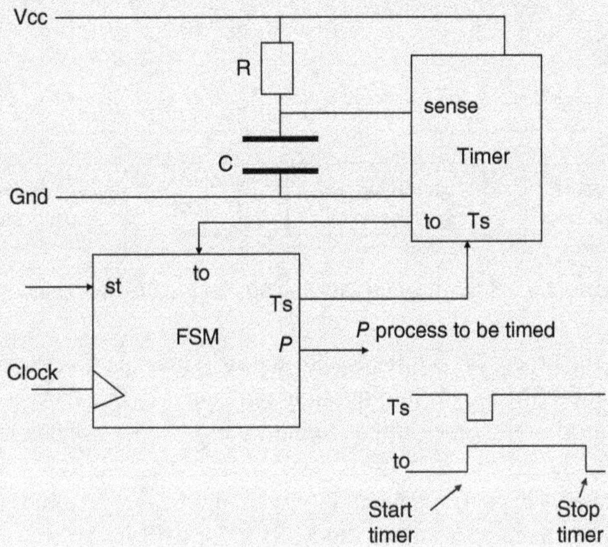


**Figure 2.3** Block diagram showing how to use the timing module with an RC time constant.

In Figure 2.3, you can see that the FSM is controlling a timer unit. This timer unit happens to be a little different from the one seen in Frame 2.1 in that it does not have a clock input. It is, in fact, a timer based around an RC charging circuit (a practical device would be the 555 timer units that are readily available). This does not alter things in any way as far as the FSM is concerned. You could, of course, have used a digital counter as a timer.

**Note that Ts from the state diagram is active low.**

The start input st starts the FSM from state s0. The idea here is to turn on the process *P* for a period of time (dictated by the RC time constant of the timer unit) then turn it back off again.

**See if you can produce the state diagram to do this. Then turn to Frame 2.3 to see if you got it right.**

### Frame 2.3

The state diagram to implement the arrangement is shown in Figure 2.4.

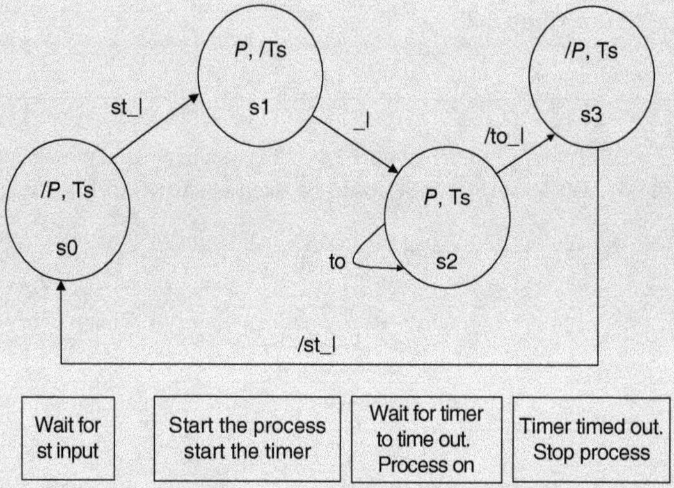


**Figure 2.4** State diagram showing how to use the timing module.

Note that the process *P* is turned on in state s1 since the timer is started in this state before the FSM moves into the time out state s2. The FSM will remain in state s2 monitoring the timer output signal to until the timer has timed out, then move to state s3.

Assume clock frequency is much higher than time out period of the timer.

Using this arrangement allows the FSM to be held in the wait state for any length of time, depending only on the value of the resistor capacitor time constant. It also allows for outputs to be held 'active' for this particular time as well.

If you got this wrong, you might need to go back and re-read Frames 2.1 and 2.2.

Please turn to **Frame 2.4**.

### Frame 2.4

Now that you have seen how to control an outside world device like a timer, you are ready to see how other outside world devices can be controlled.

This is really what FSMs are all about: controlling outside world devices.

Look at how an FSM can be used to control an ADC and/or a memory device:

- An analogue-to-digital converter (ADC).
- A memory device.

### Controlling an analogue-to-digital converter (ADC)

ADCs are used to convert analogue data into digital form. They are needed to allow digital computers to be able to process data from the 'real' world, which is essentially analogue in nature. In most systems that use an ADC, the ADC will be controlled from a microprocessor (or microcomputer). However, it is often the case that a system (or part of a system) will be implemented using a customized chip design, or a programmable logic device (PLD).

In these cases, the ADC (and indeed other devices) may be controlled by an FSM.

Consider the ADC shown in Figure 2.5.

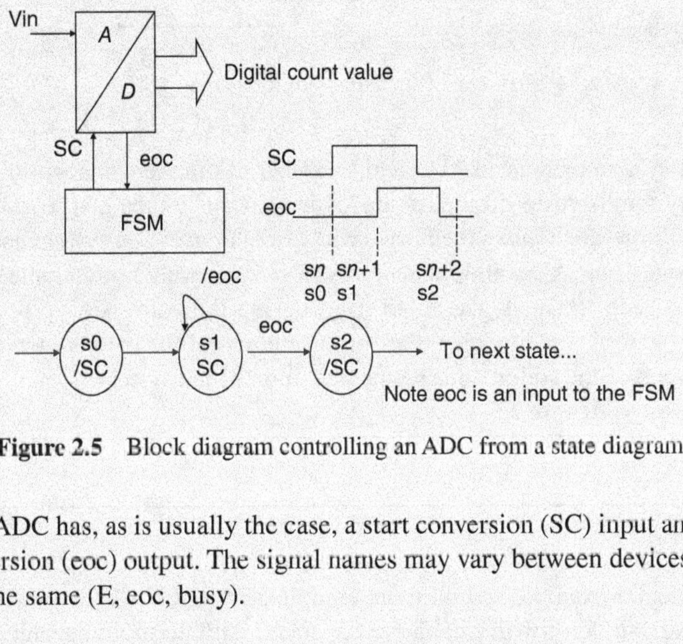


**Figure 2.5** Block diagram controlling an ADC from a state diagram.

This ADC has, as is usually the case, a start conversion (SC) input and an end of conversion (eoc) output. The signal names may vary between devices, but the idea is the same (E, eoc, busy).

- The digital outputs are connected to the external circuit and are not part of the FSM.
- The FSM can read selected inputs and control data.

Now consider the small data acquisition system shown in Figure 2.6.

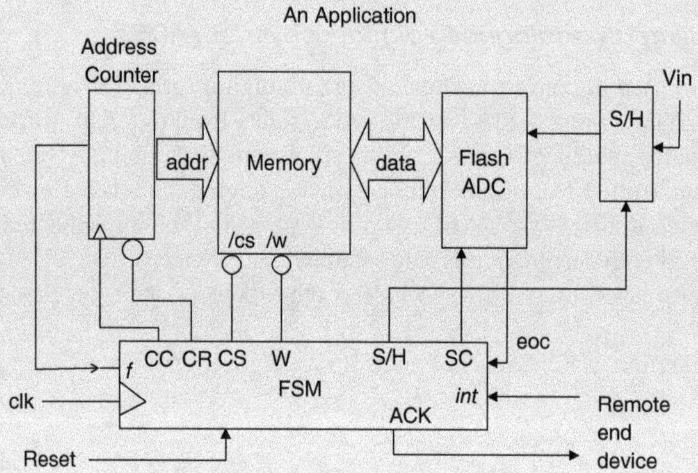


**Figure 2.6** Block diagram for a small DAS.

In this system there is an ADC and a number of other outside world devices.

This system is designed to take analogue values from the real world, digitize them, and store them into a memory device for later use. The system needs to be able to control the ADC and the memory device (as well as a 'sample and hold circuit' to keep the analogue value fixed during the conversion process). The address counter is used to keep track of the address of the memory device, so as to keep track of the address during the operation of the system.

Now turn to **Frame 2.5**.

### *Frame 2.5*

This particular example is a bit more complicated than the examples you have seen so far, but don't worry: it can be separated out into more manageable bits, as you will find out.

Figure 2.6 shows an FSM used to control a sample and hold amplifier (S/H), an ADC, a random access memory (RAM) device, and a simple binary counter.

All these outside world components allow the FSM to be able to:

- Sample ac analogue data from the outside world.
- Store the data in a RAM memory.

All under control of a remote end device, which could be a microcomputer, or perhaps another FSM.

But before attempting to produce a state diagram for this FSM you must first consider how an FSM can control the RAM and counter.

Consider Figure 2.7.

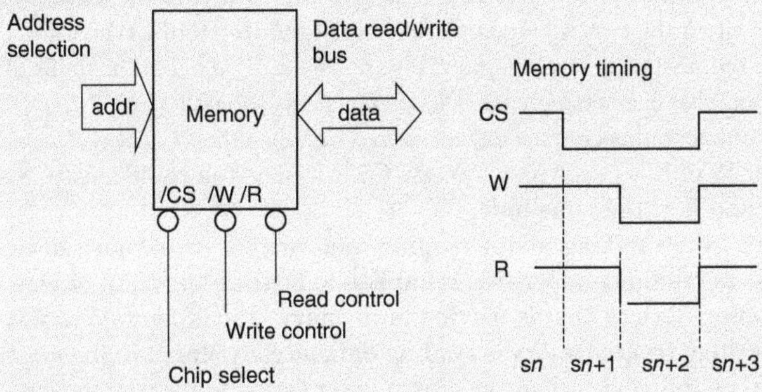


**Figure 2.7** Control of a memory device note negating circles at /CS,/W,/R.

This shows a memory device controlled by an FSM.

Note that the memory device has a collection of address input lines (commonly called the 'address bus') and a set of data lines called the 'data bus'.

If the memory is **read only** the data bus lines will be outputs. If the memory is a **random access memory** the data bus lines will be **bi-directional**. This means that the /R and /W control signals can be used to condition the data bus lines to be either inputs when /W is used to write data to the memory, or /R used to read the memory. In addition there is a **chip select** input, which in this case is **active low** and used to control the selection of the memory device. These control lines need to be applied in the correct sequence.

Now turn to **Frame 2.6**.

### *Frame 2.6*

To access the memory device the chip select line must be asserted (this means that the chip select line must be active, in this case active is logic 0, or active low), as most memory devices seem to be at the time of writing.

Then, if data are to be written into the RAM device, the chip must first be selected by bringing the chip select line low, then set the /W signal line low also. This is writing data to the memory.

If you wish to read the contents of the RAM then you must first select the chip select line by making it go low then set the /R line low. This is reading data in the memory.

In most cases you can then raise all control lines high, to dis-assert them at the same time both /CS and /W, or /CS and /R go high. It is usually at this point in the cycle that the memory device is read or write. **Note that the memory action occurs when the control signals are raised high (the 0 to 1 transition).**

The segment of timing diagram (Figure 2.7) in Frame 2.5 illustrates this process (note that each change needs its own state as shown ($sn$, $sn + 1$, $sn + 2$, and $sn + 3$). Note in this case the CS is lowered first then the W or R is lowered after, then the W or R is raised first, then the CS is raised. **You could raise CS and W or CS and R at the same time.**

**Note: when talking about reading and writing to memory devices the process of reading and writing is implied to be from the point of view of the controlling device. That is writing to memory is sending data to the memory; reading from memory is reading data already stored in the memory.**

The controlling device in a microprocessor system is the microprocessor.

In our case the controlling device is the FSM.

If you do not at this stage understand what is happening with the memory device may be you could read a book on memory devices which contains this information before continuing further, or simply re-read this section of the book.

**Most memory is controlled by a microprocessor, which has its own CS, W, and R signals. A state machine inside the microprocessor controls these. You are making your own state machine to do this.**

**The FSM must have an input signal to select either a memory read or a memory write.**

**Consider the use of an input signal W/R which when logic 1 will allow a write, and when logic 0 will allow a read to take place. However, in this example we are using separate R and W signals. During my time working with microprocessors the Intel range used separate /W and /R signals, while the Motorola processors used a single W/R signal. Other microprocessors followed their own arrangement.**

Figure 2.8 illustrates the timing for a typical memory chip. Note that in a practical system you would not want to hold both W and R active at the same time.

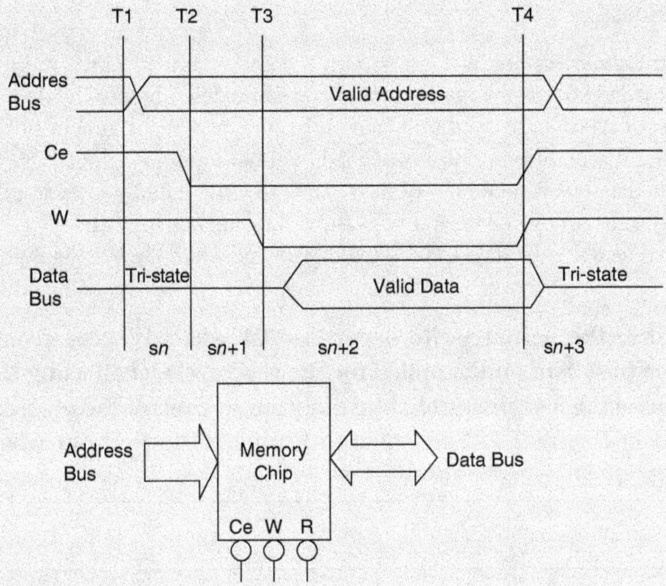


**Figure 2.8** Timing of the control of a memory device.

In Figures 2.8 and 2.9 we do not show the read signal R.

Consider Figure 2.9. The waveforms T1, T2, T3, and T4 are used to help with the correct timing of the memory device. They are often generated within a microprocessor or microcontroller.

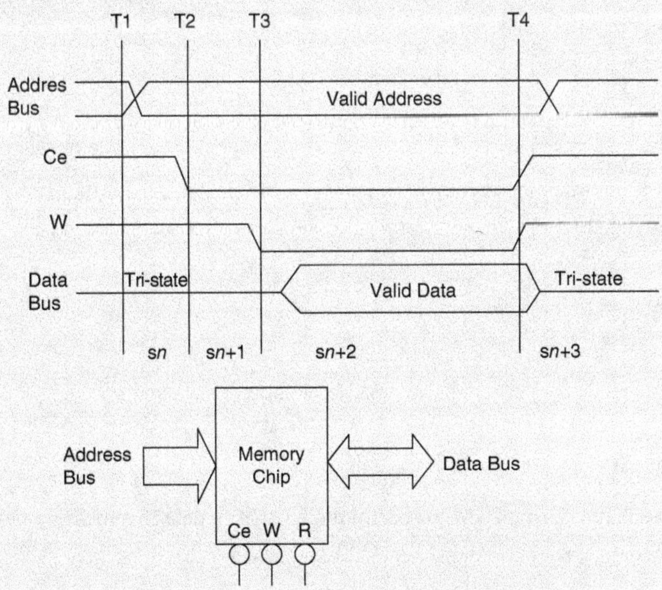


**Figure 2.9** T1 to T4 timing stages.

The periods:

- T1 is the address set-up time.
- T2 is provided so that the memory chip can settle down before the data bus is taken out of its tri-state condition (remember the data bus is generally kept in an 'open circuit' condition until it is needed to either input or output).
- At T3 the data bus is taken out of its tri-state to write data by lowering the W signal. The rest of the T3 period is to allow the data bus to settle.
- T4 is used to write the data to the memory device when the W signal goes high again.

**Therefore, the actual write occurs in T4 when W goes from 0 to 1. Remember this! The same applied for the read cycle when using R signal.**

**Try producing a segment of state diagram to control the memory device illustrated in Figure 2.7. Then turn to Frame 2.7 to find out whether you got it right.**

## *Frame 2.7*

The segment of state diagram to control the memory is illustrated in Figure 2.10.

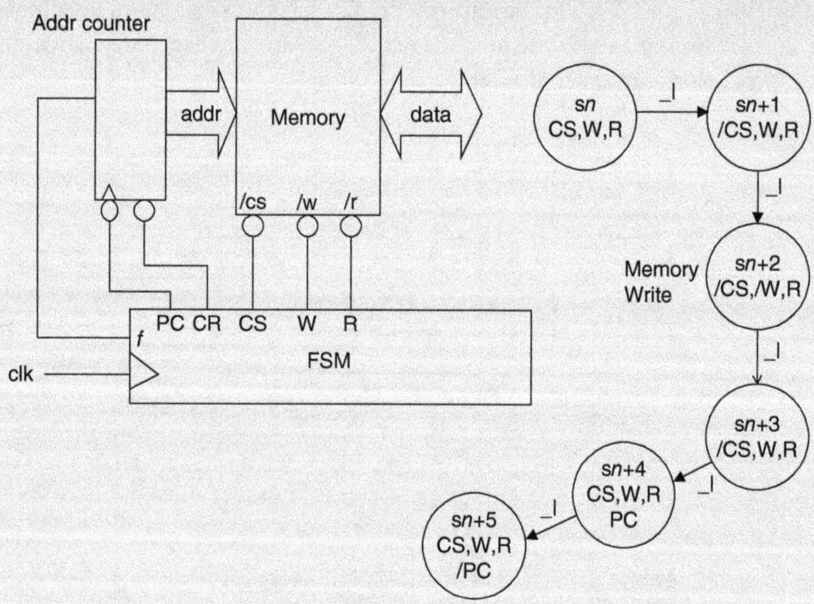


**Figure 2.10** Using FSM to control the writing of data to a memory device.

This segment of the state diagram shows the sequence to write data to a selected memory location. The address counter is used to select the memory location as it is connected to the address bus of the memory. **Note: this is not a complete sequence state diagram.**

In state $sn + 1$ the chip select line CS is lowered to select the memory chip.

In state $sn + 2$ the write line W is lowered to select the data bus.

In state $sn + 3$ W is raised so that the data can be entered to the selected memory location.

In state $sn + 4$ the chip select CS is de-selected.

In state $sn + 4$ the address counter is pulsed (PC = 1) to advance the memory address to the next memory (with the memory chip de-selected of course). Finally, in state $sn + 5$ the PC line is disabled ready for the next memory write cycle.

Note that using the R read is the same except it is the R rather than the W that goes low in $sn + 2$.

Note that initially you need to reset the address counter with CR = 0 so that it selects the first memory location 0. In all other states, CR = 1.

**The state sequence is the same for reading the data in the memory device, except that the read line R is lowered in place of the write line W in $sn + 2$.**

When you read or write to a memory device, the data transaction will occur in the memory element currently accessed by the address bus. You then need to access another memory element by selecting another address. **This is what the address counter in the diagram of Figure 2.6 is being used for.**

In this case each memory element of the memory is selected in sequence by incrementing the binary counter after accessing each memory location.

Turn to **Frame 2.8**.

### *Frame 2.8*

Now you have seen how the individual outside world devices are controlled in the data acquisition system diagram (Figure 2.6) in Frame 2.4.

**Try to produce a state diagram to implement the FSM used to control the system in Figure 2.6.**

Some hints: the FSM is to wait in state s0 until it receives an interrupt signal from the remote end device over the *int* signal line. Look at Figure 2.6.

When this occurs, the FSM is to:

- Obtain a sample of data.
- Perform an analogue to digital conversion.

- Store the converted value into the memory device.
- Increment the counter to point to the next available memory location.

The FSM should keep on doing this until the memory device is full. The FSM will know this when the *f* input (from the counter) goes high. **Note that the *int* signal and ACK works like a handshake between the FSM and the control device.**

At this point, the FSM is to send an acknowledge signal to the remote end device using the ACK signal line, then once the signal line *int* is low (remember it was asserted high at the beginning of the sequence) returns to state s0 ready for another cycle.

Also at this point the memory will be full of new data it has converted from analogue to digital values. The program could be developed further to read the memory contents to some other location (but not just yet).

**Now complete the state diagram and then turn to Frame 2.9.**

## *Frame 2.9*

The complete state diagram is illustrated in Figure 2.11. Don't worry if your state diagram does not look quite like this one. There is more than one way of drawing a state diagram and there is no one correct solution. However, this diagram is very concise.

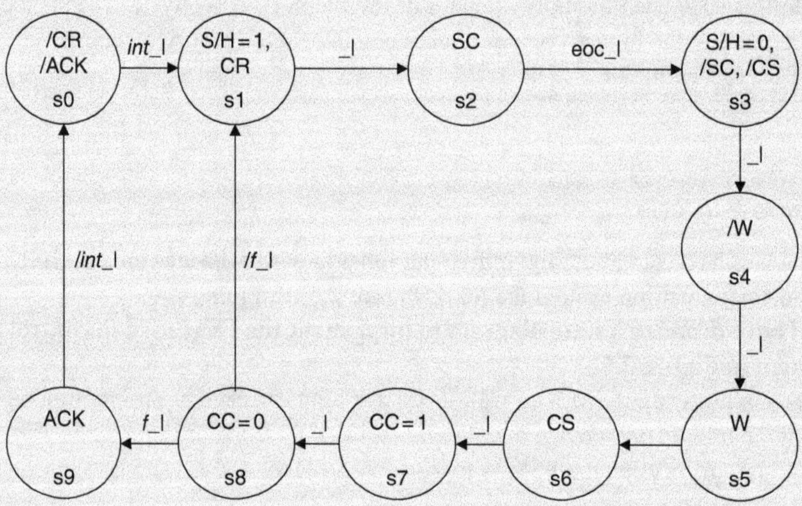


**Figure 2.11** State diagram of the DAS.

The first thing to note is that the counter reset line CR used to hold the memory address counter in the reset condition held active (asserted low) in state s0 (CR = 0). Thereafter it is held in its DIS-ASSERTED state. This **action will clear the address counter to zero so it points to the first memory location on entering the state diagram.**

Not all states show this, but the sequential nature of the state diagram implies it by making s0 state have /CR low. **Once into state s1 through to s9 CR = 1 and is disabled in these states.**

When the *int* input is asserted, the FSM moves into state s1, removing the reset from the address counter and simultaneously asserting the sample and hold amplifier (S/H = 1).

On the next clock pulse, the FSM moves to state s2 where (with the S/H still set), the start conversion signal of the ADC is asserted (SC = 1). At this point, the FSM waits for the end of conversion signal (eoc = 1) then, on the next clock pulse, moves to state s3 where the S/H signal is dis-asserted (S/H = 0) and the chip select line of the memory device is asserted (CS = 0). On the next clock pulse the FSM moves to s4 where the memory device write signal line is asserted low (W = 0).

Note that in s3 the signal S/H is dropped since it is no longer asserted and is assumed to be dis-asserted low. The next clock pulse will move the FSM from s4 into state s5, where W is dis-asserted high.

**In this case W was taken high in s5, then CS taken high in s6, but both CS and W could have been taken high in s5 at the same time (this is often done in most microprocessor systems).**

**Thus, it is the dis-asserting of W to high that causes the data at the output of the ADC to be written into the memory device. It is common to raise both CS and W in microprocessors and microcontrollers (often at the same time).**

In s9 the FSM raises ACK high to indicate to the external device (could be a microcontroller) that the FSM has completed filling the memory. In this way *int* and ACK act like a handshake between the FSM and the external device.

The FSM now moves into state s7 on the next clock pulse.

Turn to **Frame 2.10**.

### *Frame 2.10*

At this stage in the FSM cycle, the FSM is in state s7, with the CC line high.

The 0 to 1 transition on this signal line, caused by the FSM moving from state s6 to s7 is to the address counter to increment with CC = 1. Note that in state s6 we have dropped the CS and W signals since they are now dis-asserted and no longer used.

In state s8 the FSM can move either to state s1, if the signal *f* is dis-asserted low, and hence repeat the sequence s1 through to s7 again. Alternatively, if signal *f* is asserted high, the FSM can move from s8 into s9. In s8 the CC signal is returned to its low value.

The signal *f* is used to indicate whether the address counter has reached the end of the memory. If it has not reached the end of the memory ($f = 0$) then another cycle is started; otherwise, the FSM moves into state s9 to wait for the signal *int* going low.

By waiting for *int* to go low, you are assured that the FSM will be ready for the next low to high transition on *int* to start the next cycle of operations. In this system the *int* signal is controlling the operation of writing all data to the memory. This signal may come from a microprocessor/microcontroller, or from another FSM device.

The foregoing example is quite a complex one and shows how an FSM can be used to control a complex sequence to control a number of outside world devices.

Don't be too disappointed if you did not get this quite right. It takes time to develop the skills necessary to implement complex FSMs, but at least you have now seen how it is done.

**The state diagram acts as a very useful tool to allow you to see the required sequence.**

As you can see from this previous example the development of a state diagram is largely an intuitive process. However, by applying the techniques discussed using this book you can become experienced in developing your own state diagrams.

Do not worry yet about how you will convert this into a physical circuit for implementing the design. It turns out that this is largely an automated process that you can learn. Indeed, you can write a computer program to do this for you.[1] One question you may be asking is how do you generate the *f* address counter full signal from the counter in Figure 2.6 and Figure 2.11?

**This is quite simple. Consider a 4-bit counter with address values 0000 to 1111.**

**Add an extra flip-flip to the counter as 00000 where the most significant bit is the *f* bit (0000f).**

**Now addresses 00000 to 11110 will count from 0000 to 1111. An additional clock pulse will move the counter from 11110 to 00001 where the *f* bit is seen**

[1] I have a computer program in Turbo C 2 that works in Windows XL that allows the user to use a state machine to produce state equations and Verilog HDL as well as VHDL for a state diagram using *T and D* type flip-flops.

**to become 1 (address counter full). Note in the Logisim folder there is an example showing how this works.**

Some of the ideas put forward in this text are:

- Using dummy states to obtain unit distance coding (Frame 1.13).
- Using an external timer to provide a wait state (Frame 2.1).
- Using the FSM to control outside world devices like ADC and memory devices.
- The memory full signal *f*.

By the way, if you are still puzzled about how you get from a state diagram to a functional circuit, don't worry: the steps from state diagram to final circuit are very 'mechanical' and you will pick this up later on in the book (Chapter 3), if you have not already done so at a previous level.

Later, you will be provided with an example using Logisim to implement FSM. See Appendix A2 on how to use Logisim and Logisim examples.

There are still a number of points we need to consider, so move on to **Frame 2.11**.

## *Frame 2.11*

Consider the block diagram of the two *D* pulse detector FSM shown in Figure 2.12a.

In this particular FSM, a single clock pulse is required at the output *P* whenever the *d* input is asserted high twice.

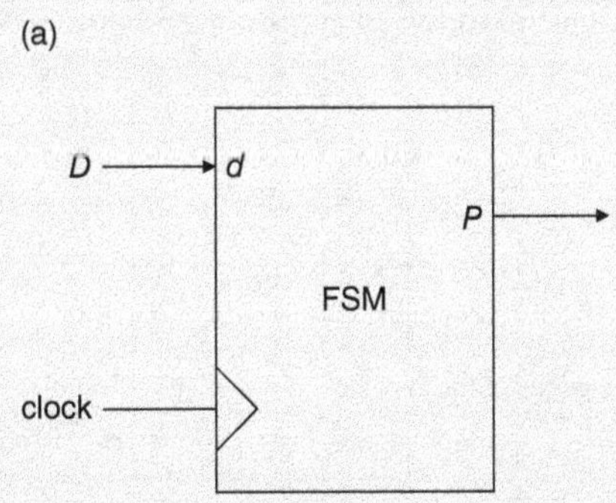


**Figure 2.12a** Block diagram for the two *D* detector FSM.

In Figure 2.12b, we see the state diagram to perform this operation.

Turn to **Frame 2.12**.

*Frame 2.12*

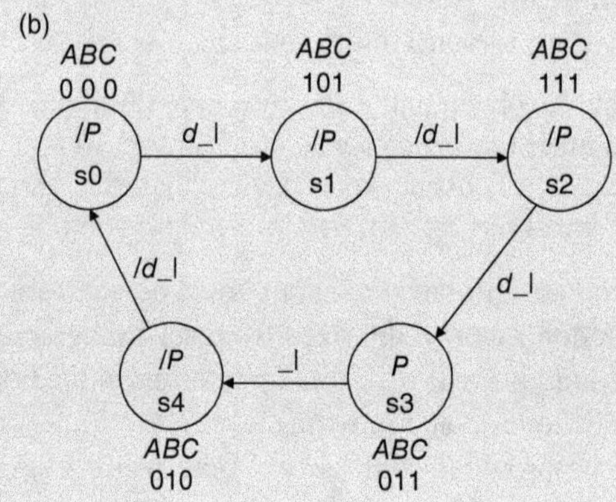


**Figure 2.12b** State diagram to detect *d* twice.

Here the state diagram is for monitoring *d* twice:

- The monitoring of *d* is very important in this example since the FSM must determine when *d* has happened twice.
- To do this you need to monitor *d* going high, then monitor *d* going low. At this point the FSM has seen *d* go high then low once.
- Then monitor *d* going high again, followed by monitoring *d* going low again. At this point the FSM has seen *d* go high then low twice before asserting output *P* in s3.

Now we look at the *P* output producing a 1010 pulse sequence.
The block diagram is illustrated in Figure 2.13.

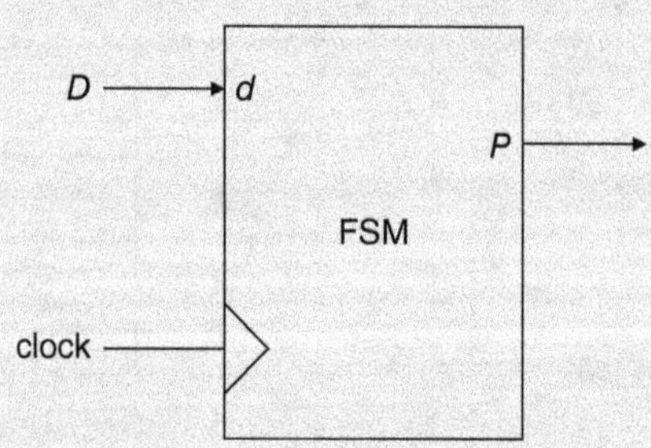


**Figure 2.13** Block diagram to detect a 1010 sequence.

In this block diagram you can see that the FSM monitors the *d* input going high, then low, then monitors *d* going high then low again.

In Figure 2.14 you can see the waveforms for this detector. From these waveforms the *d* 1010 sequence is detected on the rising edge of the clock waveform.

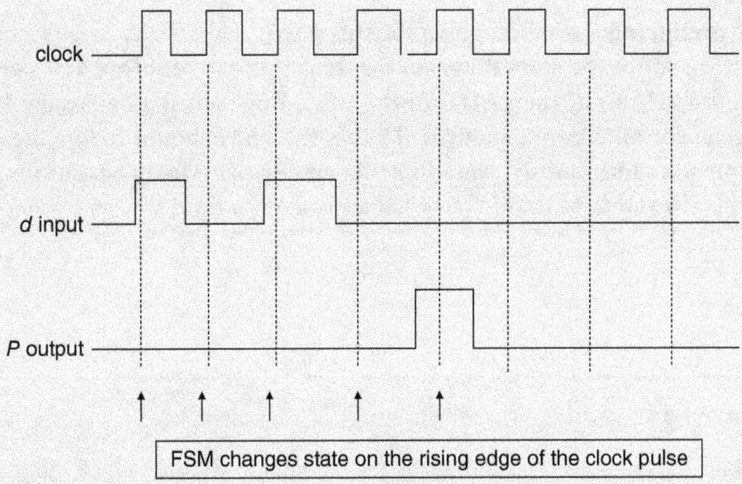


**Figure 2.14** Timing diagram showing the 1010 *d* input and its effect on the *P* output.

Turn to **Frame 2.13**.

## *Frame 2.13*

Now consider the detection of a different sequence, the 110 sequence. The block diagram for this is shown in Figure 2.15.

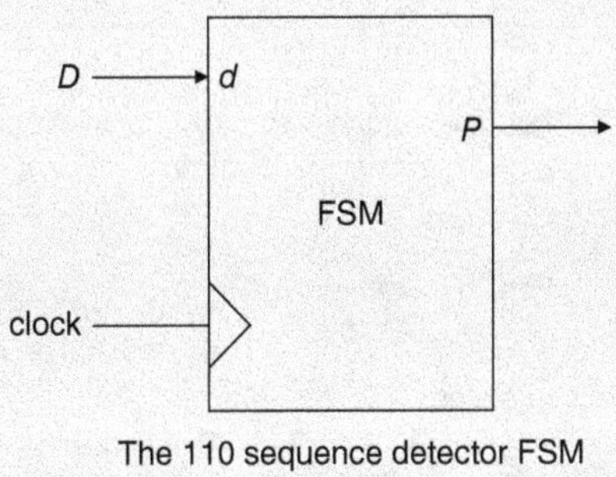


**Figure 2.15** A 110 sequence detector FSM.

Figure 2.15 shows a 110 detector where *d* is sampled.

The FSM changes state on the 0 to 1 clock transition.

Note that in the state diagram of Figure 2.16 you have *d* following a 110 sequence.

**Try producing a state diagram for this FSM.**

**Hint: produce the state diagram for detecting the required 110 sequence for Figure 2.15 first, then ADD to this state diagram the necessary transitions to cover all other sequences. That is the FSM should follow the three-bit sequence continuously. Note this diagram uses a Mealy output for *P*.**

Turn to **Frame 2.14** to find out whether you got it right.

### *Frame 2.14*

The state diagram for the 110 sequence detector for the block diagram of Figure 2.15 is shown in Figure 2.16.

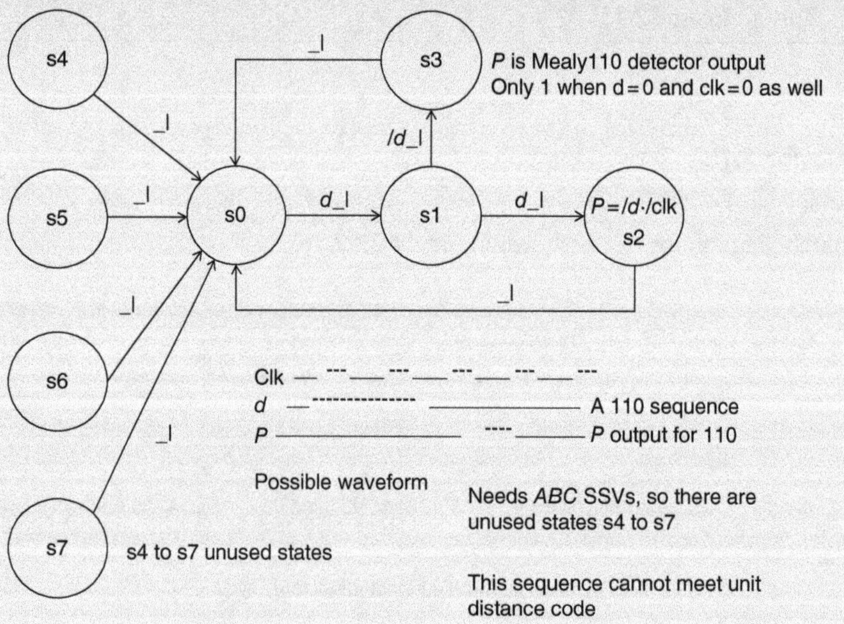


**Figure 2.16** A possible 110 detector attempt.

The sequence s0, s1, and s2 detects the 110 sequence (note that you need to assume that the *d* input is a succession of pulses with the waveform shown in Figure 2.16).

Note that in s2 Mealy output $P = /d \cdot /\text{clk}$ will set $P = 1$ when the clock is low. Hence 101, 100 etc., will **not** set $P = 1$.

The loop back terms catering for some sequences and returns the FSM back to state s0, **note they may not be unit distance.** You cannot use dummy states in this example.

However, other sequences need to be considered ensuring that after three clock pulses the FSM returns to s0. This needs to be considered as illustrated in Figure 2.16 so as to keep to the three-bit sequence.

Consider a possible state sequence:

| Sequence | | | States covered | | | | |
|---|---|---|---|---|---|---|---|
| 0 | 0 | 0 | s0, | s0, | s0 | | |
| 1 | 0 | 0 | s0, | s1, | s3, | s0 | |
| 0 | 1 | 0 | s0, | s1, | s3, | s0 | |
| 1 | 1 | 0 | s0, | s1, | s2, | s0 | $P = /d \cdot /\text{clk} = 1$ when clk is low. |
| 0 | 0 | 1 | s0, | s0, | s1 | | |
| 1 | 0 | 1 | s3, | s0, | s1 | | |
| 0 | 1 | 1 | s3, | s0, | s1 | | |
| 1 | 1 | 1 | s0, | s1, | s2, | s0 | $P = /d \cdot /\text{clk} = 0$ since $d = 1$. |
| 0 | 0 | 0 | the sequence continues.. | | | | |

On reflection, it may be better to use a shift register to capture each three-bit data pattern and compare it to the required 110 pattern.

Another way of doing this that allows any sized pattern to be detected as it appears at the input of the FSM is shown in Chapter 5.

Another problem is that Figure 2.16 uses four states. The SSVs allow eight states; therefore, there are four states that are not used. What would happen if the FSM were to fall into one of these unused states?

The answer to this question is that the FSM would not be able to get out of the unused state and the FSM would 'hang', i.e. cannot move from the unused state.

To avoid this calamity, it is common to direct all unused states back to state s0 so that the FSM can recover from this misadventure. This is shown in Figure 2.16, where states s4 to s7 are directed back to state s0 on the next clock pulse.

**Remember, if you are using *D* type flip flops (or DFFs) the FSM will not hang in an unused state, but simply fall back to s0. Using other types of flip-flops (e.g. JKFFs, TFFs, SRFFs), it will hang.**

Chapter 5 indicates how the one hot method can be used to avoid the unit distance problem when using *D* type flip-flops with a number of techniques.

You might like to take another well-earned break now, before moving on to Chapter 3 on FSM synthesis.

For more details, see Minns (1995).

# 3 Introduction to FSM Synthesis

## 3.1 INTRODUCTION

At this point in the course, you have covered the main requirements to design your own finite state machine (FSM). There are refinements and these are discussed in this chapter, and also later chapters. The process of FSM synthesis is a mechanized activity and can be performed at a number of levels. However, to start, you need to look at the different types of flip-flops used in the design of FSMs.

It is possible to develop clocked FSMs using the following flip-flop types:

- *D* type flip-flops (DFFs)
- *T* type flip-flops (TFFs)
- *JK* type flip-flops (JKFFs)

using a high-level hardware descriptive language (HDL) such as VHDL or Verilog HDL.

Such an HDL can be used to enter the state diagram directly into the computer. This can then be converted by the HDL into actual hardware on a programmable chip such as a programmable logic device (PLD) or a field programmable gate array (FPGA). These are very commonplace.

The HDL can then be used to produce a design based upon any of the above flip-flop types using one of a number of technologies.

*Digital System Design using FSMs: A Practical Learning Approach*, First Edition. Peter D. Minns.

Companion website: www.wiley.com/go/minns/digitalsystemdesign

One approach is referred to as the **behavioural method** and can be studied in Chapter 8 and Appendix D of Minns and Elliott (2008), which **will also allow you to do lots of other things with FSM systems.**

An alternative approach would be to create a state diagram and obtain the Boolean equations for your design that can be used to convert to Verilog HDL. **This is known as the 'data method' and makes use of equations to describe the circuit needed to make up the FSM. This chapter describes this method.**

Another way would be to take the information from the state diagram and convert it directly into a logic circuit and enter this information into an HDL language. This is known as the **structure method** and breaks the design down into the logic gates and flip-flops needed to make up the full circuit.

Each method has its own advantages and this detail will be revealed to you as you gain access to further information.

Using either the direct synthesis approach, or an HDL, the final design can be implemented using:

- Discrete TTL or CMOS components – build up the circuit with TTL or CMOS gates.
- Programmable logic devices and field programmable gate array (PLD/FPGA) VLSI (very large silicon integrated) chip.
- FPGAs are used quite a lot with the one hot method in Chapter 5.

Most technologies support *D* and *T* type flip-flops, and in practice these devices are used a lot in industrial designs; therefore, we will look at the way in which *T* flip-flops and *D* flip-flops can be used in a design.

We will start by looking at the implementation of FSMs using *T* type flip-flops, and then move on to look at designs using *D* type flip-flops.

Why *T* and *D* type flip-flops?

These are the most common types of flip-flop used today. The reason is that the *T* type can be easily implemented from a *D* type, and the *D* type requires only six gates, (compared with the *JK* type which requires about 10 gates). This means that *D* type occupy less chip area than *JK* types. Another reason is that the *D* type flip-flop is more stable than the *JK* flip-flop (more on this later).

The alternative arrangement for Verilog HDL using the behavioural method is described in Appendix A4.

Now on with Chapter 3: synthesizing FSM systems.

## Frame 3.1 The T Type Flip-Flop

A *T* type flip-flop can be implemented with a standard *D* type flip-flop as illustrated in Figure 3.1 using UK logic symbols (Appendix A1).

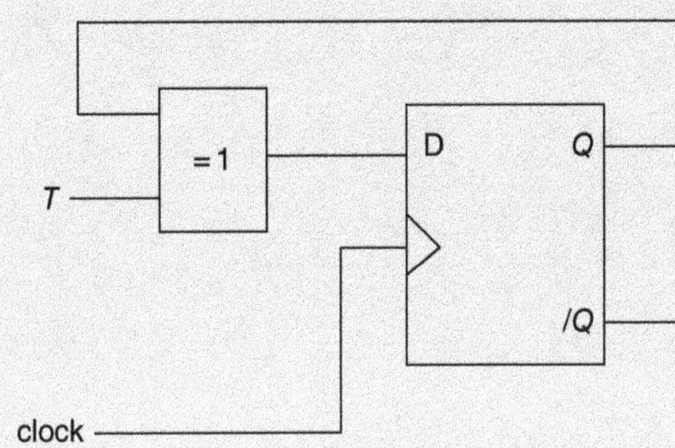


**Figure 3.1** *T* type flip-flop using exclusive OR and a *D* type flip-flop.

The table for this is:

Steering table

| *T* | Q*n* | Q*n* + 1 | |
|---|---|---|---|
| 0 | 0 | 0 | No change |
| 1 | 0 | 1 | Toggle high |
| 1 | 1 | 0 | Toggle low |
| 0 | 1 | 1 | No change |

As you can see from the diagram, the *T* flip-flop is implemented by using a *D* flip-flop with an exclusive OR gate (see Appendix A1). The table below the flip-flop shows its characteristics in a steering table.

In this table Q*n* is the **present** state output of the flip-flop (prior to a clock pulse) whilst Q*n* + 1 is the **next** state output of the flip-flop (after the clock pulse).

The table shows that the flip-flop will change state on each clock pulse provided that the *T* input is high. But if the *T* input is low, then the flip-flop will not change state.

We can therefore use the *T* input to control the flip-flop since whenever we wish the flip-flop to change state we simply set the *T* input high; otherwise, it is held low.

Please turn to **Frame 3.2**.

### Frame 3.2 The T Flip-Flop Example

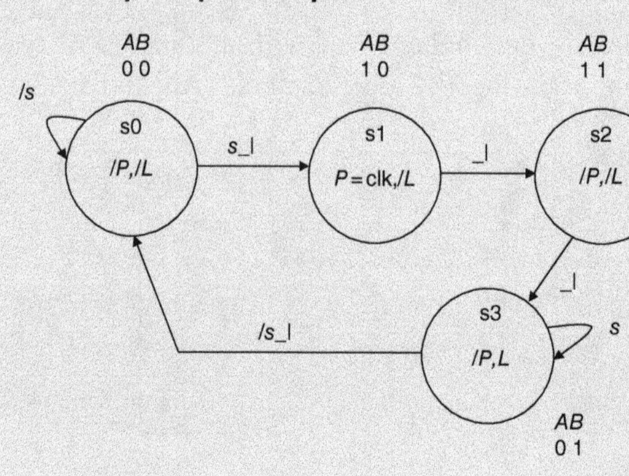


**Figure 3.2** State diagram for the single-pulse generator with memory having Mealy *P* in s1.

Consider the simple single-pulse example of Figure 3.2.

We follow the state transitions for secondary state variable *A*, and wherever there is a 0 to 1 or a 1 to 0 transition in *A* we write down the state term.

So in Figure 3.2 we have a 0 to 1 transition in *A* between states s0 and s1 and a 1 to 0 transition between states s2 and s3. Therefore, we write down:

$$A \cdot T = \mathrm{s0} \cdot \mathrm{s} + s2.$$

This equation defines the logic that will be connected to the *T* input of the flip-flop.

Whenever the FSM is in state s0 AND input *s* is high, the *T* input of flip-flop *A* will be high. As we now know, whenever the *T* input is high the flip-flop will toggle. Since in state s0 both flip-flops are reset, in state s0, when *s* goes to logic 1, the next clock pulse will cause the flip-flop *A* to go from 0 to 1.

In state s1 the *T* input to flip-flop *A* will be at logic 1 since the *T* input will hold this input high in this state. Therefore, in state s1 the flip-flop *A* will hold its *T* input low and not toggle with the clock pulse. When the FSM reaches state s2, the *T* input will go high again and the next clock pulse will cause the flip-flop to toggle back to its reset state as intended. Note that in state s3 the *T* input to flip-flop *A* will be low again so the flip-flop will not toggle on the next clock pulse in state s3.

**Note that the equation for $A \cdot T$ uses the present state conditions to set the *T* line high. This is necessary in order to ensure that the flip-flop will toggle on the next clock pulse.**

**Try producing the equation for the input logic on flip-flop $B{\cdot}T$ input.**

$$0 > 1$$

$$B \cdot T =$$

$$1 > 0$$

Since you are looking for 0 to 1 transitions and 1 to 0 transitions in *B*.

Please turn to **Frame 3.3** to see whether you got it right.

## Frame 3.3

The equation for the $T$ input of flip-flop $B$ is:

$$\begin{gathered}0>1\\ B\cdot T=\text{s1}+s3\cdot/\text{s}.\\ 1>0\end{gathered}$$

Since in state s1 we need the $B \cdot T$ input to be logic 1 so that, on the next clock pulse, the flip-flop will change from a reset state to a set state. Note that there is no outside world input condition between states s1 and s2 and between states s2 and s3.

The second term s3·/$s$ will cause the $B$ flip-flop to toggle from its set state to its reset state in state s3 when the outside world input $s = 0$ on the next clock pulse. Hence the /$s$ is used.

In summary **look for the 0 to 1 or 1 to 0 transitions in each flip-flop**.

Figure 3.3 is another example.

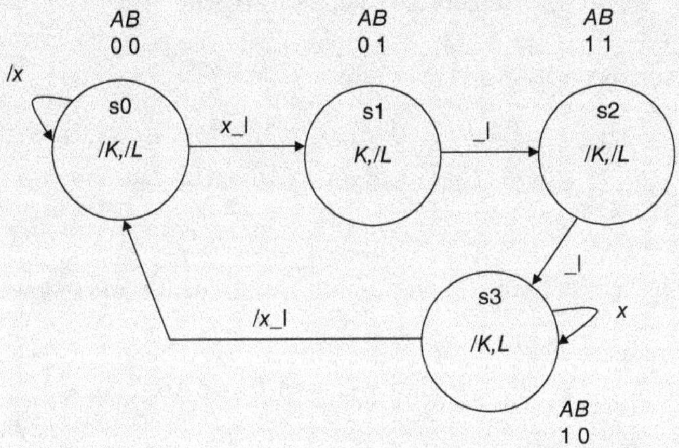


**Figure 3.3** A different state diagram using $T$ type flip-flops.

Now try the Figure 3.3 example for yourself. Produce the output equations as well for this design.

$$\begin{gathered}0>1\\ A\cdot T=\\ 1>0\\ 0>1\\ B\cdot T=\\ 1>0\end{gathered}$$

Turn to **Frame 3.4**.

## Frame 3.4

The state diagram is repeated in Figure 3.4 for convenience.

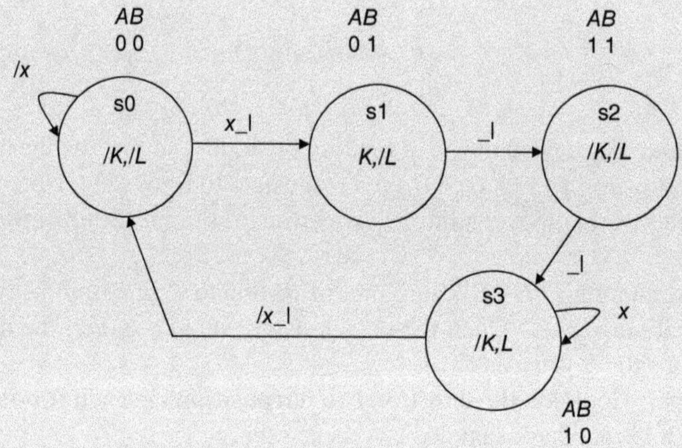


**Figure 3.4** Repeated diagram.

The equations for $A \cdot T$ and $B \cdot T$ are:

$$0 > 1$$
$$A \cdot T = \mathrm{s1} + \mathrm{s3} \cdot /x = /AB + A/B \cdot /x.$$
$$1 > 0$$

Note I have replaced the '*' with '·' for AND. Which is often done.

$$0 > 1$$
$$B \cdot T = \mathrm{s0} \cdot x + \mathrm{s2} = /A/B \cdot x + AB.$$
$$1 > 0$$

The outside world outputs are:

$$K = \mathrm{s1} = /A \cdot B.$$
$$L = \mathrm{s3} = A \cdot / B.$$

Since $K$ is high in state s1 and $L$ is high in state s3.
If you made any mistakes go back and read Fames 3.1–3.3.
Please turn to **Fame 3.5**.

### *Frame 3.5*

Attempt the following examples.

**Produce the *T* flip-flop equations, and output equations, for each of the following state diagrams. If you are not sure, re-read Chapter 3 up to here.**

**State diagram in Frame 2.3, Figure 2.4 using the following Secondary State Variables:**

#### *Time state problem*

| | *AB* |
|---|---|
| s0 | 00 |
| s1 | 10 |
| s2 | 11 |
| s3 | 01 |

**State diagram in Frame 2.7, Figure 2.10 using the following SSVs:**

#### *Memory control problem*

| | *ABC* |
|---|---|
| s0 | 000 |
| s1 | 100 |
| s2 | 110 |
| s3 | 111 |
| s4 | 011 |
| s5 | 001 |

**State diagram in Frame 2.9, Figure 2.11 using the following SSVs:**

#### *Data acquisition problem*

| | *ABCD* |
|---|---|
| s0 | 0000 |
| s1 | 1000 |
| s2 | 1100 |
| s3 | 1110 |
| s4 | 1111 |
| s5 | 0111 |
| s6 | 0011 |
| s7 | 1001 |
| s8 | 0001 |
| s9 | 0101 |

See **Frame 3.6** for the solution to these problems (but try them first, don't cheat!).

### *Frame 3.6*

Answers to the problems in Frame 3.5 are:

**For the state diagram of Frame 2.3, Figure 2.4**

#### *Timer state problem*

| | *AB* | answers: |
|---|---|---|
| s0 | 00 | $A \cdot T$ = s0 · st + s2 · /to = /*A*/*B* · st + AB·/to |
| s1 | 10 | $B \cdot T$ = s1 + s3 · /st = *A*/*B* + /*AB* · /st |
| s2 | 11 | |
| s3 | 01 | *P* = s1 + s2 = *A*, Ts (active low) = /s1 = /(*A*/*B*)<br>Ts (active high) = s0 + s2 + s3 = /*A*/*B* + AB + /*AB* = /*A* + *B*.<br>**Use aux rule.** |

**For the state diagram of Frame 2.7, Figure 2.10. This is an incomplete state diagram.**

#### *Memory control problem*

| | *ABC* | answers: |
|---|---|---|
| s0 | 000 | $A \cdot T$ = s0 + s4 = /*A*/*B*/C + /*ABC* |
| s1 | 100 | $B \cdot T$ = s1 + s4 = *A*/*B*/C + /*ABC* |
| s2 | 110 | $C \cdot T$ = s2 = *AB*/*C* (not enough terms to complete) |
| s3 | 111 | |
| s4 | 011 | /CS = /(s1 + s2 + s3) = /(*A*/C + *AB*) from /(*A*/*B*/*C* + *AB*/*C* + *ABC*) |

Here active low so invert all states used.

| | | |
|---|---|---|
| s5 | 001 | R = 1 not used here (active low so always 1 in this diagram).<br>/W = /s2 = /(*AB*/*C*) used here note active low. |

**For the state diagram of Frame 2.9, Figure 2.11**

#### *Data acquisition problem*

| | *ABCD* | Answers |
|---|---|---|
| s0 | 0000 | $A \cdot T$ = s0·int + s5 + s6 + s8 · *f*.<br>= /*A*/*B*/*C*/*D* · int + /*ABCD* + /*A*/*BCD* + /*A*/*B*/*CD* · *f*. |
| s1 | 1000 | |
| s2 | 1100 | |
| s3 | 1110 | $B \cdot T$ = s1 + s5 + s8 · *f* = *A*/*B*/*C*/*D* + /*ABCD* + /*A*/*B*/*CD* · *f*. |
| s4 | 1111 | |
| s5 | 0111 | $C \cdot T$ = s2 · eoc + s6 = *AB*/*C*/*D* · eoc + /*A*/*BCD*. |
| s6 | 0011 | |
| s7 | 1001 | $D \cdot T$ = s3 + s8 · /*f* + s9 · int. |
| s8 | 0001 = reader to do. | |
| s9 | 0101 | /R = 1 active low output not used.<br>S/H = s1 + s2 = *A*/*C*/*D*.<br>SC = s2 = *AB*/*C*/*D*.<br>/CS = /(s3 + s4 + s5) = /(*ABC* + *BCD*) active low output.<br>/W = s4 = /(*ABCD*) active low output. |

You could try using a unit distance code for this one – just for fun!

We have now covered the complete cycle of designing an FSM and synthesizing it using *T* type flip-flops.

*T* type flip-flops, as we have already seen in Frame 3.1, can be implemented from a basic *D* type flip-flop using an exclusive OR gate. A lot of PLDs support both the *D* and *T* type flip-flop so our FSM designs can be implemented using these PLDs.

***T* type flip-flops look for 0 to 1 and 1 to 0 transitions.**

A popular PLD that supports the *T* type flip-flop is the EP910. This has 48 macro cells (48 flip-flops), which can be programmed to be either *D* type or *T* type flip-flops.

However, most PLD devices support *D* type flip-flops, in particular, the cheaper PLD devices. Therefore, it is worthwhile considering how *D* type flip-flops can be used to synthesize FSMs.

As it turns out, using *D* type flip-flops to synthesize FSMs requires a little thought so you will now spend a few frames looking at the techniques required in order to make use of *D* type flip-flops in the design of FSMs. This will be time well spent since it opens up a large number of potential programmable devices, which we can use to design our FSMs.

Turn to **Frame 3.7**.

## *Frame 3.7*

Synthesizing FSM using a *D* type flip-flop is shown in Figure 3.5.

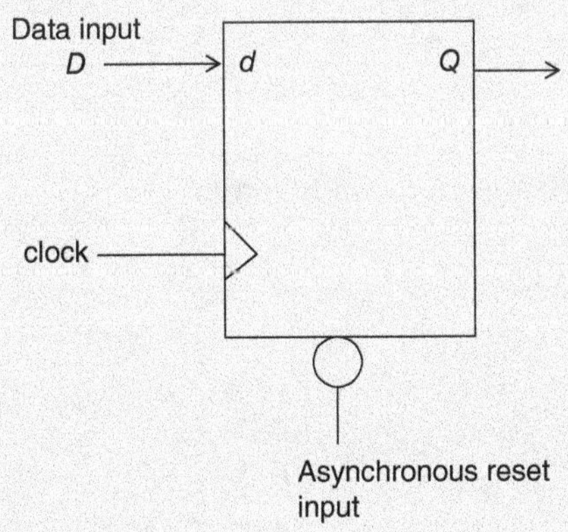


Steering Table

| Qn | Qn+1 | D |
|---|---|---|
| 0 | 0 | 0 |
| 0 | 1 | 1 |
| 1 | 0 | 0 |
| 1 | 1 | 1 |

**Figure 3.5** *D* type flip-flop with its characteristics.

Consider the basic *D* flip-flop device.

The *D* type flip-flop has a single data input (apart from the clock input).

- **The data line must be asserted high before the clock pulse for the *Q* output to be clocked high by the 0 to 1 transition of the clock.**

- **Also, for the $Q$ output to remain high, the $d$ input must be held high so that subsequent clock pulses will clock the 1 on the $d$ input into the $Q$ output.**

Those two bullet points are very important, and you should bear them in mind when you are using $D$ type flip-flops.

Consider the following waveforms applied to a $D$ flip-flop (Figure 3.6).

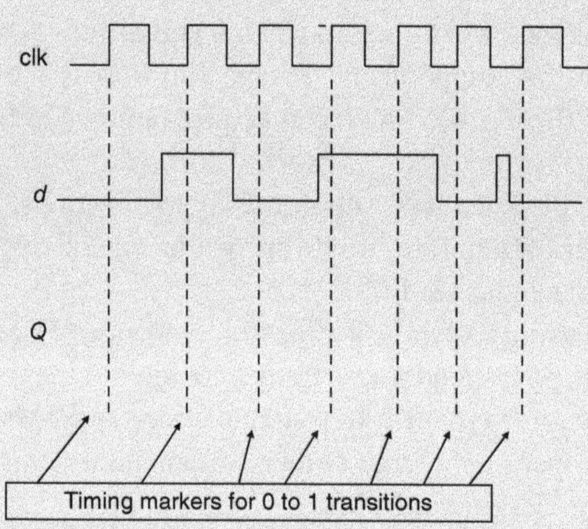


**Figure 3.6** Timing showing behaviour of $D$ type flip-flop.

**Complete the timing diagram for the output $Q$.** Hint: study the content of this frame – you will find the steering table in this frame useful.

Turn to **Frame 3.8** when you have completed the diagram to find out whether you got it right.

### *Frame 3.8*

The completed timing diagram is illustrated in Figure 3.7.

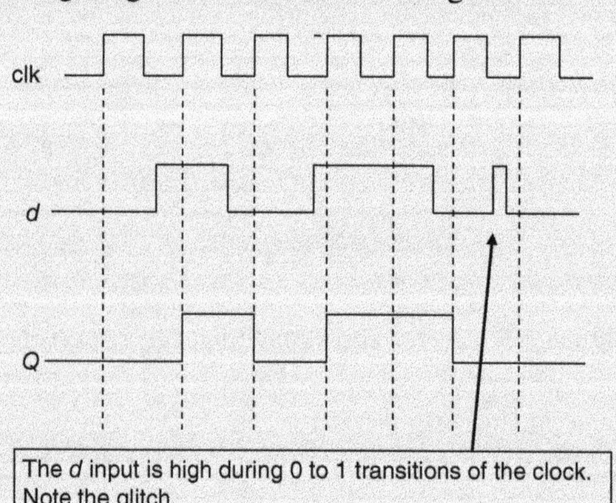


**Figure 3.7** The effect of a glitch.

The trick here is to look at the value of the *d* line whenever the clock input makes a transition from 0 to 1, and whatever the *d* line logic level is, the *Q* output level becomes.

**This is because the *D* type flip-flop sets its *Q* output to whatever the *d* input logic level is at the time of the clock 0 to 1 transition.**

Note that in the timing waveform the *d* input is held high over two clock periods. This means that the *Q* output will also be held high over the same two clock periods.

Also note the point on the timing waveform when the *d* input makes a transition to logic 1 for a brief period (between clock pulses). The flip-flop is unable to see this transition of the *d* input so the flip-flop is unable to respond.

**The flip-flop can only update its *Q* output at the 0 to 1 transition of the clock input.**

Turn to **Frame 3.9**.

### *Frame 3.9*

Now that we have covered the basics of the *D* flip-flop, consider Figure 3.8.

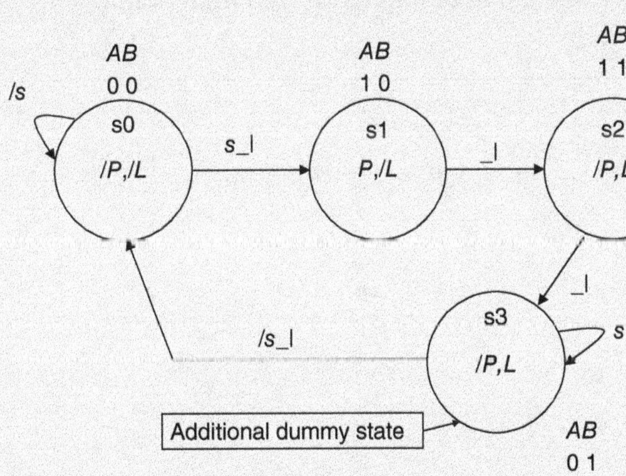


**Figure 3.8** A complete example.

It is of course the single-pulse FSM we first saw in Frame 1.13.

Let us now attempt to synthesize this using *D* type flip-flops.

The equation for flip-flop *A* is:

| 0 > 1 | 0 >1 | 1 > 1 |
|---|---|---|
| $A{\cdot}D =$ | $s0{\cdot}s +$ | $s1 = /A/B{\cdot}s + A/B = /B{\cdot}s + A/B$. Use of aux rule here (Appendix A1). |

Since we need to set the *A* line of flip-flop *A* in state s0 and hold it set over state s1.

Now consider the equation for flip-flop *B*.

| 0 > 1 | 0 > 1 | 1 > 1 | 1 > 0 > negate |
|---|---|---|---|
| *B·D* = | s1 + | s2 + | s3·*s*. |
| 1 > 1 and 0 > 1 >negate | | | |

s1 is the 0 to 1 transition, s2 is a 1 to 1 transition with the s3·*s* term as a 1 to 0 path with negated input.

In state s3 we want the *B* line to be held high if the input *s* is logic 1 since when *s* = 0, the FSM should return to state s0 (by resetting flip-flop *B*). Therefore, whilst *s* = 1, the third term in the equation for *B·D* will be high (not /*s*). When *s* = 0, this term will become logic 0 and the *B* flip-flop will reset, causing the FSM to move to state s0.

**NEGATED input term** (*s* in this case) with s3 to hold the *B* input of the flip-flop high.

**Rule 1: Whenever we have a 1 to 0 transition with an input term present along a transitional line of the state diagram, we AND the state with the NEGATED input.**

**Read this Rule 1 line again to make sure you understand it.**

Turn to **Frame 3.10**.

### *Frame 3.10*

Now consider Figure 3.9.

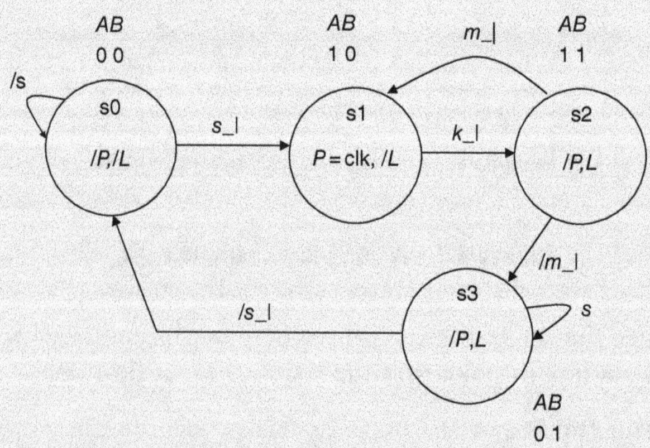


**Figure 3.9** State diagram with single/multiple-pulse FSM.

This is just a modification of the single-pulse FSM, which allows the FSM to produce multiple pulses *P* if input m = 1 and single pulses *P* if m = 0. Assuming k = 1.

So for multiple pulses *P* the state diagram moves between s1 and s2 repeatedly. While for single pulses *P* it moves from s0 to s1 to s2, and back to s0 via s3. Note that in s1 *P* = clk so that *P* can only be 1 if in s1 and the clk is 1 also; this *P* is a Mealy output.

The equation for the *A* flip-flop is:

$$A \cdot D = \text{s0} \cdot \text{s} + \text{s1} + \text{s2} \cdot m = /A/B \cdot \text{s} + A/B + AB \cdot m.$$

The first term is to set the *D* input high for the next clock pulse to set the flip-flop and cause the FSM to move into state s1.

The second term is to hold the flip-flop set between states s1 and s2.

**Note that the input term along the transitional line between states s1 and s2 (*k*) is not present in the second term. This is because it is not needed. The flip-flop is to remain set regardless of the state of the input *k*.**

**Rule 2: A term denoting a transition between two states where the flip-flop remains set does not need to include the input term.**

The third term in the equation for *A·D* is a bit more complicated.

This term is a holding term, for the two-way branch state s2. In state s2, the transitional path between states s2 and s3 is a 1 to 0 transition. Therefore, we need to apply Rule 1 (defined in Frame 3.9). The term is therefore s2·*m*. The other path, between states s2 and s1, is a 1 to 1 transition. **Note that the term denoted by the s2 to s1 transition is not present in the equation for *A·D*. This is because it is not required.** So we have a third rule.

**Rule 3: A two-way branch in which one path is a 1 to 0 transition, and the other a 1 to 1 transition will always produce a term involving the state and the 1 to 0 transition with the input along the 1 to 0 transitional line negated. The 1 to 1 transitional path will be ignored.**

**This is a very important rule and is used whenever we have a 1 to 0 path with an input along the transitional line. This needs to use the 1 to 0 negate rule.**

You need to remember this rule, as it is quite different from what you have done before.

Finally:

$$B \cdot D = \text{s1} \cdot k + \text{s2} \cdot /m + \text{s3} \cdot \text{s} = /AB \cdot k + AB \cdot /m + /AB \cdot s.$$

The term s1·*k* is the 0 to 1 transition, s2·*/m* is the two-way branch where you follow the 1 to 0 transition with a negated input *m*, and s3·*s* is a 1 to 0 negated input transition.

You will be looking at more two-way branches later.

Turn to **Frame 3.11**.

### Frame 3.11 Another Look at the D Type Flip-Flop Rules

To see why these three rules apply, look at the state diagram of Figure 3.9 again (which we reproduce in Figure 3.10).

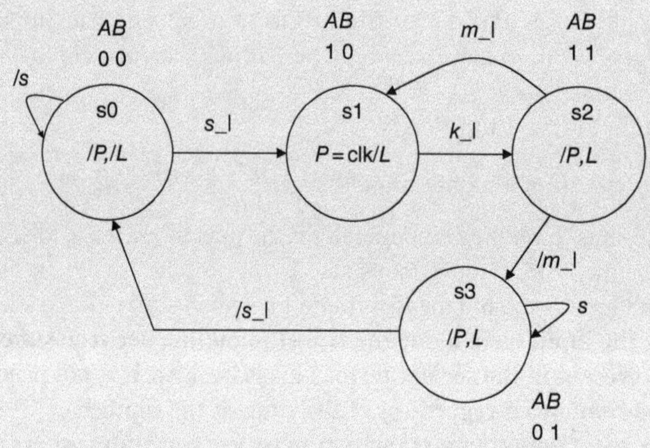


**Figure 3.10** State diagram of Figure 3.9 again.

**The rules you need to remember are:**

- **Rule 1: Whenever we have a 1 to 0 transition with an input term present along a transitional line of the state diagram, we AND the state with the NEGATED input.**
- **Rule 2: A term denoting a transition between two states with an input term in which the flip-flop remains set does not need to include the input term.**
- **Rule 3: A two-way branch in which one path is a 1 to 0 transition, and the other a 1 to 1 transition, will always produce a term involving the state and the 1 to 0 transition with the input along the 1 to 0 transitional line negated. The 1 to 1 transitional path will be ignored. If you include the 1 to 1 transitional path it is wrong – so do not.**

The equations for this example are:

$$A \cdot D = \mathrm{s0} \cdot \mathrm{s} + \mathrm{s1} + \mathrm{s2} \cdot m = /A/B \cdot \mathrm{s} + A/B + AB \cdot m.$$

$$B \cdot D = \mathrm{s1} \cdot k + \mathrm{s2}/m + \mathrm{s3} \cdot \mathrm{s} = /AB \cdot k + AB \cdot /m + /AB \cdot \mathrm{s}.$$

$$P = \mathrm{s1} \cdot \mathrm{clk} = A/B \cdot \mathrm{clk} \text{ a Mealy output.}$$

$$L = \mathrm{s3} = /AB \text{ a Moore output.}$$

**Look at this example and follow the rules used in the equations.**

Turn to **Frame 3.12**.

### *Frame 3.12*

Figure 3.11 illustrates all possible conditions for a two-way branch in a state diagram.

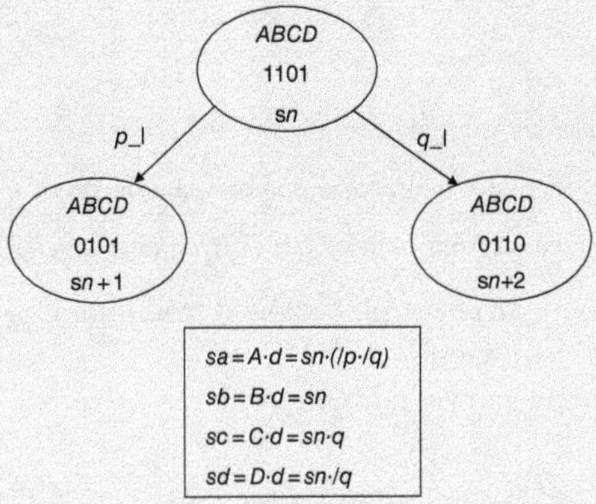


**Figure 3.11** Different *ABCD* transitions in a two-way branch.

In particular you will note that the condition for flip-flop *A* is a two-way branch with both transitions being 1 to 0. The term $A{\cdot}d = sn{\cdot}(/p{\cdot}/q)$ implies that the negation of the input term along each transitional line is being used. Only if both $p = 0$ **and** $q = 0$ will the FSM remain in state *sn*.

**Study the diagram carefully then turn to Frame 3.13.**

### *Frame 3.13*

**Look at the state diagram segment in Figure 3.12.**

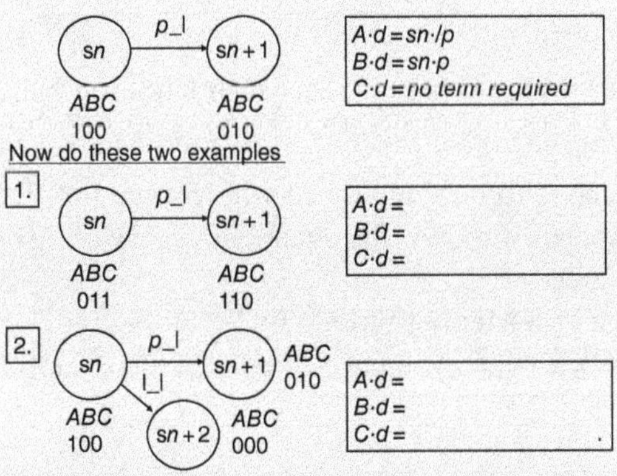


**Figure 3.12** Some more examples to try.

Now do the following examples in a similar manner.

When you have completed these examples, turn to **Frame 3.14**.

### *Frame 3.14*

The answer to the two problems in Frame 3.13 (Figure 3.12) is:

1. $A \cdot d = sn \cdot p.$

   $B \cdot d = sn.$

   $C \cdot d = sn \cdot / p.$

2. $A \cdot d = sn \cdot /p \cdot /l$ since must not leave $sn$ if $p = 0$ and $l = 0$.

   $B \cdot d = sn \cdot p$ since there is a 0 to 1 transition between $sn$ and $sn + 1$.

   $C \cdot d = 0$ $C$ is not turned on from $sn$ to $sn + 1$, or from $sn$ to $sn + 2$.

If you did not get these answers, go back and read Figures 3.11–3.13 and try to find out where you went wrong.

Now try the problem shown in Figure 3.13.

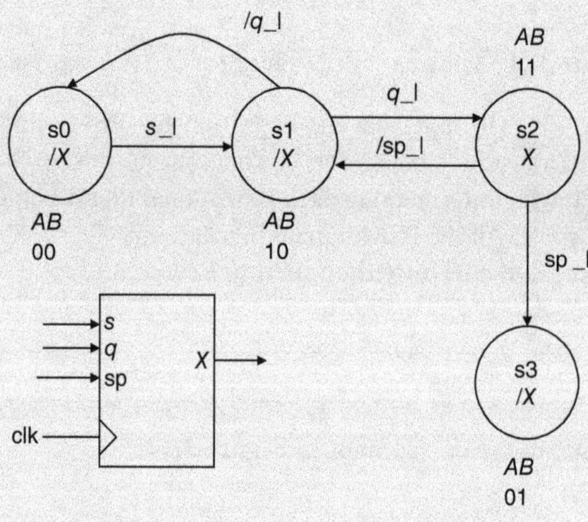


**Figure 3.13** A state diagram with two two-way branches.

This FSM, which is to be synthesized with $D$ type flip-flops, has two states with two-way branches. Produce the equations for the two $D$ flip-flops, as well as the output equation for $X$.

**Is this a Moore or a Mealy FSM? Why?**

Turn to **Frame 3.15** when you have completed this example.

### *Frame 3.15*

The solution to the problem in Frame 3.12 is:

$$A \cdot d = s0 \cdot s + s1 \cdot q + s2 \cdot / sp.$$

Since s0·*s* is a set term, s1·*q* is the 1 to 0 transition between s1 and s0 (with negated /*q* term), and s2·/sp is the 1 to 0 transition between s2 and s3. With a negated sp term.

$$B \cdot d = s1 \cdot q + s2 \cdot sp + s3.$$

Since s1·*q* is the set term, s2·sp is the 1 to 0 transition branching term between s2 and s1, and s3 is the holding term in state s3. Note that there is no way of leaving state s3 apart from using the reset input. **This is, of course, a Moore system.**

The output equation is *X* = s2, which makes the FSM a Moore FSM because the output is a function of the SSVs only.

To provide a way out of s3, and to provide an initialization mechanism, it is wise to provide a reset input to **all** FSMs. In any case you should always provide a means of initializing your FSM to a known state.

Turn to **Frame 3.16**.

### *Frame 3.16 Resetting the Flip-Flops*

If the flip-flops have synchronous reset inputs, this is easily accomplished by a common connection to all reset inputs.

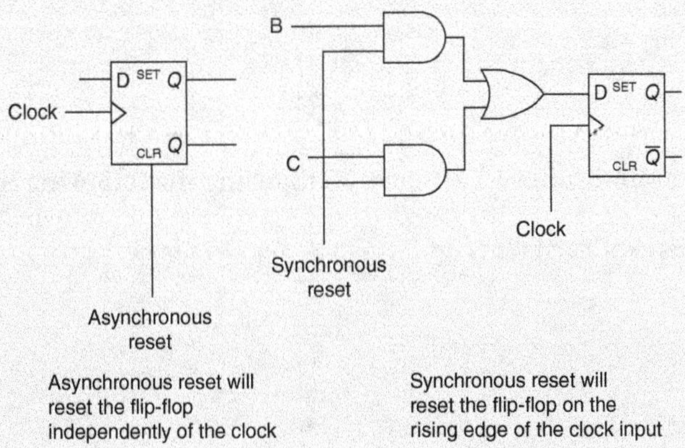


**Figure 3.14** The effect of asynchronous and synchronous resets.

Note that in Figure 3.14 the reset input is assumed to be active low. This means that the synchronous reset will be active when reset = 0. It will be active when reset = 0 and have nothing to do with the clock input.

Further, the equation $D = (B + C)\cdot/\text{reset}$ is used to reset synchronous version.

**If the flip-flops do not have an asynchronous reset input (or any reset input) then a synchronous reset can be provided by 'ANDing' a reset input to each *D* input as shown.**

In the case of the synchronous input, the reset line (which is active low) is normally held high. This enables the logic for each flip-flops *D* input. Lowering the reset line disables the AND gates and results in the *D* inputs also going low. The next clock pulse will therefore cause the flip-flops to reset. Note that flip-flops will reset on the rising edge of the clock pulse for positive edge triggered flip-flops.

Now turn to **Frame 3.17**.

## Frame 3.17

**Try producing the *D* flip-flop equations, and output equations for each of the following state diagrams. If you are not too sure, re-read this chapter before starting to do the problems.State diagram in Frame 2.4 (Figure 2.4) using the following SSVs:**

### Timer problem

| | *AB* |
|---|---|
| s0 | 00 |
| s1 | 10 |
| s2 | 11 |
| s3 | 01 |

**State diagram in Frame 2.7 (Figure 2.10), using the following SSVs:**

### Memory control problem

| | *ABC* |
|---|---|
| s0 | 000 |
| s1 | 100 |
| s2 | 110 |
| s3 | 111 |
| s4 | 011 |
| s5 | 001 |

**State diagram in Frame 2.9 (Figure 2.11), using the following SSVs:**

***Data acquisition problem***

| | *ABCD* |
|---|---|
| s0 | 0000 |
| s1 | 1000 |
| s2 | 1100 |
| s3 | 1110 |
| s4 | 1111 |
| s5 | 0111 |
| s6 | 0011 |
| s7 | 1001 |
| s8 | 0001 |
| s9 | 0101 |

See **Frame 3.18** for the solution to these problems (but try them first – don't cheat!).

## *Frame 3.18 Solutions to the Problems in Frame 3.17*

**State diagram in Frame 2.3 (Figure 2.4) using the following SSVs:**

***Timer problem***

| | AB | Answers |
|---|---|---|
| s0 | 00 | $A \cdot D = \text{s0} \cdot s + \text{s1} + \text{s2} \cdot \text{to} = /B \cdot s + A/B + B \cdot \text{to}$. |
| s1 | 10 | $B \cdot D = \text{s1} + \text{s2} + \text{s3} \cdot s = A + B \cdot s$. |
| s2 | 11 | |
| s3 | 01 | $P = \text{s1} + \text{s2} = A$, Ts = /s1 = /(*A*/*B*) active low output. |

**State diagram in Frame 2.7 (Figure 2.10), using the following SSVs:**

***Memory control problem***

| | *ABC* | Answers |
|---|---|---|
| s0 | 000 | $A \cdot D = \text{s0} + \text{s1} + \text{s2} = /B/C + AB/C$. |
| s1 | 100 | $B \cdot D = \text{s1} + \text{s3} + \text{s2} = A/C + AB$. This is $A/B/C + AB/C + ABC$. |
| s2 | 110 | $C \cdot D = \text{s2} + \text{s3} + \text{s4} = AB/C + ABC + /ABC$. Do not have a term for reset *C*. |
| s3 | 111 | |
| s4 | 011 | /CS = /(s1 + s2 + s3) = /(*A*/*C* + *AB*) active low output. |
| s5 | 001 | R = 1; /W = /s2 = /(*AB*/*C*); PC = s4 = /*ABC* ; |

**State diagram in Frame 2.9 (Figure 2.11), using the following SSVs:**

### *Data acquisition problem*

| *ABCD* | Answers | |
|---|---|---|
| s0 | 0000 | $A{\cdot}D$ = s0·int + s1 + s2 + s3 + s6 + s8·/*f*. |
| s1 | 1000 | $A{\cdot}D$ = /*A*/*B*/*C*/*D*·int + *A*/*B*/*C*/*D* + *AB*/*C*/*D* + *ABC*/*D* + /*A*/*BCD* + /*A*/*B*/*CD*·/*f*. |
| s2 | 1100 | |
| s3 | 1110 | $B{\cdot}D$ = s1 + s2 + s3 + s4 + s8·*f* = *A*/*C*/*D* + *ABC* + /*A*/*B*/*CD*·*f*. |
| s4 | 1111 | |
| s5 | 0111 | $C{\cdot}D$ = s2·eoc + s3 + s4 + s5 = *AB*/*C*/*D*·eoc + *ABC* + *BCD*. |
| s6 | 0011 | |
| s7 | 1001 | $D{\cdot}D$ = s3 + s4 + s5 + s6 + s7 + s8·*f* + s9·*int*. |
| s8 | 0001 | Reader to complete. |
| s9 | 0101 | |
| | | /CR = /(/*A*/*B*/*C*/*D*); S/H = s1 + s2 = *A*/*C*/*D*; |
| | | SC = s2 = *AB*/*C*/*D*; |
| | | /CS = /(s3 + s4 + s5) = /(*ABC* + *BCD*); |
| | | /W = /s4 = /(*ABCD*); |
| | | CC = s7 = *A*/*B*/*CD*. |

**This could be simplified.**

**Now that you have completed these examples, try taking the single-pulse example of Frame 3.2 (Figure 3.2) and produce the *D* type flip-flop equations and output equations. Finally, produce a circuit diagram for the FSM using *D* type flip-flops with asynchronous reset inputs and other logic gates. Note that *L* = s2 + s3.**

When you have done this, turn to **Frame 3.19**.

## *Frame 3.19*

The complete design for the single-pulse FSM is given below.

### *The design equations*

*A* ·*D* = s0 · *s* + s1 = /*A*/*B* · *s* + *A* /*B* = /*B* · (/*As* + *A*) = /*B* · (*s* + *A*)= **/*B* · *s* + *A*/*B*.**

*B* · *D* = s1 + s2 + s3 · *s* = ***A* + *B* · *s*.**

*P* = s1 · clk = *A*/*B*·clk and /*L* = /s0 = /(/*A*/*B*) or *L* = s2 + s3 = *B*. Hence *L* should go through an OR gate. However, with *A* and *B* inputs (note via *B* output as in Figure 3.15, this would make *L* high in states s2 and s3, *L* = *B*).

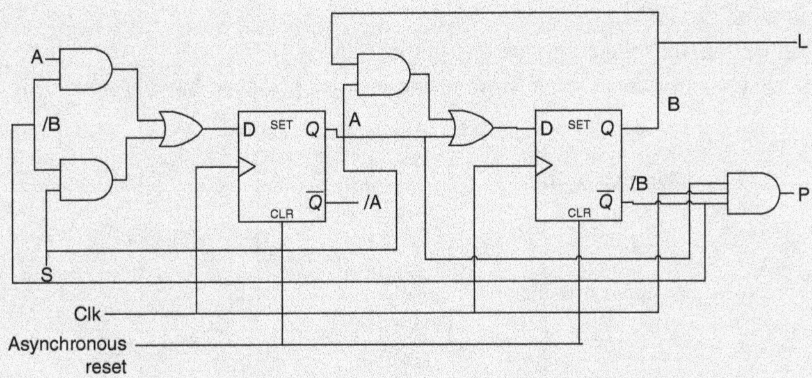


**Figure 3.15** Complete design circuit for the FSM.

In this circuit diagram you can see the memory elements (flip-flops), input decoding logic ($A \cdot D$ and $B \cdot D$ logic), and output decoding logic. Note a feed-forward input from clock to output decoding logic for clock pulse outputs could be added to make $P = A/B \cdot$ clk (Mealy output if required), as shown in Figure 3.15.

If flip-flops with asynchronous reset inputs are not available, a synchronous reset can be used, ANDed with the $A \cdot D$ and $B \cdot D$ logic, as illustrated in Figure 3.16.

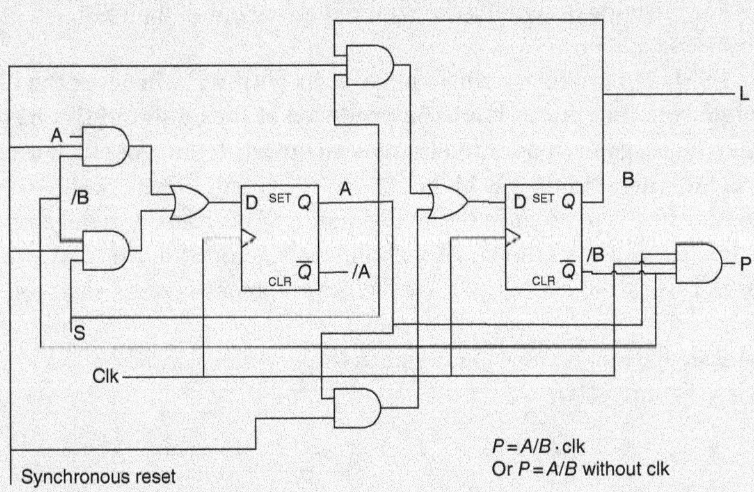


**Figure 3.16** Repeat of completed circuit showing *P* output with and without clk.

In Figure 3.16 the reset line is connected to the AND logic of each *D* input. Note the addition of the extra AND gate to the input logic of $B{\cdot}D$ so that when RESET = 0, $B{\cdot}D = 0$ also.

Turn to **Frame 3.20**.

## *Frame 3.20*

So now you have covered all aspects of designing FSMs. From initial specification, to construction of the state diagram, to synthesizing the circuit used to implement the FSM.

Let us consider the complete design process.

If we consider these steps for the Single-pulse FSM we have Figure 3.17.

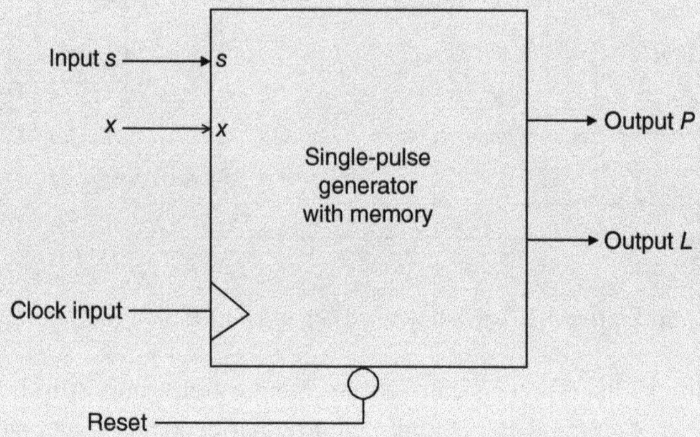


**Figure 3.17** This example has an $x$ input for Mealy output $L$.

### *The specification*

The block diagram showing inputs and outputs could be supplemented with a written specification describing the required behaviour of the FSM.

The specification is:

**The FSM is to produce a single pulse at its output *P* whenever the input *s* goes high. No other pulse should be produced at the output until *s* has gone low, then high again. In addition there is an output *L* that goes high after the *P* pulse, but only if input *x* is high.**

The next step is to produce the state diagram. This is not a trivial step since it requires us to use a number of techniques developed during this program of work. This is the skilled part of the development process that you need to make.

The state diagram is shown in Figure 3.18.

Turn to **Frame 3.21**.

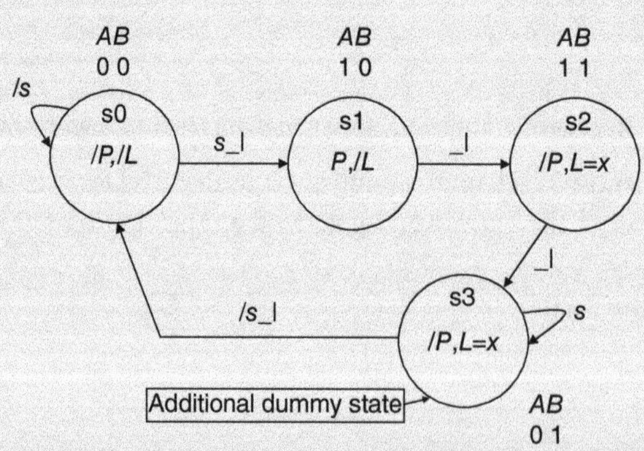


**Figure 3.18** Use this to complete the design equations.

### Frame 3.21

Now assign SSVs to the state diagram in order to continue with the synthesis of the FSM.

Next, develop the equations for the flip-flops and output logic.

### The design equations

$A \cdot D = (\text{s0} \cdot s + \text{s1}) \cdot \text{reset} = (/B \cdot s + A/B) \cdot \text{reset} = (/B \cdot s + A/B) \cdot \text{reset}.$

$B \cdot D = (\text{s1} + \text{s2} + \text{s3} \cdot s) \cdot \text{reset} = (A + B \cdot s) \cdot \text{reset}.$

$P = \text{s1} = A/B.$

$L = \text{s2} \cdot x + \text{s3} \cdot x.$ A Mealy output.

$L = B \cdot x.$

And finally, the circuit is produced from the equations (Figure 3.19).

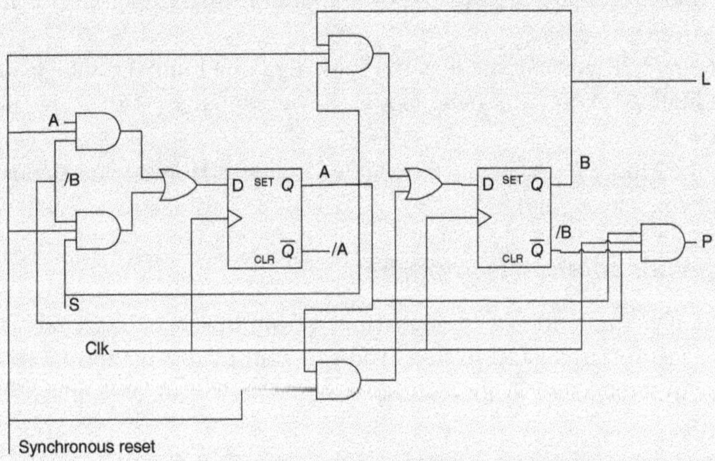


**Figure 3.19** The circuit with lower AND gate to allow reset of flip-flop *B*.

The design can then be simulated to ensure that it is functioning according to the original specification.

**You might like to add to Figure 3.19 the AND gate output to *L* with the clock input *x*.**

**When you have looked at Figure 3.19, you might like to draw the circuit using the Logisim Simulator (it is free to download). See Appendix A2 to find out how to use the Logisim Simulator.**

Now we will look at the timing arrangement of the system.

Figure 3.20 shows the timing diagram for Figure 3.19 in which you can see the operation of the system of Figure 3.18 with the clock input.

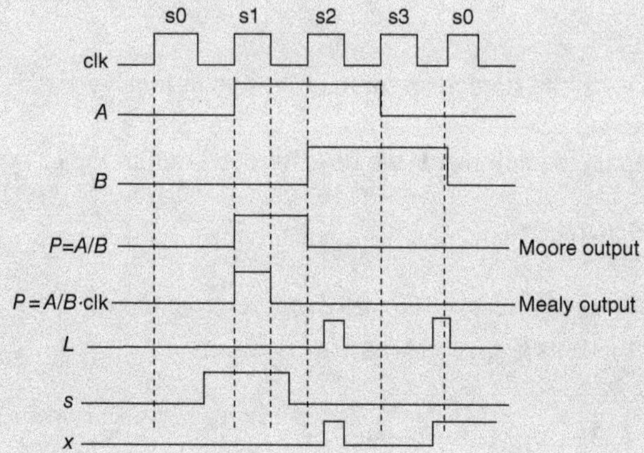


**Figure 3.20** The effect of Moore/Mealy outputs for *P* and *L*.

Now look at Figure 3.20 that shows a partial timing diagram for the state diagram of Figure 3.18.

In Figure 3.20 you can see how the *P* Mealy output only goes high when the clock is high. Also output *L* only goes high when input $x = 1$ in s2, but not in s3 due to $x = 0$.

Note the Moore *P* output and the Mealy *P* outputs. How do you create the test sequence for a state diagram?

## *Testing your solution is important*

Note that the simulation can be determined from the state diagram by looking at each transition from state to state. You will start by making reset = 0, so reset is active. Now make $s = 1$, the simulator should stay in s0. Now make reset = 1 and $s = 0$.

The simulator should stay in s0 also. Now make $s = 1$, and on the next clock pulse the simulator would move to s1.

Now two clock pulses should cause the simulator to move to s1, then s2, then s3, and should remain in s3 until you set $s = 0$, on which the simulator should move to s0 again and stop.

**You can see that the state diagram provides the necessary sequence.**

**If your state diagram contains a two-way branch, you need to test your circuit both ways during the simulation to make sure it works correctly.**

**Note how $L = x$ in Figure 3.20 the Mealy output.** $L = 1$ only, while $x = 1$, and only when in states s2 or s3.

Turn to **Frame 3.22**.

## Frame 3.22

One thing we need to consider is how to represent active low outputs in the state diagram.

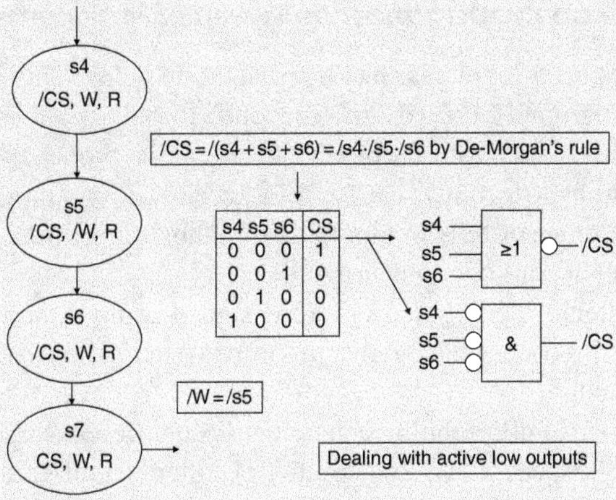


**Figure 3.21** CS and W outputs using De-Morgan's rule.

Consider Figure 3.21.

In state s4 the CS signal is low (and stays low in states s5 and s6). This can be written as /(s4 + s5 + s6) or as (/s4·/s5·/s6).

The table illustrates the NOR gate version. Likewise W = /s5 (active low).

Turn to **Frame 3.23**.

## Frame 3.23

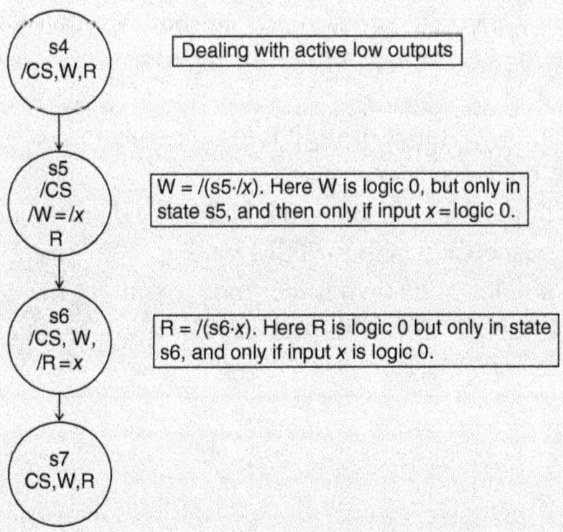


**Figure 3.22** How W and R are used.

In Figure 3.22 you can see how the signals W and R are written.

Note that as both are in terms of state and inputs, they are Mealy outputs, and so when we write them down in equations we need to define them in terms of both the state and the input.

### *Another note on propagation delays in FSM systems*

In logic systems each logic gate has a propagation delay. This could be of the order of a microsecond ($1 \times 10^{-6}$ of a second) in older gate circuits. In more modern logic gates, such as TTL or CMOS gates, the propagation delay might be of the order of nanoseconds ($1 \times 10^{-9}$ of a second). In standard TTL gates this might be of the order of tens to hundreds of nanoseconds (ns). While in some fast CMOS gates it could be in nanoseconds.

In the past digital logic systems were designed using a number of different standard logic chips connected together to form a system. These could use standard TTL or CMOS chips.

If you make use of different logic gates between different chips such as standard TTL and high-speed TTL (low-speed TTL) then the difference between the propagation delay in different gates on different chips could be quite large (i.e. LS TTL 7 ns, standard TTL 50 ns). This can lead to problems in our FSM-based systems.

However, in today's world use is made of PLDs with perhaps hundreds or even thousands of logic gates. In these circumstances, the gate delays between different gates can be very close, to the order of a few nanoseconds (1 ns to a few nanoseconds). Therefore, under these conditions our FSM-based system will be easier to control.

Most FSM-based systems make use of complex programmable logic devices (CPLDs) with a few hundreds or even thousands of logic gates, and some field programmable gate arrays (FPGAs) can contain many thousands of logic gates so that quite large logic systems can be developed to work on a single CPLD or FPGA.

You can look at any number of digital logic books to see how this works, as this is beyond this book's scope

As well as Section 3.1 in this book on the binary data serial transmitter FSM there is a random sequence generator FSM similar to the one in Minns (1995) in the Web-based examples in its own folder that you might like to look at.

**Now on to tutorials for Chapter 1, 2, and 3 covering the clock-driven FSM.**

## 3.2 TUTORIALS COVERING CHAPTERS 1, 2, AND 3

### 3.2.1 Binary data serial transmitter FSM

This is a basic FSM system that allows the user to transmit binary data sequentially via a shift register and detect when it has all been sent via the detector 'det' block. This is shown in Figure 3.23.

Each binary four-bit code is loaded into the parallel inputs of the shift register then serially sent via the FSM to line via the TX output.

The RE output of the shift register indicates to the FSM that the four-bit data have been sent.

The state diagram consists of seven states (s0 to s6) and is responsible for the sequence of operations required. After st is raised high, the reset RC is raised to allow the binary counter to operate in s1. In s2 the load input LD is lowered to allow the binary count value to be loaded into the shift register parallel loading inputs to receive the binary count value. Then the FSM cycles through s3, s4, s5, s1, etc. to pulse each binary bit to line repeatedly.

This continues until all binary four-bit counter words have been sent indicated by the Det output to the done input of the FSM.

The st input can be used to send another set of four-bit binary data to line.

Looking at Figure 3.24, you can see that not all states have a unit distance coding. Figure 3.25 looks at how to overcome this problem.

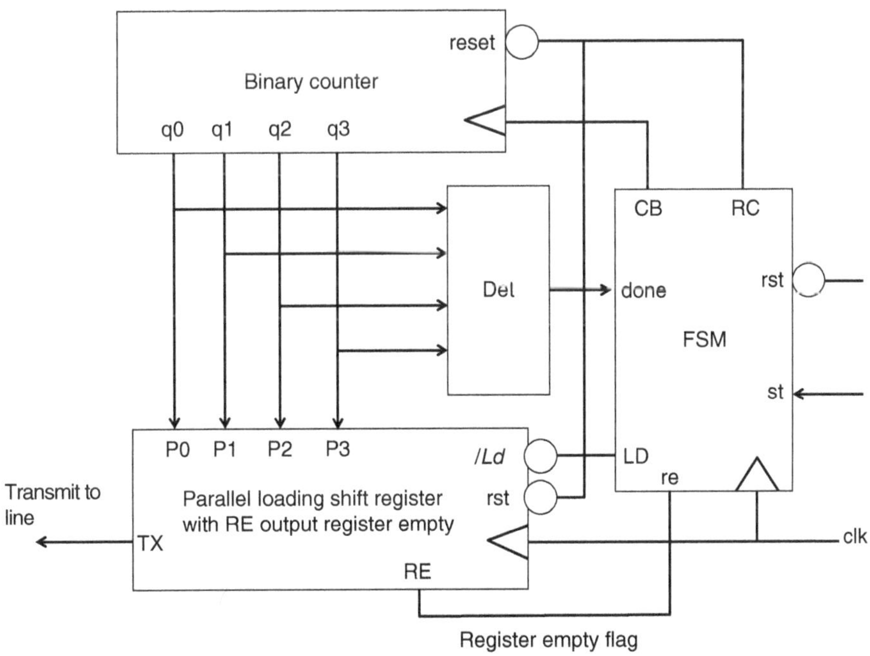


**Figure 3.23** Block diagram of the binary data serial transmitter.

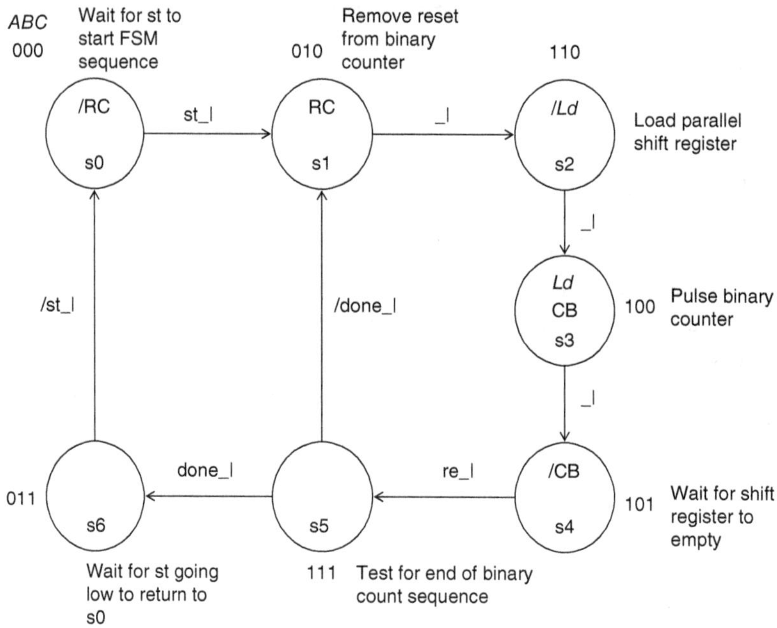


**Figure 3.24** State diagram for binary data serial transmitter.

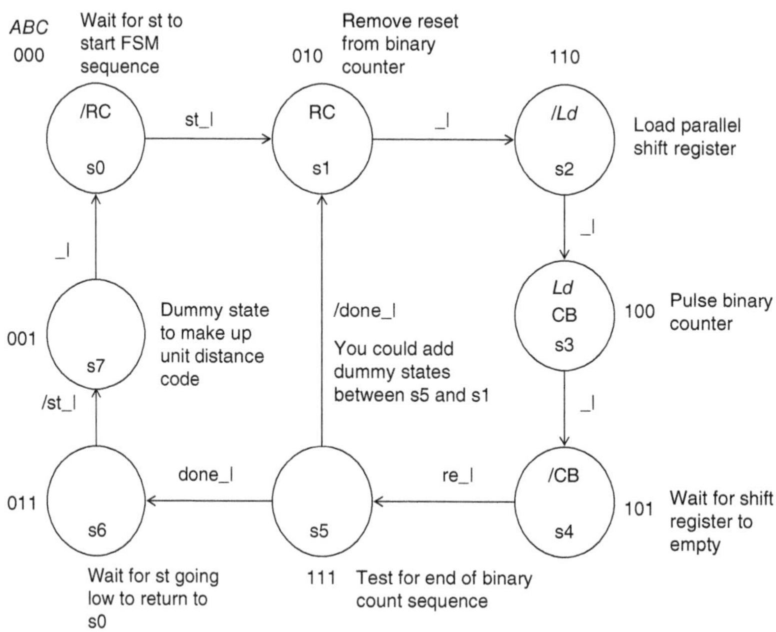


**Figure 3.25** With additional dummy state s7.

**In the state diagram you could add a dummy state between s6 and s0 to keep to the unit distance code. This dummy state would be called 's7' and would have the *ABC* value of *ABC* = 001.**

You could add additional dummy states between s5 and s1 as well; however, this would need an extra SSV *D*. This would add to the complexity of the state diagram. This would be OK using *D* type flip-flops that reset to s0 if the FSM lands in an unused state. This is not considered.

Now on with the equations for the state diagram of Figure 3.25:

$A \cdot d = \text{s1} + \text{s2} + \text{s3} + \text{s4} + \text{s5} \cdot \text{done} = /AB/C + AB/C + A/B/C + A/BC + ABC \cdot \text{done}.$

$B \cdot d = \text{s0} \cdot \text{st} + \text{s1} + \text{s4} \cdot \text{ren} + \text{s5} \cdot \text{done} = /A/B/C \cdot \text{st} + /AB/C + A/BC \cdot \text{ren} + ABC \cdot \text{done}.$

$C \cdot d = \text{s3} + \text{s4} + \text{s5} \cdot \text{done} + \text{s6} = A/B/C + A/BC + ABC \cdot \text{done} + ABC.$

I leave you to work out the outputs RC, LD, and CB.

The equations can be obtained from the state diagram using *D* type flip-flops as described in this chapter.

Also once this has been done, the output equations for the LEDs can also be obtained from the state diagram.

So now we have all the information needed to create the FSM system.

Figure 3.26 shows simulation of the system.

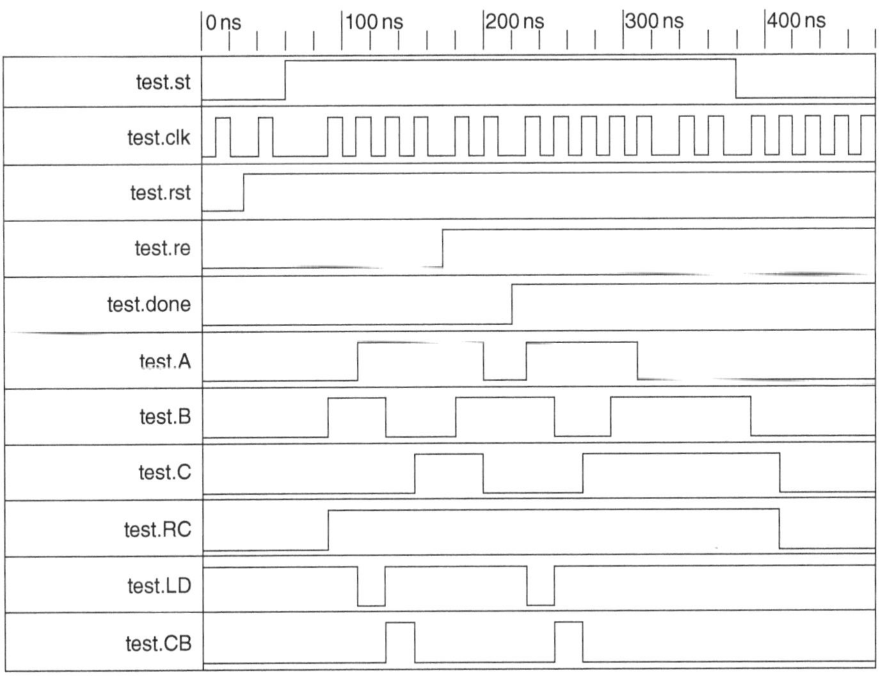


**Figure 3.26** Simulation of the binary data serial transmitter.

This is using the Verilog HDL file **binserfsm.v**

```
// Binary serial transmission FSM simulation:
// This is the FSM developed in Chapter 3 tutorials.
// Author: PDM
//
// Define DFF
module D_FF(q,d,clk,rst);

  output q;
  input d,clk,rst;
  reg q;
  always @ (posedge clk or negedge rst)
   if (rst==0)
      q<=1'b0;
   else
      q<=d;
endmodule

// The FSM module..
module fsm(rst,clk,st,done,re,A,B,C,RC,LD,CB);

 input clk,rst,st,done,re;
 output A,B,C,RC,LD,CB;

 wire ad, bd, cd;

 D_FF FFA(A,ad,clk,rst);
 D_FF FFB(B,bd,clk,rst);
 D_FF FFC(C,cd,clk,rst);

 //Assign canonical form.

 ad=~A&B&~C | A&B&~C | A&~B&~C | A&~B&C | A&B&C & done
 bd=~A&~B&~C.st | ~A&B&~C | A&~B&C&ren | A&B&C&done,
 cd=A&~B&~C | A&~B&C | A&B&C&done | A &B&C,

 RC=~(~A&~B&~C),
 LD=~(A&~B&~C),
 CB=A&~B&~C;

endmodule

// test bench module

module test;

  reg st,clk,rst,re,done;
```

```
fsm binserfsm(rst,clk,st,done,re,A,B,C,RC,LD,CB);

 initial
   begin
   $dumpfile("bintx.vcd");
   $dumpvars;
   rst=0;
   clk=0;
   st=0;
   done=0;
   re=0;
   #10 clk=~clk;
   #10 clk=~clk;         //check that fsm stays in s0
   #10 rst=1;
   #10 clk=~clk;
   #10 clk=~clk;         //still in s0.
   #10 st=1;
   #10 done=0;
   #10 re=0;
   repeat (8)
     #10 clk=~clk;
   //clk=~clk;
   #10 re=1;
   #10 clk=~clk;
   #10 clk=~clk;
   #10 clk=~clk;
   #10 clk=~clk;
   #10 done=1;
   repeat (10)
     #10 clk-~clk;
   #10 st=1;
   #10 clk=~clk;
   #10 clk=~clk;
   #10 clk=~clk;
   #10 clk=~clk;
   #10 st=0;
  repeat (10)
    #10 clk=~clk;
  $finish(200);
   end

endmodule
```

**Listing 3.1** Binary data serial transmission.

You can run this on the SynaptiCAD software to see it in operation.

### 3.2.2 The high low FSM system

In this example the FSM controls an analogue-to-digital converter (ADC) and checks its value against a hi word limit and a lo word limit. If the ADC exceeds the hi word limit, or is below the lo word limit, the FSM sounds a siren and lights an appropriate HL or LL indicator.

It will remain in this hi or lo state until the system reset is pressed. In this way the FSM can be left monitoring the ADC until an alarm is given.

From the above specification and the block diagram of Figure 3.27, the state diagram for this system can be produced.

The state diagram for the high low system is shown in Figure 3.28.

After some consideration of the specification and the block diagram the state diagram is constructed in Figure 3.28.

In this state diagram use has been made of Mealy outputs for LL and HL to connect to the LEDs. This was a decision made at the time. We could have just used a Moore output. Using Mealy outputs has helped to reduce the number of states in the state diagram.

Note that De-Morgan's rule has been applied to the terms for /(hi + lo) to produce /hi·/lo.

Also HL = /(s3·/hi) and LL = /(s3·/lo). Applying De-Morgan's rule.

These are active low to allow the LEDs to operate.

The equations from the state diagram are:

$$A \cdot d = \mathrm{s0} \cdot \mathrm{st} + \mathrm{s1} + \mathrm{s2} \cdot /(\mathrm{lo} + \mathrm{hi}) = /A/B \cdot \mathrm{st} + A/B + AB \cdot /\,\mathrm{hi} \cdot /\,\mathrm{lo}.$$

Simplifying to:

$$A \cdot d = /B \cdot \mathrm{st} + A/B + B \cdot /\,\mathrm{lo} \cdot /\,\mathrm{hi}.$$

$$B \cdot d = \mathrm{s1} \cdot \mathrm{eoc} + \mathrm{s2} \cdot (\mathrm{lo} + \mathrm{hi}) + \mathrm{s3}.$$

$$B \cdot d = A/B \cdot \mathrm{eoc} + A \cdot \mathrm{lo} + A \cdot \mathrm{hi} + /AB.$$

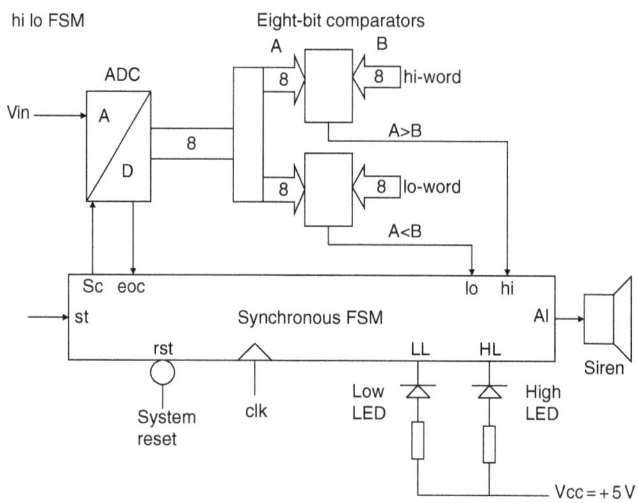


**Figure 3.27** The high low FSM system.

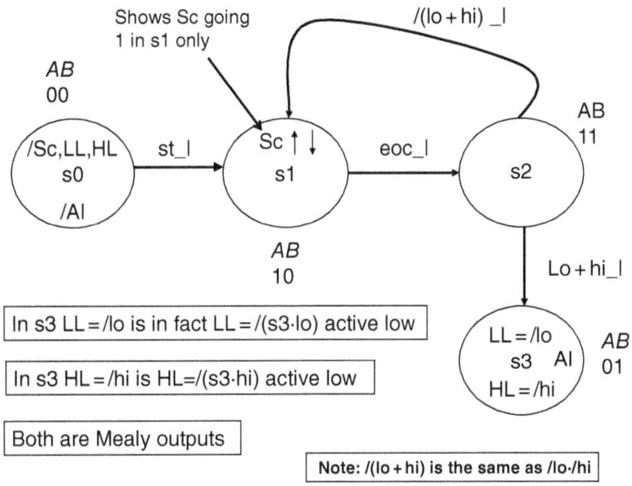


**Figure 3.28** State diagram of the high-low system.

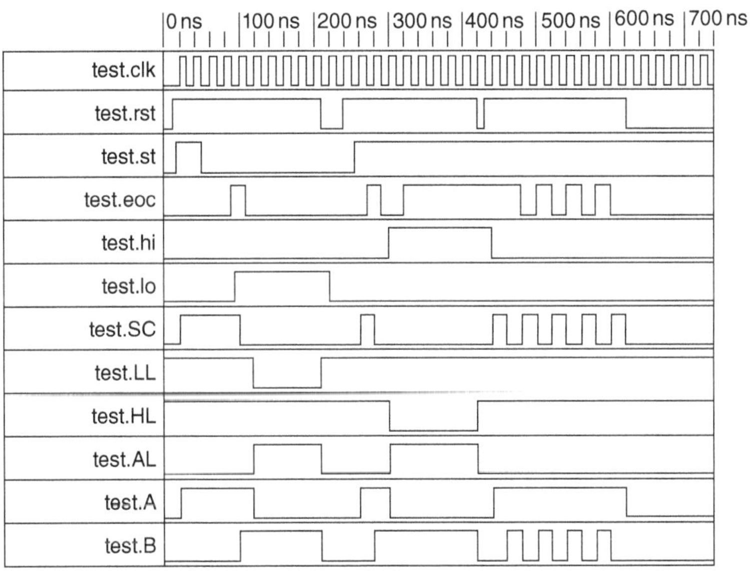


**Figure 3.29** Simulation of the high-low system.

Note that aux rule applied to s2(lo + hi) term.

$$B \cdot d = A \cdot \text{eoc} + A \cdot \text{lo} + A \cdot \text{hi} + /AB.$$

The remaining Moore outputs are Sc = s1 = $A/B$ and siren Al = s3 = $/AB$.
Note that Al is the same as AL in Figure 3.29.
The simulation waveform for this FSM is shown in Figure 3.29.
See Verilog HDL for High and Low inputs and outputs.
Listing 3.2 is taken from **hiloinsm.v file**.

```
////////////////////////////////////////////////////
//  HiLo indicator FSM from Chapter 3.
//  This shows both Lo and Hi actions.
//  Author: PDM.
///////////////////////////////////////////////////

//define the dff..
module D_FF (q,d,clk,rst);

  output q;
  input d,clk,rst;
  reg q;
  always @ (posedge clk or negedge rst)
   if (rst==0)
     q<=1'b0;
   else
     q<=d;
endmodule


//////////////////////////////////////////
module fsm(st,eoc,lo,hi,rst,clk,Sc,LL,HL,AL,A,B);

 input clk,rst,st,eoc,lo,hi;
 output Sc,LL,HL,AL, A, B;
 wire da,db;

 D_FF FFA(A,da,clk,rst);
 D_FF FFB(B,db,clk,rst);
 // if stuck do in canonical form:
 assign
 da=~B&st | A&~B | A&~hi&~lo,
 db=A&~B&eoc | A&lo | A&hi | ~A&B,
 Sc=A&~B,
 LL=~(~A&B&lo),
 HL=~(~A&B&hi),
 AL=~A&B;

endmodule
//////////////////////////////////////////



module test;
reg clk, rst, st, eoc,hi, lo;
// instantce the fsm..
 fsm hilofsm(st,eoc,lo,hi,rst,clk,Sc,LL,HL,AL,A,B);
```

```
initial
  begin
   $dumpfile("hilofsm.vcd");
   $dumpvars;
   rst=0;
   clk=0;
   st=0;
   lo=0;
   hi=0;
   eoc=0;

   #10 rst=1;
   repeat(180)
    #10 clk=~clk;
   end
   initial
    begin
    #15 st=1;
    #35 st=0;
    #38 eoc=1;
    #5 lo=1'b1;
    #15 eoc=0;
    #1 lo=1'b1;
    #100 rst=0;
    #10 lo=0;
    #20 rst=1;

    #15 st=1;
    #15 eoc=1;
    #20 eoc=0;
    #10 hi=1;
    #20 eoc=1;
    #20 eoc=1;
    #80 rst=0;
    #10 rst=1;
    #10 hi=0;
    #20 eoc=1;
    #20 eoc=0;
    #20 eoc=1;
    #20 eoc=0;
    #20 eoc=1;
    #20 eoc=0;
    #20 eoc=1;
    #20 eoc=0;
    #20 rst=0;
    #120 $finish;
    end
endmodule
```

**Listing 3.2** High-low FSM.

There is a Logisim file called **hiloalarmFSM.circ** you can download from: www.wiley.com\go\minns\digitalsystemdesign.

### 3.2.3 The clocked watchdog timer FSM

Most microprocessors these days have a built-in watchdog timer (WDT). The WDT is an addressable device to visit on a regular basis. The idea is that the timer is usually a down counter that can be regularly written to with a known value. Between writes the counter will be clocked towards zero. If the microprocessor does not write to the WDT between countdown periods then the counter will reset to zero and this action will be used to reset the microprocessor.

The WDT thus acts as a safeguard to prevent the microprocessor from running out of control (jumping to an instruction that is not part of the program sequence), providing a means of determining a microprocessor system failure.

Figure 3.30 illustrates the block diagram with its corresponding waveform timing.

The system is made up from an FSM that controls the counter. When the chip select (Ce) becomes active (see Figure 3.30), the T2 clock of the FSM moves the FSM to s1 (see Figure 3.31), then when it detects the /iow signal going low it moves to s2 (see Figure 3.31) on the T3 clock, where it brings the WDT signal low, and on

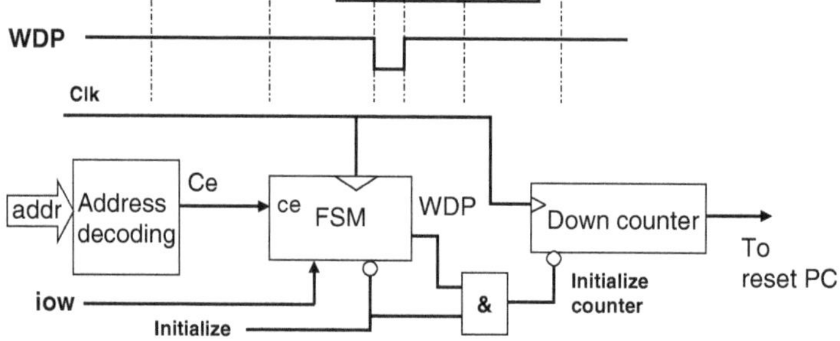


**Figure 3.30** The clocked watchdog timer.

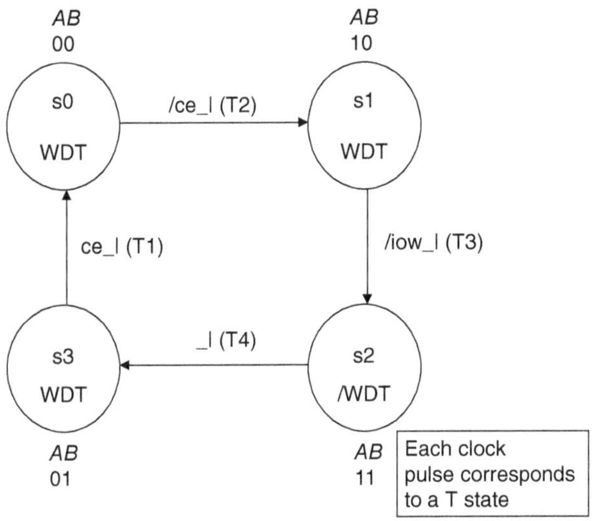


**Figure 3.31** State diagram of the watchdog timer.

the next clock pulse (T4) it moves the FSM through s3 back to s0 via T1 (two clock pulses) to s0.

From Figure 3.31 the equations can be obtained.

### 3.2.3.1 FSM equations

$$A \cdot d = \text{s0} \cdot /\,\text{ce} + \text{s1} = /A/B \cdot /\,\text{ce} + A/B = /B \cdot /\,\text{ce} + A/B.$$

$$B \cdot d = \text{s1} \cdot /\,\text{iow} + \text{s2} + \text{s3} \cdot /\,\text{ce} = A \cdot /\,\text{iow} + AB + B \cdot /\,\text{ce}.$$

$$\text{Output is WDP} = /\text{s2} = /\left(AB\right).$$

The address of the ce used in this design needs an address code. For example, if the address code is 300h (11 0000 0000 binary) then the address would be:

$$\text{ce} = /\left(\text{a9} \cdot \text{a8} \cdot /\text{a7} \cdot /\text{a6} \cdot /\text{a5} \cdot /\text{a4} \cdot /\text{a3} \cdot /\text{a2} \cdot /\text{a1} \cdot /\text{a0}\right).$$

The timing diagram in Figure 3.32 illustrates the working of the WDT in action.

There is a Logisim file called **watchdogtimer.circ** you can download from: www.wiley.com\go\minns\digitalsystemdesign.

As you can see, the ce pulse, followed by an iow low, results in the generation of WDP pulse output from the FSM in Figure 3.32. Note the action of the SSVs *A* and *B* following the four-state sequence of the FSM.

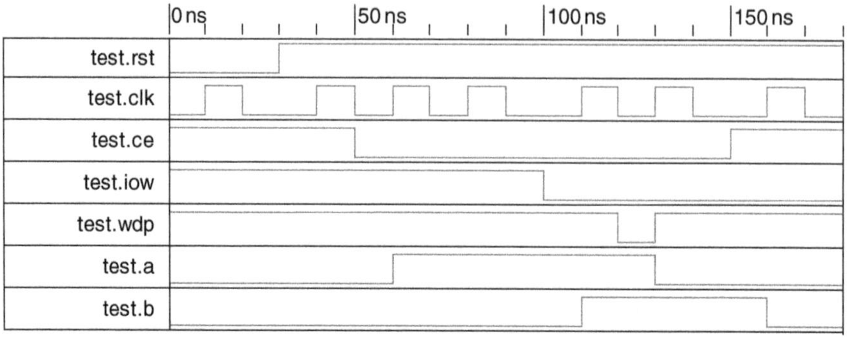


**Figure 3.32** Simulation of the watchdog timer.

There could be additional qualifiers, i.e. in the I/O in the PC using the I/O memory map such as to identify between dynamic memory access (DMA) and I/O access (**aen** signal). Also, the /iow signal would be needed to identify a write cycle.

This would change the equation for ce to:

$$ce = /\left(a9 \cdot a8 \cdot /a7 \cdot /a6 \cdot /a5 \cdot /a4 \cdot /a3 \cdot /a2 \cdot /a1 \cdot /a0 \cdot /\mathbf{aen}\right).$$

The signal aen is the address enable.

The down counter is defined in Appendix A3 and would result in the down counter equations of:

$$\begin{aligned}
&Q0 \cdot t = 1. \\
&Q1 \cdot t = /q0. \\
&Q2 \cdot t = /q0 \cdot /\, q1. \\
&Q3 \cdot t = /q0 \cdot /\, q1 \cdot /\, q2. \\
&Q4 \cdot t = /q0 \cdot /\, q1 \cdot /\, q2 \cdot /\, q3.
\end{aligned}$$

The above equations relate to the eight-state down counter using T type flip-flops.

Note that the counter needs an asynchronous initialization signal connected to each *T* type flip-flop to form the parallel input logic.

Note that the FSM will not respond if there is no ce input, and so it only works with addressed signals to the WDT.

You can find the watchdog system FSM at **wdtfsm.v**.

Finally, the Verilog HDL code for this example is shown in Listing 3.3.

```
////////////////////////////////////////////////
// Clocked Watchdog Timer FSM.
// Author: PDM.
///////////////////////////////////////////////
```

```
module D_FF (q,d,clk,rst);
 output q;
 input d,clk,rst;
 reg q;
 always@(posedge clk or negedge rst)
   if (rst==0)
     q=1'b0;
   else
     q=d;
endmodule

module fsm(rst,clk,ce,iow,wdp,A,B);
  input rst,clk,ce,iow;
  output A,B,wdp;
  wire ad,bd;

  D_FF ffa(A,ad,clk,rst);
  D_FF ffb(B,bd,clk,rst);

  // Could place into canonical form.
  assign
   ad  =~B&~ce | A&~B,
   bd  =A&~iow | A&B | B&~ce,

   wdp =~(A&B);
endmodule

module test;
  reg rst,clk,ce,iow;
  fsm uut(rst,clk,ce,iow,A,B);

    initial
    begin
    $dumpfile("fsm.vcd");
    $dumpvars;
    clk=0;
    rst=0;
    ce=1;
    iow=1;
    // so
    #10 clk=~clk;
    #10 clk=~clk;
   // still in s0.
   #10 rst=1;
   #10 clk=~clk;
```

```
    #10 clk=~clk;
    // still in s0.
    #10 ce=0;
    #10 clk=~clk;
    #10 clk=~clk;
    //move to s1.
    #10  clk=~clk;
    #10  clk=~clk;
    // stay in s1.
    #10 iow=0;
    #10 clk=~clk;
    #10 clk=~clk;
    // move into s2
    #10 clk=~clk;
    #10 clk=~clk;
    // then s3.
    #10 iow=1;
    #10 ce=1;
    #10 clk=~clk;
    #10 clk=~clk;
    #10 $finish;
    end
endmodule
```

**Listing 3.3** Clocked watchdog timer.

### 3.2.4 The asynchronous receiver system clocked FSM

This is quite a complex example that makes use of not just the FSM but also a number of different digital devices for its operation.

There is a shift register, a data latch and a divide by 11 counter in the design.

Appendix A3 describes some of these in some detail, which you might want to read about as you go further into the design. This is a useful appendix that is very worthwhile looking at as it contains many different aspect of FSM design used in this book.

This is a rather unusual design. The designer can dictate the baud rate and serial protocols, so as to control the device. In this example, the serial data input is encapsulated into an asynchronous data packet with selected start (st) and stop (sp) bits that have been added to the serial transmission packet. These are used to provide a means of identifying the data packets as they arrive. This allows the data packet to arrive at any time and at any selected baud rate.

The problem with receiving data is that it is necessary to ensure that the shift register used in the design is clocked with the correct data bits. To do this, the FSM clock is used to drive the FSM to create a shift register clock RXCK in the middle of the data bit time period. This RXCK clock pulse can be seen in Figure 3.33 as the arrow pulses occur every forth FSM clock (denoted by the arrows).

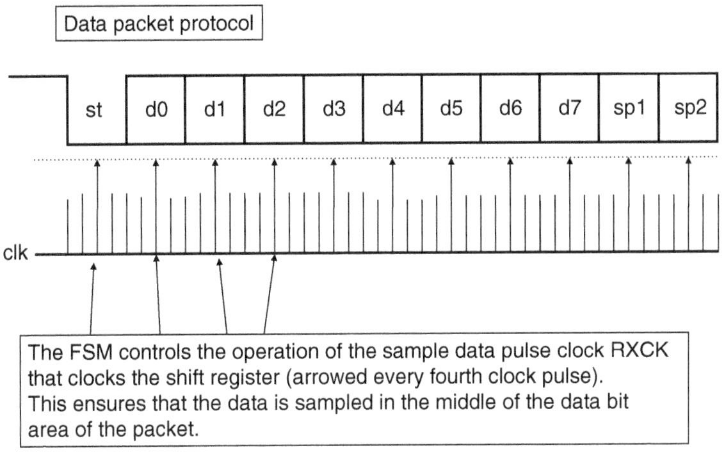


**Figure 3.33** Data packet protocol.

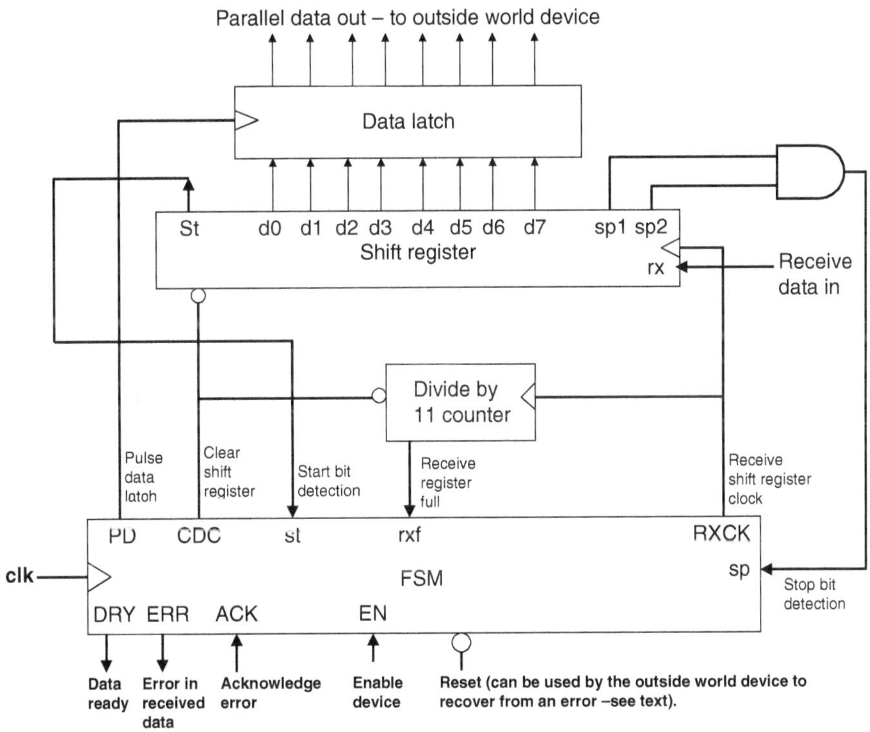


**Figure 3.34** Block diagram of the asynchronous receiver.

**Also note that in this example there are four clock pulses between each sample point.**

As you can see in Figure 3.34 there is a shift register, a data latch and a divide by 11 counter, all of which are described in detail in Appendix A3.

The Verilog HDL code for this example requires an FSM module, shift register module, a data latch module and a divide by 11 counter module that will run in parallel in the Verilog HDL system.

### 3.2.4.1 Brief note on the development of the test bench generator

The test bench generator needed for this example is complex. The one in Listing 3.4 is simple as it tests out the FSM on its own. Note the whole design including the shift register and the divide by 11 counter.

The skills required to do this can be challenging and may require the use of the Verilog repeat statement to keep the Verilog code looping to wait for the divide by 11 counter. Such code development is beyond the scope of this book but can be obtained by studying Verilog code. You could try developing your own code to do this.

### 3.2.4.2 The state diagram

The state diagram for an asynchronous receiver is shown in Figure 3.35.

OK, so this design is a bit unusual. However, it is an acceptable design.

The _|1, and other FSM clock pulses describe the pulses the FSM needs to run through in order to look for the middle pulse it needs to find. This is shifted into the shift register.

Thereafter, the FSM moves around states s5, s6, s7, and s8 as it picks up the protocol packet bits of the data packet. When it has all the data packets, it looks at the stop bit and looks for any errors (ERR bit should be 1).

Any errors cause the FSM to move to s10 and stop (no point in going on if there are errors!). Otherwise, it pulses the data latch and raises the data ready (DRY) bit to indicate to the outside world that all is OK at this point. The outside world issues an acknowledge pulse (ACK) to indicate it has seen this and the FSM lowers ACK to continue on with the next packet.

Note that the bit PD is used to pulse the data latch to place the shift register contents into the data latch ready for the outside world to access.

To keep this system simple, it has four clock pulses in each data packet. In practice you may use around 16 clock pulses between each data bit in the packet. This would make the FSM appear quite large.

The reader should look at the state diagram carefully (and perhaps also look at Appendix A3).

**Note: state s3 is missing from the state diagram that moves from s2 to s4. This is OK, as we do not have an s3 state (missed out!) and allows the equations to line up with the state diagram.**

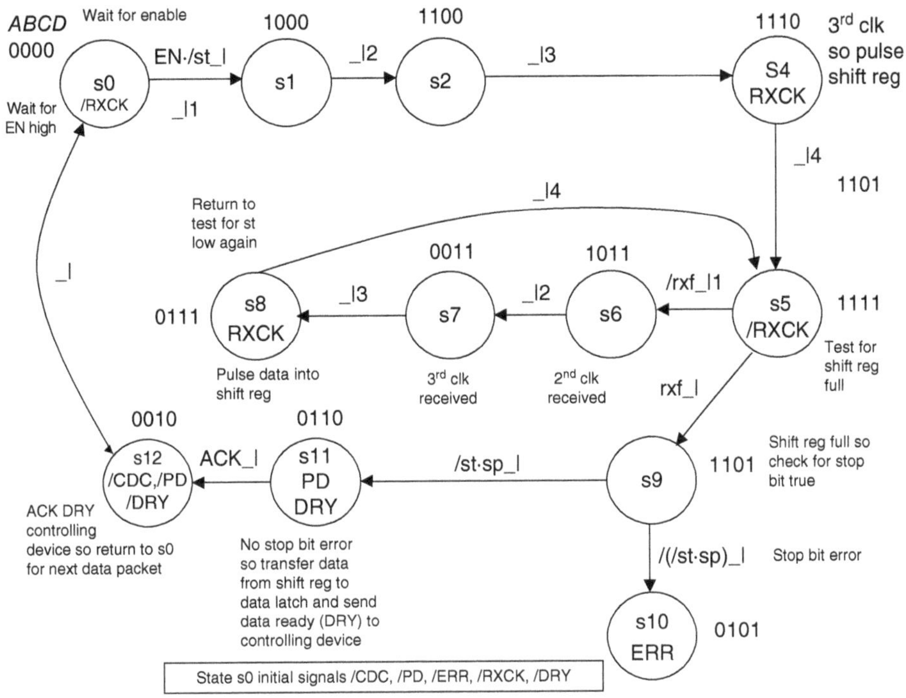


**Figure 3.35** The state diagram for an asynchronous receiver.

### 3.2.4.3 The state diagram equations

$$A \cdot d = s0 \cdot EN \cdot / st + s1 + s2 + s4 + s5 + s8.$$

$$B \cdot d = s1 + s2 + s4 + s5 \cdot rxf + s7 + s8 + s9 + s10 + s11 \cdot / ACK.$$

$$C \cdot d = s2 + s4 + s5 \cdot / rxf + s6 + s7 + s8 + s9 \cdot (/st \cdot sp) + s11.$$

$$D \cdot d = s4 + s5 + s6 + s7 + s8 + s9 \cdot / (/st \cdot sp) + s10.$$

**You can solve these in canonical form.**

### 3.2.4.4 The outputs

$$\text{RXCK} = s4 + s8 = ABC/D + /ABCD.$$

$$\text{PD} = \text{DRY} = s11 = /ABC/D.$$

$$\text{ERR} = s10 = /AB/CD.$$

$$\text{CDC} = \sim(\mathbf{s12} + \mathbf{s0}) = \sim(/\boldsymbol{A}/\boldsymbol{BC}/\boldsymbol{D} + \sim \boldsymbol{A} \sim \boldsymbol{B} \sim \boldsymbol{C} \sim \boldsymbol{D}).$$

The reader might like to complete the $A{\cdot}d$ to $D{\cdot}d$ equations to obtain the final details of the FSM.

**Note: you can also obtain the equations from the serfsm.v file listed below.**

```
//////////////////////////////////////////////////////
// Asynchronous serial receiver system.
//
// Complete system simulation:
// This system is developed from the block diagram
// of the asynchronous receiver with two stop bits.

// The design has not used arrays for buses so that
// the details of all the connections can be seen.
//
// Note the acknowledge and ready signal handshake
// (ACK, and DRY) at the end of the simulation.
//
// Author: PDM
/////////////////////////////////////////////////////
//-------------------------------------------
// Define DFF
module D_FF(q,d,clk,rst);
  output q;
  input d,clk,rst;
  reg q;
  always @ (posedge clk or negedge rst)
   if (rst==0)
     q<=1'b0;
   else
     q<=d;
endmodule
//-------------------------------------------
// define TFF
// Needed for the divide by 11 asynchronous
// counter.
//-------------------------------------------
module T_FF (q,t,clk,rst);
  output q;
  input t,clk,rst;
  reg q;
  always @ (posedge clk or negedge rst)
    if (rst==0)
      q<=1'b0;
   else
     q<=t^q;
endmodule
```

```
//-----------------------------------------------
// define DFF with asynchronous preset and clear.
// Needed by the shift register.
// Could have just made all FF's this type and
// dispensed with the basic DFF with clear only.
//-----------------------------------------------
module D_FF_pc(q,d,clk,rst,prs);
  output q;
  input d,clk,rst,prs;
  reg q;
  always @(posedge clk or negedge rst or negedge prs)
    if(~rst & prs)
      q<=1'b0;
    else
      if(~prs & rst)
      q<=1'b1;
      else
        if(clk)
          q<=d;
        else;

endmodule

//------------------------------------------
// Divide by 11 counter.
// This is a simple asynchronous counter
// which will count up to 1011 and stop.
// It is cleared by the FSM.
//------------------------------------------
// Now define the counter..
module divide by11(clk,rst,rxf);

input clk, rst;            //clk and rst are inputs.
output rxf;                //,q0,q1,q2,q3; // all q/s outputs.
wire t0,t1,t2,t3,
stop,q0,q1,q2,q3;          //all ts are interconnecting wires.
// need to define instances of each TFF defined earlier.
T_FF ff0(q0,t0,clk,rst);
T_FF ff1(q1,t1,clk,rst);
T_FF ff2(q2,t2,clk,rst);
T_FF ff3(q3,t3,clk,rst);

// now define the logic connected to each t input.
// we use an assign for this.
assign
```

```
t0=1'b1&stop,               //this is just following the technique
t1=q0&stop,                 //for binary counter design.
t2=q0&q1&stop,              //will generate AND gates..
t3=q0&q1&q2&stop,
stop=~(q0&q1&~q2&q3),       //to detect 11 clock pulse.
rxf=~stop;
endmodule                   //end of the module counter.
//------------------------------------------
// Data Latch: Used to provide a storage for
// the received data for user to collect.
//-------------------------------------------------------------
// define data latch.
// This module does not require a reset so this
// input is kept high (disabled).
//-------------------------------------------------------------
module latch(ld,d0,d1,d2,d3,d4,d5,d6,d7,q0,q1,q2,q3,q4,q5,q6,q7);
  input ld,d0,d1,d2,d3,d4,d5,d6,d7;
  output q0,q1,q2,q3,q4,q5,q6,q7;
  D_FF dl0(q0,d0,ld,1'b1);
  D_FF dl1(q1,d1,ld,1'b1);
  D_FF dl2(q2,d2,ld,1'b1);
  D_FF dl3(q3,d3,ld,1'b1);
  D_FF dl4(q4,d4,ld,1'b1);
  D_FF dl5(q5,d5,ld,1'b1);
  D_FF dl6(q6,d6,ld,1'b1);
  D_FF dl7(q7,d7,ld,1'b1);
endmodule
//-----------------------------------------
// define shift register
// The shift register clock is rxclk which
// is controlled by the fsm.
// The protocol bits (st, sp1, and sp2) are
// shifted into their own FF's.
// Note that the flip-flop for st must be
// initialised to 1 (hence the preset FF).
//------------------------------------------
module shifter(rst,clk,din,qst,q0,q1,q2,q3,q4,q5,q6,q7,qsp1,qsp2);

 input clk,rst,din; //Note din is really the rx input data to
the shift register.
 output qst,q0,q1,q2,q3,q4,q5,q6,q7,qsp1,qsp2;

 wire dst,d0, d1, d2, d3, d4, d5, d6, d7, dsp1, dsp2 ;
```

```
 D_FF_pc qstd(qst,dst,clk,rst,1'b1);  //preset to 1 for st=1
on reset.
 D_FF qod(q0,d0,clk,rst);
 D_FF q1d(q1,d1,clk,rst);
 D_FF q2d(q2,d2,clk,rst);
 D_FF q3d(q3,d3,clk,rst);
 D_FF q4d(q4,d4,clk,rst);
 D_FF q5d(q5,d5,clk,rst);
 D_FF q6d(q6,d6,clk,rst);
 D_FF q7d(q7,d7,clk,rst);
 D_FF qsp1d(qsp1,dsp1,clk,rst);
 D_FF qsp2d(qsp2,dsp2,clk,rst);

 assign
 dst=d0,
 d0=d1,
 d1=q2,
 d2=q3,
 d3=q4,
 d4=q5,
 d5=q6,
 d6=d7,
 d7=dsp1,
 dsp1=qsp2,
 dsp2=din;

endmodule



// the Asynchronous serial receiver module controller.
// This FSM module does not cater for parity errors
//----------------------------------------------------------------------------
module ser_fsm(rst,clk,st,sp,rxf,en,ACK,a,b,c,d,PD,CDC,RXCK,E
RR,DRY);
  input rst,clk,st,sp,rxf,en,ACK;
  output a,b,c,d,PD,CDC,RXCK,ERR,DRY;
  wire ad,bd,cd,dd;

  D_FF ffa(a,ad,clk,rst);
  D_FF ffb(b,bd,clk,rst);
  D_FF ffc(c,cd,clk,rst);
  D_FF ffd(d,dd,clk,rst);
```

```
  // un-minimised equations in Canonical form:
  assign
    ad=~a&~b&~c&~d&en&/st  | a&~b&~c&~d | a&b&~c&~d |
a&b&c&~d | a&b&c&d | ~a&b&c&d,
    bd=a&~b&~c&~d | a&b&~c&~d | a&b&c&~d | a&b&c&d&/rxf |
~a&~b&c&d | ~a&b&c&d | a&b&~c&d
        |~a&b&~c&d | ~a&b&c&~d&~ack,

    cd=a&b&~c&~d | a&b&c&~d | a&b&c&d&~rxf | a&~b&c&d &~st&sp |
~a&~b&c&d | ~a&b&c&d | a&b&~c&d & ~st&sp | ~a&b&c&~d ,

    dd=a&b&c&~d | a&b&c&d | a&~b&c&d | ~a&~b&c&d | ~a&b&c&d |
a&b&~c&d&~(~st & sp) | ~a&b&~c&d,

    ERR=~a&b&~c&d,
    PD=~a&b&c&~d,
    DRY=~a&b&c&~d,
    RXCK=a&b&c&~d | ~a&b&c&d,
    CDC=~(~a&~b&~c&~d | ~A&~B&C&~D);

endmodule

//---------------------------------------------------
// The top level module makes use of all the above
// modules to form the complete system.
// This is to be connected to the test bench module.
//---------------------------------------------------
module asynch_receiver (rst,en,din,ACK,clk,DRY,ERR,A,B,C,D,OQ
0,OQ1,OQ2,OQ3,OQ4,OQ5,OQ6,OQ7);
  input rst,clk,en,din,ACK;
  //output DRY,ERR,a,b,c,d,oq0,oq1,oq2,oq3,oq4,oq5,oq6,oq7;
  output DRY,ERR,A,B,C,D,OQ0,OQ1,OQ2,OQ3,OQ4,OQ5,OQ6,OQ7;
  wire wst,wsp,wsp1,wsp2,wpd,wcdc,wrxf,wrxck,wq0,wq1,wq2,wq3,
wq4,wq5,wq6,wq7;

  ser_fsm controller(rst,clk,wst,wsp,wrxf,en,ACK,A,B,C,D,wpd,
wcdc,wrxck,ERR,DRY);
  shifter rec_shift_reg(wcdc,wrxck,din,wst,wq0,wq1,wq2,wq3,wq4,
wq5,wq6,wq7,
  wsp1,wsp2);
  latchreceive_reg(wpd,wq0,wq1,wq2,wq3,wq4,wq5,wq6,wq7,OQ0,OQ1,
OQ2,OQ3,OQ4,
  OQ5,OQ6,OQ7);
 divideby11 packet_counter(wrxck,wcdc,wrxf);
 and g1(wsp,wsp1,wsp2);
```

```
endmodule
//==============================================================
// end of design.
// a basic attempt to show how the FSM might be tested - see
state diagram.
module test; //The Test Bench..
// This Test Bench covers the FSM testing and not the shift
register and divide by 11 //counter.  You do not need the
shift reg or divide by 11 counter.  Din is rx to shift reg.
  reg rst,clk,din,en,ack;
  asynch_receiver
  uut(rst,en,din,ACK,clk,DRY,ERR,A,B,C,D,OQ0,OQ1,OQ2,OQ3,OQ4,
OQ5,OQ6,OQ7);
  initial
  begin
  $dumpfile("asynch_rx.vcd");
  $dumpvars;
  rst=0;                        //reset the system.
  clk=0;                        //baseline level for clock
                                signal.
  st=1; sp=1;                   //st and sp set.
  en=0;                         //initially disabled.
  ACK=0;                        //not ACK from user yet.
  //din=1;                      //initial value of serial data
                                input line dealt with later.
  //------------------------------------- //start FSM Test.
  #10 rst=1;                    //release reset from system.
 #10 en=1;                      //enable system.
 #10 clk=~clk;
 #10 clk=~clk;                  //issue a single clock pulse.
 #10 clk=~clk;
  #10 clk=~clk;                 //to s0
  #10 clk=~clk;
  #10 clk=~clk;                 //to s0
  #10 clk=~clk;
  #10 clk=~clk;                 //to s0 - st not 0 yet.
  st=0;
  #10 clk=~clk;
  #10 clk=~clk;                 //s1
  #10 clk=~clk;
  #10 clk=~clk;                 //s2
  #10 clk=~clk;
  #10 clk=~clk;                 //s4, RXCK pulse
  #10 clk=~clk;
  #10 clk=~clk;                 //s5, /RXCK
```

```
  //din=0; dealt with later in the whole system.
  #10 rxf=0
    //din=1; Dealt with later..
  repeat(8)
  begin                         // receive the next data bits
into shift register.
   #10 clk=~clk;                // 8 times around the s5 to
                                s8 loop.
   #10 clk=~clk;                //s6
   #10 clk=~clk;
   #10 clk=~clk;                //s7
   #10 clk=~clk;
   #10 clk=~clk;                //s8
   #10 clk=~clk;
   #10 clk=~clk;                //s5
   //#10 din=~din;  Dealt with later.
  end
  #10 rfx=1;                    // to allow FSM into s9.
  #10 clk=~clk;
  #10 clk=~clk;                 //s9
  #10 ACK=0;                    //raise ACK in response to
data ready signal from user.
  #10 clk=~clk;
  #10 clk=~clk;                 //s11
  #10 clk=~clk;
  #10 clk=~clk;
  #10 clk=~clk;                 //stay in s11 until ACK
                                arrives from user.
  #10 ack=1;

  #10 clk=~clk;                 //s12
  #10 clk=~clk;
  #10 clk=~clk;                 //s0.
  #10 clk=~clk;
  #10 clk=~clk;
  #10 ack=0;
  #10 clk=~clk;
  #10 clk=~clk;
 // more tests needed to test out more of system.
  #10 $finish(10);
  end
endmodule
//////////////////////////////////////////////////////////
```

**Listing 3.4** Asynchronous serial receiver system.

This Verilog HDL contains all the different modules that make up the complete asynchronous serial system but only tests out the FSM part of the system for each transition of the state diagram.

### 3.2.4.5 Verilog HDL simulation of the completed system

Finally the completed system is illustrated in Figure 3.36. Note that OQ0 to OQ7 are not shown updated in this waveform. You could try to simulate this yourself.

For more details, see Minns (1995).

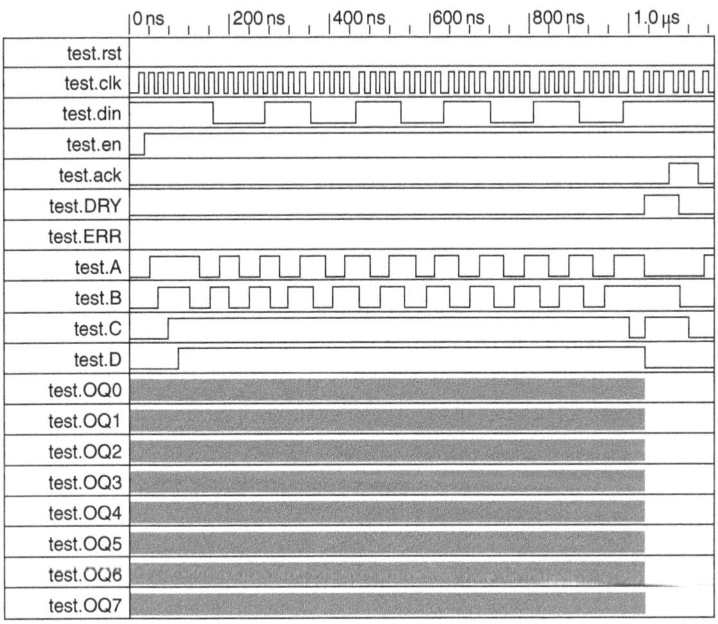


**Figure 3.36** Simulation of the asynchronous receiver.

# 4

# Asynchronous FSM Methods

## 4.1 INTRODUCTION TO ASYNCHRONOUS FSM

In the previous chapters the designs are all based on synchronous (clocked) systems.

This chapter introduces the idea of asynchronous FSM-based designs.

In asynchronous-based designs there is no need for a system clock. Instead, the transition from one state to the next takes place when the input becomes available. This allows transitions between states to occur when the input (an event) takes place. As a result, the transitions can be much faster than those in a clocked system since they do not have to wait for the clock.

Asynchronous FSM systems are sometimes referred to as being **event driven** since the transition from one state to the next can take place whenever an input event occurs.

Because there is no need for a clock, event-based systems can run at lower power levels than clocked-based systems. This is because in a clocked-based system the transistors (CMOS FETs) need to turn on and off at the clock rate, while in event-based systems the transistors only need to turn on or off when the input signals (events) take place. Recall that in clocked-based systems the higher the clock speed, the higher the power drain.

This chapter introduces the concept of event-driven systems, the event cell, and the design methods used to develop event-based systems. A number of real-world examples are looked at and the practical realization and limitations discussed.

If you have not used Boolean algebra for some time, it might be prudent to look over Appendix 1 at the end of this book.

*Digital System Design using FSMs: A Practical Learning Approach*, First Edition. Peter D. Minns.

Companion website: www.wiley.com/go/minns/digitalsystemdesign

### *Frame 4.1*

Answer the following question then turn to **Frame 4.2**.

In an event-based system the FSM will move to the next state when:

- **The input event occurs.**
- **The input event and the clock occurs.**
- When the clock signal occurs.

### *Frame 4.2*

The answer to the question is, of course, **when the input event occurs**, since there is no clock in an event-driven FSM.

As there is no clock in an event-based system, clocked flip-flops such as the *T* type or *D* type cannot be used. Instead a special **event cell** must be used. There will be as many event cells as there are secondary state variables (SSVs) needed to implement the event-driven state diagram describing the system.

**In a five-state event-based FSM system, how many of the following will there be?**

**A. There will be five *D* type flip-flops.**
**B. There will be five event cells.**
**C. There will be five *T* type flip-flops.**
**D. There will be three event cells.**
**E. There will be three *D* type flip-flops.**
**F. There will be three *T* type flip-flops.**

Turn to **Frame 4.3**.

### *Frame 4.3*

The answer is D, since five states requires $2^3$ states; therefore, three event cells are needed to identify each of the states in the state diagram.

Consider the two block diagrams and state diagrams in Figure 4.1.

**Which diagram is a clocked system, and which is an event system?**

**Note: the block diagram on the right has the same inputs as the one on the left.**

When you have decided turn to **Frame 4.4**.

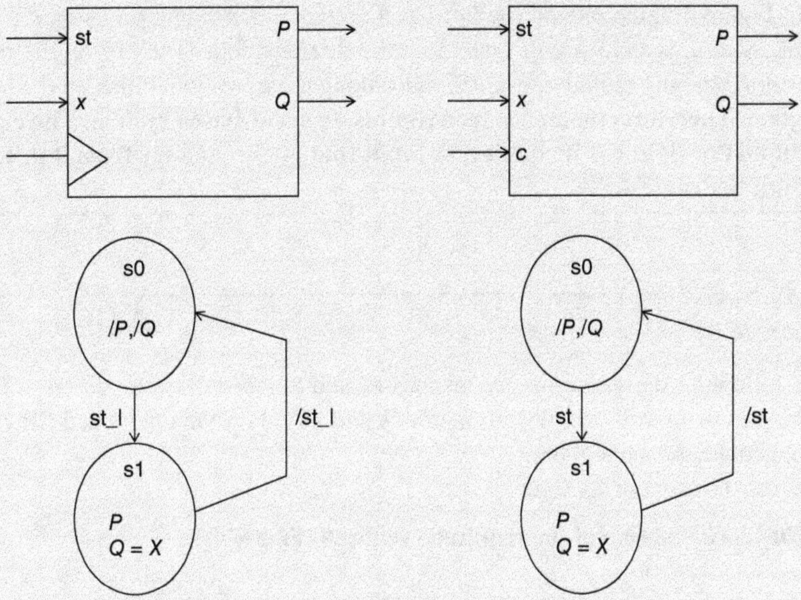


**Figure 4.1** Event and clocked block diagram and state diagram.

***Frame 4.4***

The answer is shown in Figure 4.2.

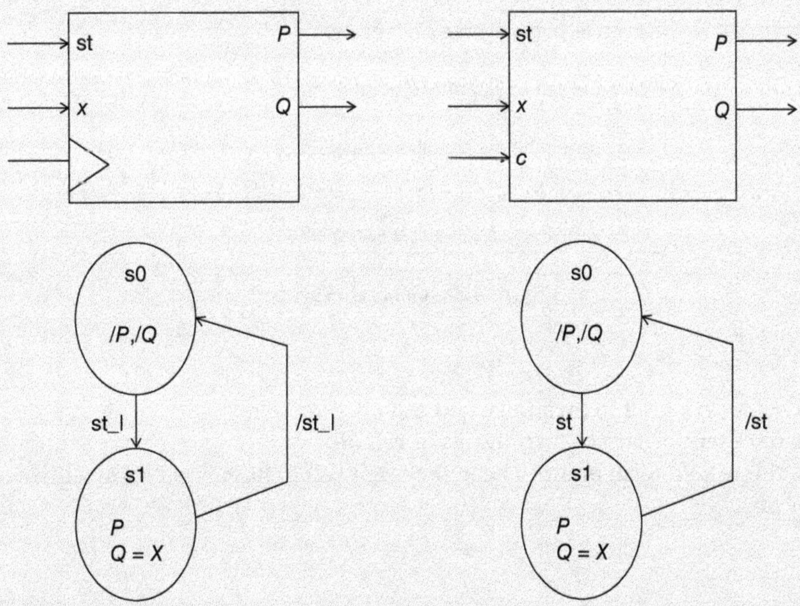


**Figure 4.2** Event and clocked block diagram and state diagram.

In Figure 4.2, the one on the left has a system clock signal on its block diagram, as well as a clock pulse next to the state transitions. The one on the right does not, but only the input event signals along the transitional lines in the state diagram. **Therefore, the one on the right is an event-based system. The clock input on the right *c* is just an event input that goes regularly from 0 to 1.**

Turn to **Frame 4.5**.

### *Frame 4.5*

The following derivation is my own work and has been developed from first principles using my own design methods and based on the unclocked SR flip-flop, using Boolean algebra.

What the event cell will consist of:

- **The block diagram of the required event cell (Figure 4.3).**

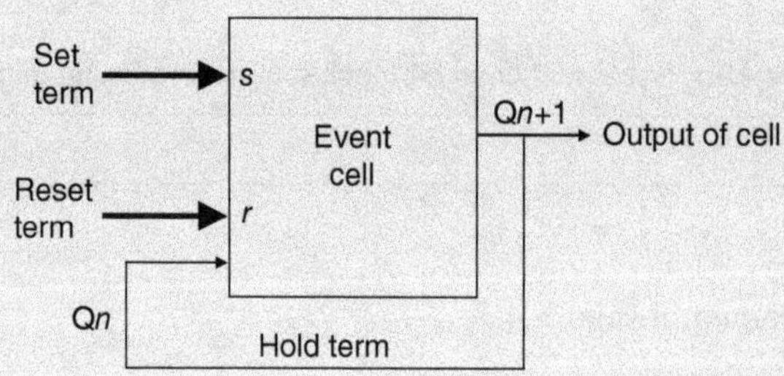


*Qn* is cell output before *s* or *r* event.

*Qn* + 1 is cell output after *s* or *r* event.

This implies the event cell is controlled by *s* or *r* events.

'Hold term' implies the ability to remember.

**Figure 4.3** General event cell.

This has:

- **A set term *s* used to turn the event cell on.**
- **A reset term *r* used to turn the event cell off.**
- **A single hold term input to keep the event cell in its current state (either on or off).**

- **A single output ($Qn$) from the cell that is either logic 0 (cell turned off) or logic 1 (cell turned on). The inverted cell output can be added if necessary.**

In the development of the event cell there will initially be only one set input and one reset input.

Turn to **Frame 4.6**.

## *Frame 4.6*

Figure 4.4 shows the event cell.

Specification of Event Cell

| Row | s | r | Qn | Qn+1 | |
|---|---|---|---|---|---|
| 1 | 0 | 0 | | | no change |
| 2 | 0 | 0 | | | no change |
| 3 | 0 | 1 | | | reinforced reset |
| 4 | 0 | 1 | | | reset |
| 5 | 1 | 0 | | | set |
| 6 | 1 | 0 | | | reinforced set |
| 7 | 1 | 1 | | | not allowed |
| 8 | 1 | 1 | | | not allowed |

Each row corresponds to a possible particular state of the event cell.
This can be used to determine the characteristic equation for the event cell.

**Figure 4.4** Inputs to the event cell.

In the figure we have eight rows defining the different possible states that the event cell can occupy.

Note that column $Qn$ refers to the cell output at the same time that the cell inputs are applied. This is referred to as the 'current state of the event cell'. The column $Qn + 1$ refers to the cell output some brief time after the cell inputs take up their values in the same row. It is the **next state** of the event cell.

In row 1 the set term $s$ is 0, as is the reset term $r$. The present state of the cell output is 0. The cell must therefore remain in the 0 state as there is no requirement to set the cell on.

In row 2 the cell output is at logic 1 but its set $s$ and reset $r$ inputs are both 0. The cell should therefore remain in the set state.

In row 3 the cell output is in its 0 state and its reset input is at logic 1. This means that the cell must remain reset. This is referred to as a 'reinforced reset condition'.

**Try to complete the table and then move on to Frame 4.7.**

## *Frame 4.7*

The completed table is illustrated in Figure 4.5.

Specification of Event Cell

| Row | s | r | Qn | Qn+1 | |
|---|---|---|---|---|---|
| 1 | 0 | 0 | 0 | | no change |
| 2 | 0 | 0 | 1 | | no change |
| 3 | 0 | 1 | 0 | | reinforced reset |
| 4 | 0 | 1 | 1 | | reset |
| 5 | 1 | 0 | 0 | | set |
| 6 | 1 | 0 | 1 | | reinforced set |
| 7 | 1 | 1 | 0 | | not allowed |
| 8 | 1 | 1 | 1 | | not allowed |

Each row corresponds to a possible particular state of the event cell.
This can be used to determine the characteristic equation for the event cell.

**Figure 4.5** Inputs and present state output of event cell.

Here are the results for the $Qn + 1$ column:

| Row | $s$ | $r$ | $Qn$ | $Qn + 1$ |
|---|---|---|---|---|
| 1 | 0 | 0 | 0 | 0 |
| 2 | 0 | 0 | 1 | 1 |
| 3 | 0 | 1 | 0 | 0 |
| 4 | 0 | 1 | 1 | 0 |
| 5 | 1 | 0 | 0 | 1 |
| 6 | 1 | 0 | 1 | 1 |
| 7 | 1 | 1 | 0 | $X$ |
| 8 | 1 | 1 | 1 | $X$ |

Check this against your own results.

Note that in row 4 the cell output $Qn$ is 1 and the reset input is active at logic 1. This must turn the cell off as indicated by the 0 in the $Qn + 1$ output.

In row 5 the cell output $Qn$ is 0 and the set input of the cell is 1, indicating that the cell $Qn + 1$ should be set. This is the turn on state of the event cell, hence $Qn + 1 = 1$.

Row 6 illustrates the cell already set ($Qn = 1$) with a set input at 1. This is a reinforced set condition.

The rows 7 and 8 indicate the impossible condition of the event cell in which both the set and reset cell inputs are at logic 1. Clearly, they cannot be allowed to occupy this condition since it would imply that the cell should both set and reset at the same time. In our design we will assume that these two states cannot happen and assume that rows 7 and 8 are **don't care** states. This simplifies the design of the event cell.

Rows 7 and 8 can therefore be used as **don't care** states in our design equations for the event cell.

**This may not seem a good idea – but as you will find out soon we tend to use this idea in practice to avoid having unnecessarily complicated logic in our designs.**

Turn to **Frame 4.8**.

### Frame 4.8

The tables in Frame 4.7 can be used to derive the equation describing the behaviour of the event cell.

Look at columns *s*, *r*, Q*n*, and Q*n* + 1 in the final table in Frame 4.7.

In this table consider the Q*n* + 1 column as the expected output (next state) and columns *s*, *r*, and Q*n* as the present state. An equation for Q*n* + 1(*s*, *r*, Q*n*) can be obtained.

Try writing this equation out from the table: to help here are the first two terms:

$$Qn+1 = /s/rQn + s/r/Qn + ..$$

Turn to **Frame 4.9**.

### Frame 4.9

Here it is:

$$Qn+1 = /s/rQn + s/r/Qn + s/rQn + (sr/Qn) + (srQn).$$

The last two terms in parenthesis are **don't care** terms.

Now apply the rules of Boolean algebra to this equation to simplify:

$$Qn{+}1 = /s/rQn + s/r/Qn + s/rQn + \mathrm{sr}(/Qn + Qn).$$
$$Qn{+}1 = /s/rQn + s/r/Qn + s/rQn + \mathrm{sr} \text{ since } (/Qn + Qn) = 1.$$
$$Qn{+}1 = /s/rQn + s/r(/Qn + Qn) + \mathrm{sr} \text{ and the } (/Qn + Qn) = 1,$$

so:

$$Qn+1 = /s/rQn + s/r + \mathrm{sr} = /\mathrm{sr}Qn + s(r + /r) \text{ and with } (r + /r) = 1,$$

we obtain:

$$Qn+1 = /s\,/\,rQn + s.$$

Now apply the aux rule to this to obtain:

$$Qn+1 = /rQn + s.$$

Re-arrange to obtain:

$$Qn+1 = s + Qn \cdot /\,r. \qquad \text{Characteristic equation (4.1)}$$

Turn to **Frame 4.10**.

### *Frame 4.10*

The characteristic equation describes the behaviour of the event cell so far derived.

$$Qn+1 = s + Qn \cdot /\,r. \qquad \text{Characteristic equation.}$$

The terms on the right-hand side define the condition of the event cell in its **present state**.

The Q$n$ + 1 term on the left defines the **next state** that the cell can occupy.

Set up an initial condition for the cell, and then use the characteristic equation to calculate the next state of the event cell. For example:

*If* $s = 1, r = 0$, and $Qn = 0$ then $Qn+1 = 1 + 0 \cdot /\,0 = 1 + 0 \cdot 1 = 1 + 0 = 1.$
So $Qn+1 = 1.$

This shows that the event cell has set.

**Try to predict the other possible event cell states and then turn to Frame 4.11.**

### *Frame 4.11*

There are three other possible conditions:

Make $s = 0$, $r = 1$, while Q$n$ = 1. Then Q$n$ + 1 = 0 + 1·/1 = 0 + 1· 0 = 0 + 0 = 0. So Q$n$ + 1 = 0.

This indicates that the event cell has reset.

Now let $s = 0$, $r = 0$, and Q$n$ = 1. Then Q$n$ + 1 = 0 + 1·/0 = 0 + 1·1 = 0 + 1 = 1. The event cell remains set when $s$ and $r$ are both zero.

Finally, try $s = 0$, $r = 0$, and Q$n = 0$. This results in Q$n + 1 = 0 + 0 \cdot / 0 = 0 + 0.1 = 0 + 0 = 0$. The event cell remains reset when $s$ and $r$ are zero.

Clearly, the event cell appears to work as expected.

So what does the characteristic equation circuit look like?

**Try to draw it from the characteristic equation** then turn to **Frame 4.12**.

### *Frame 4.12*

The characteristic equation is:

$$Qn + 1 = s + Qn \cdot / r.$$

The circuit is shown in Figure 4.6.

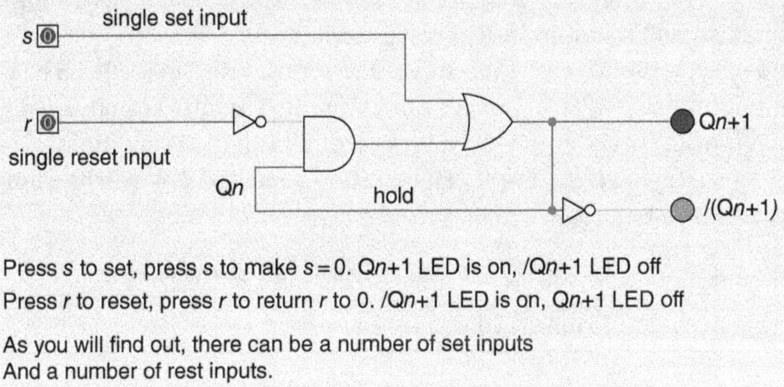


**Figure 4.6** Circuit of event cell using characteristic equation.

This circuit has a single set input $s$ and a single reset input $r$.

Figure 4.6 has been implemented using the Logisim Simulator and can be used to try out the design. It is filed as **fig4.6 event_cell.circ** and you can try it out using Logisim.

Appendix A2 gives a tutorial on how to use Logisim. **There are other examples in Chapter 4 as well as Chapter 6 that give good demonstrations of the method.** This is a very good example showing how the event cell works.

Now, let us continue with Figure 4.6.

Note in the event-driven cell the output Q$n$ + 1 is the same signal as Q$n$. This is because Q$n$ is the **present state** of the signal and Q$n$ + 1 is its next state. Note that in Figure 4.6 the AND gate output represents the Q$n$ output which can be seen better in Figure 4.17, but for now just accept this.

**From now on let $Qn + 1 = Qn = Q$.**
The characteristic equation now becomes:

$$Q = s + Q \cdot / r. \qquad \text{Event cell characteristic equation (4.2)}$$

Looking at this equation, it becomes clear that the equation is **recursive**, implying that the output is on both sides of the **equal** sign (=).

Turn to **Frame 4.13**.

### *Frame 4.13*

Up to this point the event cell has two inputs: *s* and *r*.

But in a real event-driven FSM there could be more than one set input and more than one reset input. These will, however, only occur one at a time as the asynchronous FSM moves from one state to the next. This implies a **summation** of set terms *s* and a **summation** of reset *r* terms.

Sum of set terms s1 + s2 + .. + sn i.e. Σsn terms; also Sum(sn) can be used.

Sum of reset r1 + r2 + .. + rn i.e. Σrn terms; also Sum(rn) can be used.

Therefore, Σ/(rn) is simply the inversion of all summed conditions.

**Try to write down the event cell equation repeated below with sum of set and reset terms:**

$$Q = s + Q \cdot / r. \qquad \text{Event cell characteristic equation (4.2)}$$

Now move on to **Frame 4.14**.

### *Frame 4.14*

Here is the complete equation.

$$Q = \Sigma\mathbf{sn} + Q \cdot \Sigma/\mathbf{rn}. \qquad \textbf{Sequential equation (4.3)}$$

This is called the **sequential equation** and defines the **general equation** that can be used to obtain the required equations for event cells. Its use will be looked at more closely in Frame 4.15.

Equation (4.3) has been adapted from a similar arrangement that used *JK* flip-flops and NAND gates.

Source: *Problems and Solutions in Logic Design* (Zissos 1979) equation 1.2, p. 19. Adapted with permission of the Licensor (Oxford Publishing Limited) through PLSclear.

However, my equation has been developed from first principles using *D* type flip-flops and modern technology (Minns 1995).

Turn to **Frame 4.15**.

### Frame 4.15

To see how the sequential equation will be used in designs consider the state diagram shown in Figure 4.7. This example is used to introduce a number of potential problems using the event-driven method.

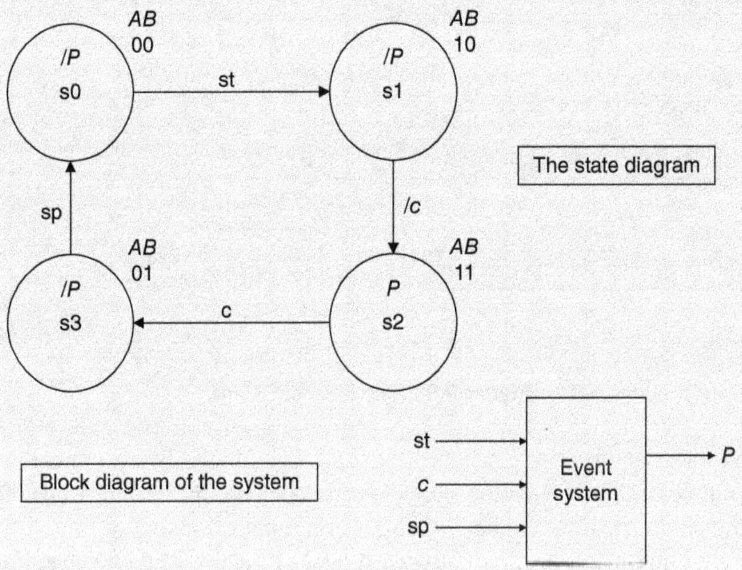


**Figure 4.7** An example of the complete method.

The sequential equations for this system would look like:

$$A = \Sigma\text{sa} + A \cdot \Sigma/\text{ra} \quad \text{for event cell } A.$$
$$B = \Sigma\text{sb} + B \cdot \Sigma/\text{rb} \quad \text{for event cell } B.$$

They could also be written as:

$$A = \text{Sum}(\text{sa}) + A \cdot \text{Sum}(/\text{ra}).$$
$$B = \text{Sum}(\text{sb}) + B \cdot \text{Sum}(/\text{rb}).$$

Turn to **Frame 4.16**.

### *Frame 4.16*

The best way to do this is to look at a practical design and work through it. Consider the state diagram in Figure 4.7, which is repeated in Figure 4.8.

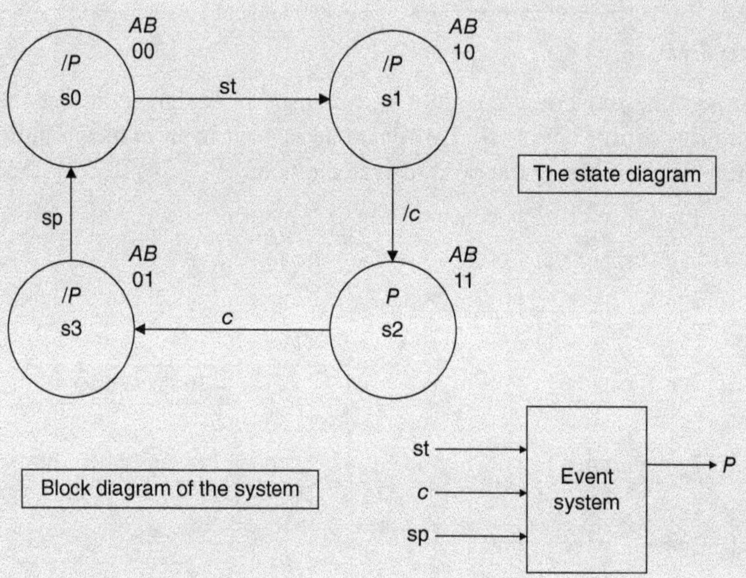


**Figure 4.8** The example again.

Here the state diagram consists of four states and uses two event cells: *A* and *B*. Take note of the fact that the state diagram does not have a clock (_|) between state transitions as it moves between states on input events.

For example, the transition between s0 and s1 occurs when the event input st goes to 1. There is no need to wait for a clock pulse. In s1 the input on the signal *c* must be zero for the machine to move to state s2. When signal *c* goes to logic 1 the machine will move to s3 where it will wait for signal sp to become 1 before moving to s0 again.

Note that if in s0 st becomes 1 (or is still 1) and signal *c* is zero, the machine will move from state s0 to s1 and then directly on to state s2, where it will wait for signal *c* to become 1.

Now let us develop the equations for the event cell *A*. The sequential equation tells us that we need to look for the set term:

$$\begin{aligned} & \quad\ \ 0>1 \qquad\ 1>0 \\ A = & \ \Sigma \mathrm{sa} + A \cdot \Sigma/\mathrm{ra}. \end{aligned}$$

Note we are looking for 0 to 1 in the sa term(*s*), and 1 to 0 in the /ra term.

The term(s) Σsa is the sum of the set inputs to event cell *A* and represents a 0 to 1 transition on an st input. In the state diagram this occurs on the s0 to s1 transition when input st goes to 1. There are no other 0 to 1 transitions between any other states in the state diagram.

$$\Sigma \text{sa} = \text{s0} \cdot \text{st}.$$

The other term /ra, is the reset of event cell *A* (negated) and represents the condition when a 1 to 0 transition occurs between two states in the state diagram. This occurs on the transition between s2 and s3.

Now /ra = /(s2·*c*). Note that the term is inverted due to the /ra:

$$\text{So now}: A = \text{s0} \cdot \text{st} + A \cdot /(\text{s2} \cdot c) \text{ as the event equation for cell } A.$$

**Can you now write this down in terms of the SSVs for states s0 and s2? Turn to Frame 4.17.**

***Frame 4.17***

Your equation should look like:

$$A = \text{s0} \cdot \text{st} + A \cdot /(\text{s2} \cdot c) = /A\,/B \cdot \text{st} + A \cdot /(AB \cdot c).$$

However, there is a problem with this equation.

Consider the equation for *A* = /*A*/*B*·st + *A*·/(*AB*·*c*).

In state s0 with *A* = *B* = 0 the st input going to logic 1 will cause the event cell *A* to go from 0 to 1. However, this can only happen when *A* and *B* are both 0 since the turn on set of *A* is /*A*/*B*·st.

When *A* becomes 1 then the set term becomes 0. If the event cell for *A* has not latched on then the cell could revert back to its reset state. This is a timing issue. To prevent this from happening, an additional set term is added.

*A* = s0·st + **s1·st** shows the set term with s1 using the same input as s0.

This makes sure that the event cell still has a set term when it reaches state s1 (see state diagram in Figure 4.8 in Frame 4.16). Now, add the next state but use the present state input condition.

There is a similar problem when the event cell *A* needs to reset.

In this case the reset term is /(s2·*c*), hence another term is added to make /(s2·*c* + **s3·*c***) to ensure that the reset term is still present when the state machine reaches state s3, thus adding the next state but using the present state input condition.

The sequential equation for event cell $A$ is now:

$$A = \Sigma \text{sa} + A \cdot \Sigma/\text{ra}.$$

$$A = \text{s0} \cdot \text{st} + \text{s1} \cdot \text{st} + A \cdot /(\text{s2} \cdot c + \text{s3} \cdot c). \quad \text{Complete equation}$$

Replacing the state numbers with the SSVs:

$$A = /\boldsymbol{A}/B \cdot \text{st} + \boldsymbol{A}/B \cdot \text{st} + A \cdot /(\boldsymbol{A}B \cdot c + /\boldsymbol{A}B \cdot c). \quad \text{Full event cell equation}$$

From this you should be able to see that the highlighted variables are complemented and that using the logical adjacency rule (**see Appendix A on Boolean algebra if you are a little rusty on this**):

$$A = (/A + A)B \cdot \text{st} + A \cdot /([A + /A] \cdot B \cdot c)$$

results in:

$$A = /B \cdot \text{st} + A \cdot /(B \cdot c) = /B \cdot \text{st} + A/B + A/c. \quad \text{Reduced equation}$$

A little thought will show that this will always be the case.

- **In the set term for $A$, the $/A$ term will be removed.**
- **In the reset term for $A$, the $A$ term will be removed.**

And in general:

- **In the set term for event cell $x$, the $/x$ term will be removed.**
- **In the reset term for event cell $x$, the $x$ term will be removed.**

Note that if st and sp are both active, the FSM will continue to produce $P$ pulses. Therefore, to produce a single $P$ pulse st pressed and released followed by sp pressed and released.

**Try now to produce the event cell equation for $B$ in the state diagram of Frame 4.16 (Figure 4.8).**

**When you do this task, produce the full event cell equation then reduce it, as was done in this frame for $A$.**

When you have completed this, turn to **Frame 4.18**.

***Frame 4.18***

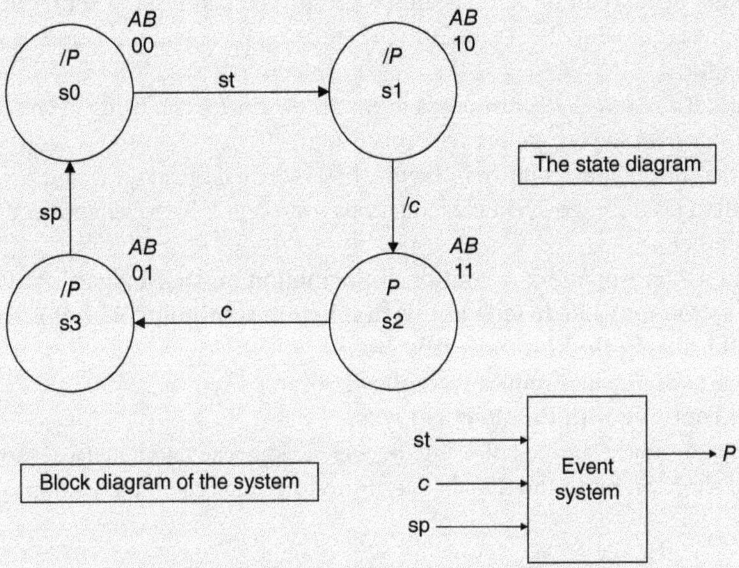


**Figure 4.9** Event cell to help form $B$ equation.

From the state diagram (Figure 4.9), for $B$:

$$B = \Sigma \text{sb} + B \cdot \Sigma / \text{rb} = \text{s1} \cdot /c + B \cdot /(\text{s3} \cdot \text{sp}).$$
$$B = \text{s1} \cdot /c + B \cdot /(\text{s3} \cdot \text{sp}) = \text{s1} \cdot /c + \text{s2} \cdot /c + B/(\text{s3} \cdot \text{sp} + \text{s0} \cdot \text{sp}).$$
$$B = A/B/c + AB \cdot c + B/(/AB\text{sp} + /A/B\text{sp}).$$
$$B = (A/B + AB) \cdot /c + B/(/AB + /A/B)\text{sp}.$$
$$B = A \cdot /c + B/(/A\text{sp}) \text{ De-Morgan's rule to obtain:}$$
$$B = A \cdot /c + AB + B/\text{sp}.$$

Therefore, the final equation becomes:

$$B = A \cdot /c + B \cdot /(/A \cdot \text{sp}) = A \cdot /c + AB + B/\text{sp}.$$

Turn to **Frame 4.19**.

### *Frame 4.19 Using the Short Cut Rule*

You do not need to go through all this work, as you can use the short cut rule. This implies removing the *A* and */A* in the equation for *A* and the *B* and */B* in the equation for *B*.

Note that Zissos (1979) just drops terms all the time when using event-driven FSM designs. Now you can see why this is done.

Use the state diagram shown in Figure 4.10 (which has had the *A* and *B* values changed) to produce the event cell equations for *A* and *B* in the same way as was shown earlier.

**As an aside: Appendix A contains information on De-Morgan's rule and its use, so you may like to look at this first before continuing with this frame. You will be using De-Morgan's rule often.**

This is used in the examples that follow.

**Let's continue with the short cut rule.**

As already stated, instead of using the logical adjacency we can use a short cut rule and simply remove the *A* and */A* terms to form the required equations.

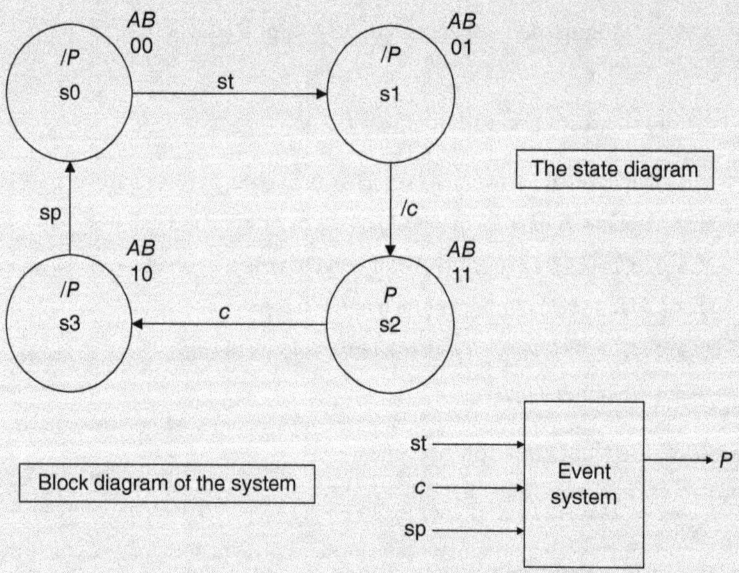


**Figure 4.10** Using the short cut method.

**Obtain the equations for *A* and *B*.**

When you have done this, turn to **Frame 4.20**.

### *Frame 4.20*

From the state diagram: note the highlighted parts where we have used the short cut rule (and De-Morgan's rule) are used.

$$A = \Sigma \text{sa} + A{\cdot}\Sigma \,/\, \text{ra}.$$
$$A = \text{s1} \cdot /c + A \cdot /(/\text{s3} \cdot \text{sp}).$$
$$A = B \cdot /c + A \cdot /(/B \cdot \text{sp}), \text{ so applying De-Morgan's rule:}$$
$$A = B \cdot /c + AB + A\,/\,\text{sp}.$$
$$B = \Sigma \text{sb} + A \cdot \Sigma \,/\, \text{rb}.$$
$$B = \text{s0} \cdot \text{st} + B \cdot /(\text{s2} \cdot c).$$
$$B = /\,A \cdot \text{st} + B \cdot /(A \cdot c). \text{ now using De-Morgan's rule :}$$
$$B = /\,A \cdot \text{st} + /\,AB + B \cdot /c.$$

If you made a mistake, go back to Frames 4.16 and read through to Frame 4.19 to see where you went wrong before continuing with the work. **Remember to take into account the short cut rule and take care with the use of De-Morgan's rule.**

**Finally, if you keep *P* = s2 = *AB*, you will have a single-pulse FSM (but since st = 1 at any time of /*c* the *P* pulse can vary in width).**

Turn to **Frame 4.21**.

### *Frame 4.21*

Please note that it would be a good idea to replace input sp with /st so that the user can make use of st only to produce a single-pulse *P* at the output. This has been done in Figure 4.11. Note that we have also changed *AB* values in Figure 4.11.

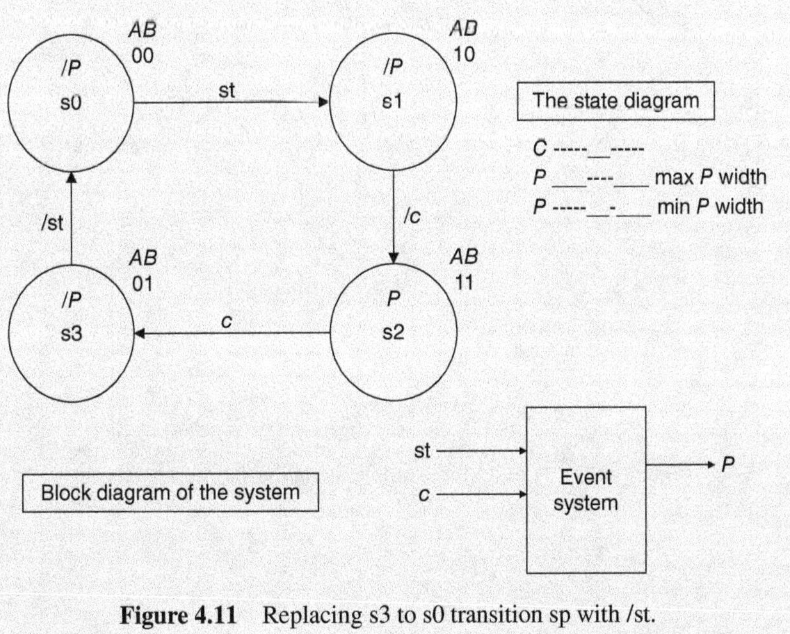


**Figure 4.11** Replacing s3 to s0 transition sp with /st.

Looking at Figure 4.11, you can see the possible maximum and minimum $P$ pulse width.

Taking Figure 4.11 to obtain:

$$A = \mathrm{s0} \cdot \mathrm{st} + A \cdot /(\mathrm{s2} \cdot c) = /B \cdot \mathrm{st} + A \cdot /(B \cdot c).$$

$$A = /B \cdot \mathrm{st} + A \cdot (/B + /c).$$

$$\mathbf{A = /B \cdot st + A \cdot /B + A \cdot /c.}$$

$$B = \mathrm{s1} \cdot /c + B \cdot /(\mathrm{s3} \cdot /\mathrm{st}) = A \cdot /c + B \cdot /(/A \cdot /\mathrm{st}).$$

$$B = A \cdot /c + B \cdot (A + \mathrm{st}).$$

$$\mathbf{B = A \cdot /c + AB + B \cdot st.}$$

$$\mathbf{P = s2 = AB.}$$

This can be converted into a circuit and then drawn with Logisim. The Logisim circuit can then be simulated to show how it works. This will be left to the reader.

Turn to **Frame 4.22**.

## *Frame 4.22*

Now consider the state diagram of Figure 4.12.

This is similar to Figure 4.11 but with different inputs.

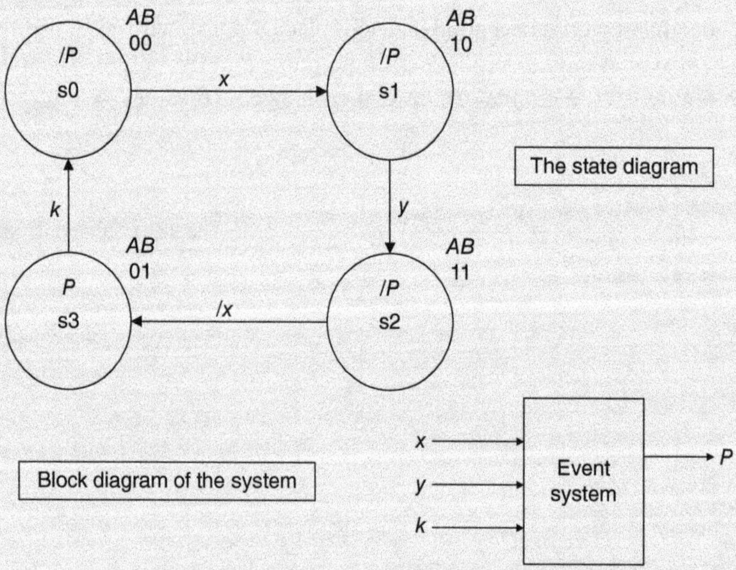


**Figure 4.12** A different example using the short cut method.

In this case the short cut method is used.

$$A = \Sigma \text{sa} + A \cdot \Sigma\, /\text{ra} = \text{s0} \cdot x + A \cdot /\left(\text{s2} \cdot /\, x\right) = /B \cdot x + A \cdot /\left(B \cdot /\, x\right).$$

$$A = /B \cdot x + A \cdot /\left(B \cdot /\, x\right) = /B \cdot x + A/B + Ax \text{ using De-Morgan's rule.}$$

$$B = \Sigma \text{sb} + B \cdot \Sigma\, /\text{rb}.$$

$$B = \text{s1} \cdot y + B \cdot /\,(\text{s3} \cdot k) = A \cdot y + B \cdot /\,(/A \cdot k).$$

Now apply De-Morgan's rule to equation for $B$:

$$B = A \cdot y + B \cdot /\left(/A \cdot k\right) = A \cdot y + B \cdot \left(A + /k\right) = A \cdot y + AB + B \cdot /\, k.$$

The outputs for the event state machine follow in the same way that outputs for synchronous clock-driven FSM did.

Therefore, in the state diagram in Figure 4.12, the output is $P = \text{s3} = /A \cdot B$.

Turn to **Frame 4.23**.

### *Frame 4.23 With Mealy output R*

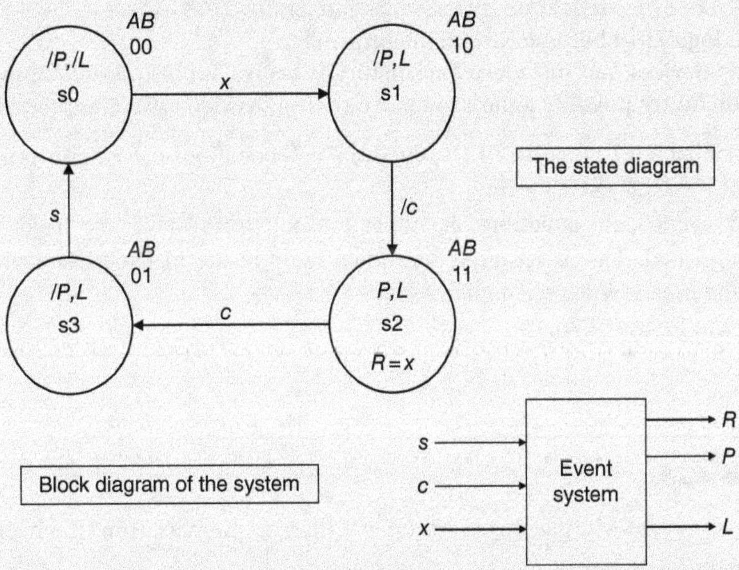


**Figure 4.13** Example with Mealy output $R$.

Note that in Figure 4.13 we have: in s2 $P = 1$, and in s1, s2, s3, $L = 1$. This makes $L = /\text{s0} = /(/A \cdot /B)$ active low. Note that $R = 1$ in state s2, but only when $x = 1$.

**Using the state diagram in Figure 4.13, produce the output equations.**

When you have completed this turn to **Frame 4.24**.

### Frame 4.24

The outputs in the state diagram of Figure 4.13 are:

$$P = s2 = AB.$$

$$L = s1 + s2 + s3 = A\,/\,B + AB + /AB = A(/B + B) + B(A + /A) = A + B.$$

$$R = s2 \cdot x(\text{a Mealy output}).$$

$$R = AB \cdot x.$$

Turn to **Frame 4.25**.

### Frame 4.25

In most situations, our event-driven FSMs will be implemented using:

- **Programmable logic devices (PLDs) or complex programmable logic devices (CPLDs) or even field programmable gate arrays (FPGAs).**
- **TTL logic (will become rare in the future).**
- **Relay devices (not just electromechanically relays but also static relays).**
- **In the future possibly nano-switched devices (Drexler 1991, Chapter** 13).

When using PLDs and CPLDs, the event cell equations can be converted into 'sum of product' (SOP) form.

The event cell equations in their present form have the reset terms complemented. This is required additional logic in the PLD/CPLD device. As indicated in the sequential equations.

Turn to **Frame 4.26**.

### Frame 4.26

You can use De-Morgan's rule to remove the complement from the event cell equations.

**Remember: Appendix A contains information on De-Morgan's rule.**

Here's a quick reminder: using the **De-Morgan's rule** $/(A \cdot B) = /A + /B$. Here you just **change everything**.

Then $/(A \cdot B)$ becomes $/A + /B$. If you had $/(/A \cdot B)$ De-Morgan's rule would produce $//A + /B = A + /B$.

Turn to **Frame 4.27**.

### *Frame 4.27*

And now a different problem that looks for an error input from an analogue-to-digital converter (ADC). Remember, ADCs are used in Chapter 3.

This problem was first looked at in the tutorials section of Chapter 3 as a clock-driven FSM system.

**This example contains a lot of detail, so you may want to skip it until you have had more experience. You could move on to Frame 4.31.**

Look at the state diagram shown in Figure 4.14. Note that the state diagram is the same as Figure 3.28 (the hi lo alarm FSM state diagram) but this time we use an event-driven FSM.

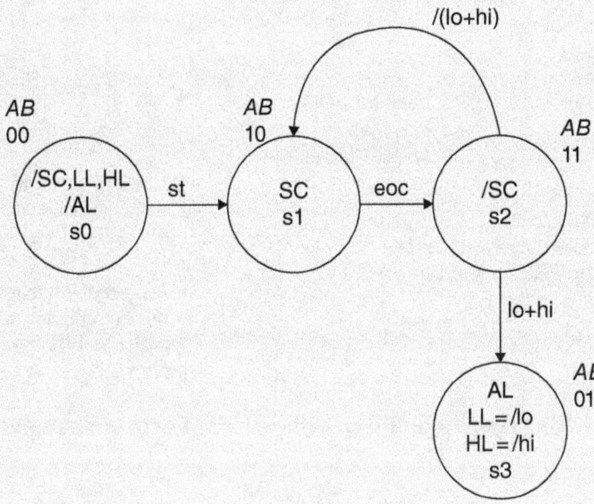


**Figure 4.14** State diagram of high-low alarm FSM.

In this example an ADC is repeatedly used to test an input analogue value. If it is within the high (hi) and low (lo) limits it just continues testing. If it is higher or lower than the hi/lo limits, the FSM moves to s3 stops (waiting for reset) and lights the high LED (HL) or low LED (LL).

Here there is a feedback term from state s2 back to state s1. The output signal SC indicates that this output goes high (logic 1) when in state s1, then returns low when the state diagram moves from s1 to s2. Also note the Mealy outputs in state s3. Output LL is only high in s3 and only if the input lo is at logic 0. Output HL is only 1 in state s3 if input hi is logic 0.

At this stage you are going to be introduced to the idea of adding a sample clock to the event-driven design.

Turn to **Frame 4.28**.

***Frame 4.28***

In Figure 4.15a a sample pulse is added to the FSM in the form of 'sp' (sample pulse).

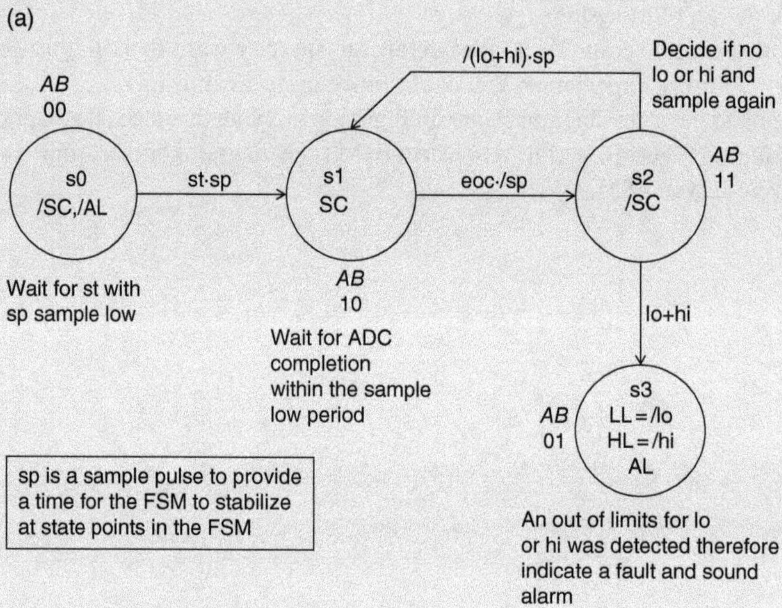


**Figure 4.15a** The event-driven high-low FSM with sampling input sp.

In this event version we have accessed a 'sample pulse sp', during which the low part of the pulse is used to allow the ADC conversion time to occur. This sp signal is simply an event input that is going from 0 to 1 repeatedly:

sp ---______----______---

where the lower period determines the ADC conversion time. The sp is just an event input that happens to be going from 0 to 1 regularly.

The high part of the pulse sp allows another cycle.

**The 'sp sample pulse' helps to keep the event waveform easier to follow.**

The event cell equation for $A = \Sigma\text{sa} + A\cdot\Sigma\ /\text{ra}$.

This is an event-driven example, not a clock-driven example, so you need to take very special care when working out the event-driven cell values for $A$ and $B$.

You can see how it works as follows:

$$A = \text{s0} \cdot \text{st} \cdot \text{sp} + A \cdot /(\text{s2} \cdot /[\text{lo}+\text{hi}]) = /B \cdot \text{st} \cdot \text{sp} + A \cdot /B(/[\text{lo}+\text{hi}])$$
$$= /B \cdot \text{st} \cdot \text{sp} + A/(B[/\text{lo} \cdot /\text{hi}]).$$
$$A = /B \cdot \text{st} \cdot \text{sp} + A \cdot /(B \cdot /\text{lo} \cdot /\text{hi}) = /B \cdot \text{st} \cdot \text{sp} + A \cdot (/B + \text{lo} + \text{hi})$$
$$= /B \cdot \text{st} \cdot \text{sp} + A/B + A \cdot \text{lo} + A \cdot \text{hi}.$$

Resulting in:

$$A = /B \cdot \text{st} \cdot \text{sp} + A/B + A \cdot \text{lo} + A \cdot \text{hi}.$$

If you did not get this right, look through it carefully then try to do the event cell *B* equation yourself. You have to make a lot of use of De-Morgan's rule.

When you have done this turn to **Frame 4.29**.

### *Frame 4.29*

The event cell equation for $B = \Sigma\text{sb} + B\cdot\Sigma/\text{rb}$.

$$B = \text{s1} \cdot \text{eoc} \cdot /\text{sp} + B \cdot /(\text{s2} \cdot /[\text{lo}+\text{hi}] \cdot \text{sp}).$$
$$B = A \cdot \text{eoc} \cdot /\text{sp} + B \cdot /(A \cdot /\text{lo} \cdot /\text{hi} \cdot \text{sp}) \text{ applying De-Morgan's rule.}$$
$$B = A \cdot \text{eoc} \cdot /\text{sp} + B \cdot (/A + \text{lo} + \text{hi} + /\text{sp}) \text{ and applying}$$

De-Morgan's rule again to the second term.

$$B = A \cdot \text{eoc} \cdot /\text{sp} + /AB + \text{Bio} + \text{Bhi} + B/\text{sp}.$$

Note that s3 is an invariant state that the FSM remains in and can only be returned to s0 via a reset.

The complete event-driven FSM cells are:

$$A = /B \cdot \text{st} \cdot \text{sp} + A/B + A \cdot \text{lo} + A \cdot \text{hi} \quad (4.4)$$

$$B = A \cdot \text{eoc} \cdot /\text{sp} + /AB + B \cdot \text{lo} + B \cdot \text{hi} + B \cdot /\text{sp}. \quad (4.5)$$

Not shown in the *B* equation is the s3 term.

The *B* term after getting to s3 needs to be reset; otherwise, the *B* event equation never terminates (you need to use the reset input to make the *B* event cell reset).

Now produce the output equations from the state diagram – note the Mealy outputs.

When you have done this turn to **Frame 4.30**.

### *Frame 4.30*

From Frame 4.28 the output equations are:

SC = s1 = $A/B$.

LL = $/(\mathrm{s3} \cdot \mathrm{lo})$ a Mealy output meaning that LL is 1 only
in s3 and then only if input lo is 0.

HL = $/(\mathrm{s3} \cdot \mathrm{hi})$ a Mealy output that is 1 only in s3 and then only when input hi is 0.

Therefore, HL = /(/*AB*·/hi) active low and LL = /(/*AB*·/lo) active low.
Al = s3 = /*AB*.

The reader can now produce the Logisim circuit for this system. **The completed Logisim circuit is called hiloAlarmwithSample3.circ.** See link details at end of Contents sheet.

My attempt is show in Figure 4.15b.

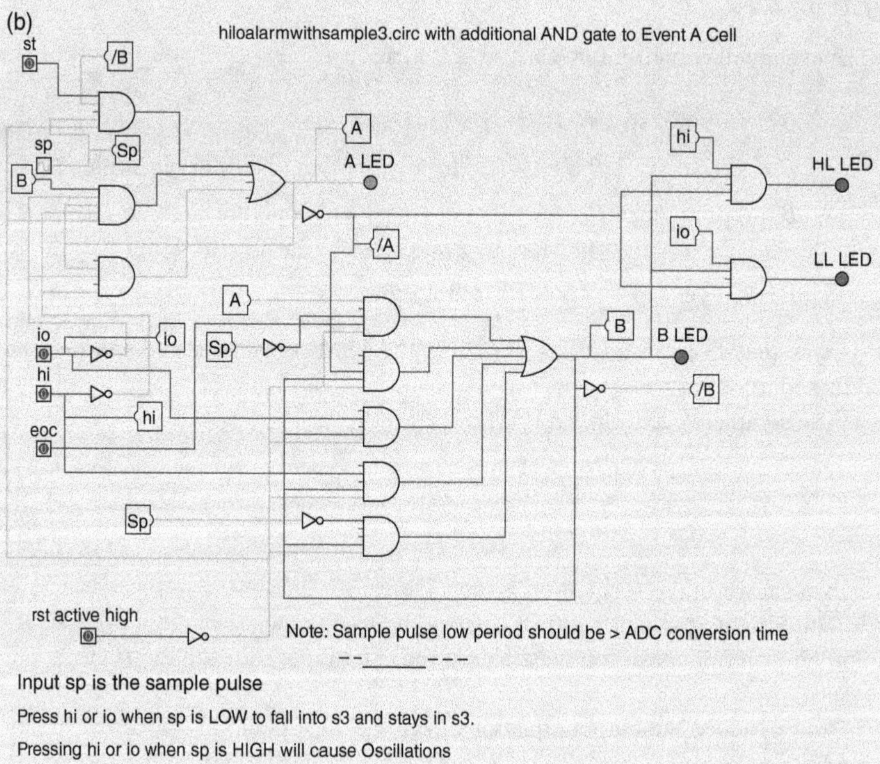


**Figure 4.15b** Logisim circuit for the high-low alarm FSM.

**Note that pressing either hi or lo when sp is LOW FSM will fall into s3 and hold.**

**If you press either hi or lo when sp is HIGH, the FSM will fall into s3 and oscillate.** It will, however, fall into state s3 and hold.

You could modify the FSM so that you can only fall into s3 with /sp.

Note that the equations used for the event cells in Figure 4.15b are:

$$A = /B \cdot \text{st} \cdot \text{sp} + AB \cdot / \text{ lo} \cdot /\text{hi} + A \cdot /\text{lo} \cdot /\text{hi}.$$

$$B = A \cdot \text{eoc} \cdot \text{sp} + B \cdot \text{lo} + B \cdot \text{hi} + B \cdot /\text{sp}.$$

**You could try using the derived event cell** Equations **4.4** and **4.5 to see what you get.**

This example was more complicated than the ones looked at before because of the more complex turn-off terms. Check this example over again if you are still unsure and make sure you understand the method before going on any further. **You will meet up with this kind of thing again in some of the examples**

Up to this point the work in this chapter has focused on how to obtain the event and output equations from an event-driven state diagram.

- **But what does the final circuit look like?**
- **How can the resulting event cells be initialized?**
- **Also, what about obtaining the state diagram in the first place?**

Turn to Frame 4.31.

### *Frame 4.31*

If you look at the examples considered so far you will see that the SSVs (the *A* and *B* event cells) follow a unit distance coding.

You may remember from the work on synchronous (clock-driven) systems in Chapter 3 that you were encouraged to try to obtain a unit distance coding in your state diagrams. The reason given was to avoid some potential glitches in the outputs.

In the design of asynchronous (event-driven) systems it is **mandatory** to follow a unit distance coding of the state diagrams; otherwise, the designs **simply will not work**. These are malfunctions.

**To see why this is so, turn to Frame 4.32.**

### *Frame 4.32*

Please note it is possible to get maloperation in an event-driven design, depending upon a number of conditions. There are three conditions:

- **Possible maloperation between primary inputs.**
- **Maloperation between secondary inputs.**
- **Maloperation between primary inputs and secondary inputs.**

Looking at the first two below:

Primary inputs maloperation is caused when two or more primary inputs are tested to happen at the same time. In practice, the delay path through the FSM will vary between gates and you cannot guarantee that the FSM will detect the arrival of two or more primary inputs happening 'at the same time', owing to the different propagation delays through the logic gates.

To avoid this do not attempt to detect the arrival of such primary inputs 'at the same time' in an FSM as you cannot predict the effect of other delays in the logic of the FSM.

Turn to **Frame 4.33**.

### *Frame 4.33*

Secondary inputs maloperation is when the designer has not followed a unit distance coding for the secondary inputs of the event cell, for example SSVs. The use of a none unit distance coding can result in the FSM falling into a state different to the one intended as a result of unequal propagation delay between event cells.

**To avoid this type of condition, do not try to look for more than one input changing at the same time by using a unit distance coding for the secondary inputs.**

To look at the third maloperation, see Frame 4.36.

Next, turn to **Frame 4.34**.

### *Frame 4.34*

A brief note: there are more details to be found on event-driven designs in the book *Problems and Solutions in Logic Design* by Zissos, published by Oxford University Press, ISDN 0-19-859359-7. This is a somewhat older book that was published back in 1979.

There is also the paper by Duncan and Zissos (1971) that you may find interesting as it goes into detail on the development of gate tolerances in sequential circuits.

Turn to **Frame 4.35**.

### *Frame 4.35*

A common question asked is: 'How do you go about obtaining a unit distance code in state diagrams where you need to connect between states that do not follow a unit distance coding?'

Figure 4.16 illustrates how this might be done.

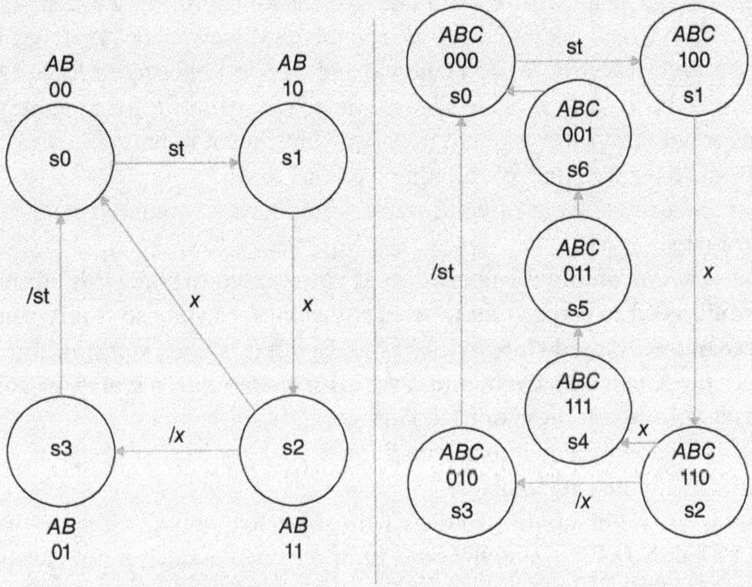


**Figure 4.16** How to keep to unit distance coding.

The state diagram on the left has a transition between states s2 and s0. Clearly this cannot be done without violating the unit distance code rule.

The state diagram on the right illustrates how to overcome the problem by adding another event cell *C*.

Now additional **dummy** states can be added so that each transition can follow a unit distance coding. **Note in some cases you may need to add extra event cells, as illustrated in Figure 4.16**. In the figure states s4, s5, and s6 are dummy states.

The output signals in the new dummy states should be the same as the output values in the state that the dummy states are going to – in this case s0. Essentially, when input signal $x$ is 1 in state s2, the FSM is required to move to s0 (via states s4, s5, and s6).

As event-driven FSM state diagrams become more complex, it can be tedious to obtain unit distance coding. This is an ongoing area of study.

**In most cases for the type of applications commonly encountered, it should be possible to obtain the correct unit distance codes. If your design is complex, perhaps you should consider using a clock-based design.**

**Now we need to consider a potential problem that could occur.**

Consider the right-hand state diagram in Figure 4.16.

In this state diagram, we have added SSV $C$. This creates a state diagram that can have a potential eight states, and we are using seven. If the FSM were to fall into the unused state ($ABC$ = 101) due to a random spike, or a glitch perhaps, then the FSM would fall into this state and the FSM would stop. This means that if we got into state $ABC$ = 011 (state s5) and then fell into state $ABC$ = 101 the FSM would be locked into state s7 (assume $ABC$ = 101 = s7, for example).

This is not acceptable as there is no way out except to press the reset input. This would have the effect of resetting the FSM to s0.

An FSM using six out of eight states would have two unused states and so forth for larger FSMs.

**This is not an automatic operation. If you wanted to make this automatic, you would need to connect all unused states back to state s0 (there could be more than one unused state in an FSM). In other words, you would need to connect the Σ /ra, and Σ /rb, and Σ /rc with these unused states, or connect the reset point to all these unused states.**

This is no doubt an additional condition to look for in your design if you are concerned with this problem.

**This is an unfortunate problem with asynchronous designs (as well as using TFF and *JK*FF in synchronous FSM designs). Also, it is not considered in Zissos (1979).**

Asynchronous FSM designs that use all states in a given design (like four states with two SSVs) will not suffer from this problem, same for three SSV in an eight-state FSM, and so forth. This means that using $2^{\wedge n}$ states in an $n-1$ state FSM would be fine.

You will not consider this potential problem for the remainder of this chapter, but you need to be mindful of it. Note that this problem is not considered in the previous Wiley book (Minns and Elliott 2008).

Turn to **Frame 4.36**.

## *Frame 4.36*

At this point the maloperation between primary inputs and secondary inputs needs to be considered.

Essentially, this is a maloperation between the primary inputs and secondary inputs to the FSM and its event cell operation. This is caused when the propagation delay through to the primary input path to set the cell is greater than the secondary delay path to keep it in that state. It can result in the cell's maloperation.

The primary delay path Tp in Figure 4.17 is from a set input, through the single NAND gate to the output NAND gate. This is through one gate.

The secondary delay path Ts is through the output NAND gate back through the feedback path to the lower NAND gate. This is through two gates.

**To prevent this type of malfunction from occurring, ensure that the primary delay Tp is less than the secondary delay Ts.**

That is:

$$\mathrm{Tp} < \mathrm{Ts}. \tag{4.6}$$

and

$$\mathrm{Tpmax} / \mathrm{Tsmin} < 1. \tag{4.7}$$

Source: *Problems and Solutions in Logic Design* (Zissos 1979), equation 1.1, p. 19. Reproduced with permission of the Licensor (Oxford Publishing Limited) through PLSclear.

This follows the 33.3% rule as detailed in Zissos (1979), which states that: 'sequential circuits are hazard-free when implemented with gates whose maximum speed tolerance is +/– 33.3%. If for example, the nominal propagation time of a gate is 10n secs, the fastest gate can be switched in 6.7n secs and the slowest in 13.3n secs without the circuit misoperating.'

Source: *Problems and Solutions in Logic Design* (Zissos 1979), section 1.14, p. 21. Reproduced with permission of the Licensor (Oxford Publishing Limited) through PLSclear.

**For most designs Tp will be smaller than Ts since the topology of the event cell provides two gates for Ts and one gate for Tp, as can be seen in Figure 4.17.**

Using:

$\boldsymbol{Q = \Sigma \mathrm{sn} + Q \cdot \Sigma/\mathrm{rn}}$. **Sequential equation (4.3) as in Frame 4.14**

$Q = \mathrm{Sum}(\mathrm{sn}) + Q \cdot \mathrm{Sum}(/\mathrm{rn})$ is same equation.

Applying De-Morgan's rule to this results in:

$$//Q = //\left(\text{Sum}[\text{sn}] + Q \cdot /\text{Sum}[\text{ra}]\right).$$

$$//Q = /(/\text{Sum}[\text{sn}] \cdot /\left(Q \cdot /\text{Sum}[\text{rn}]\right).$$

So:

$$Q = //Q = /(/\mathbf{Sum}[\mathbf{sn}] \cdot /\left(Q \cdot /\mathbf{Sum}[\mathbf{rn}]\right). \quad \textbf{NAND sequential equation (4.8)}$$

Source: *Problems and Solutions in Logic Design* (Zissos 1979), equation 1.3, p. 19. Adapted with permission of the Licensor (Oxford Publishing Limited) through PLSclear.

Equation 4.6 is obtained from Equation 4.3 using Boolean algebra.

The event cell can be drawn as shown in Figure 4.17.

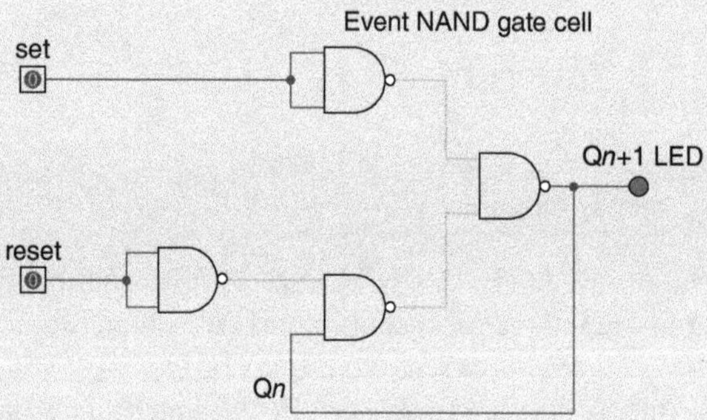


Press set on to make set = 1, Qn+1 LED is on, press set = 0. Qn+1 LED is still on.

Press reset to make reset = 1, Qn+1 LED is off press reset to make rest = 0, Qn+1 LED is off

**Figure 4.17** An event cell using NAND sequential equation.

Note you can also simulate this with Logisim and Logisim-Evolution. See Appendix A2.

From the event cell structure in the Figure 4.17, you can see the gate tolerances between these two delays.

**Generally, in most circuits produced using sequential equations the gate delays will be close to each other, particularly when implemented using PLD, CPLD, and FPGA devices.**

From Figure 4.17 you can see that Q*n* + 1 and Q*n* are the same wire.

Turn to **Frame 4.37**.

***Frame 4.37***

The next few frames go through a complete design example making use of what you have learnt up to this point. The chapter goes on to show how the design can be converted into a Verilog HDL module and tested using a test bench.

The event-based FSM controller shown below is to be used to control a motor. There is a start switch st to start the motor. Turning off the st input can stop the motor.

A sensor (possibly a hall effect device) is used to monitor the motor current and this provides (via its own interface electronics) logic 1 signal, if the motor current exceeds a given value.

There is also a test signal *t* that can be used to test the system when no over-current is present.

There is an LED indicator *L* (active low) that is to operate whenever an over-current level is detected. Once lit, this LED cannot turn off until the start input is released. Ignore the effect of the *t* input for the moment as it can be added later.

Now look at Figure 4.18.

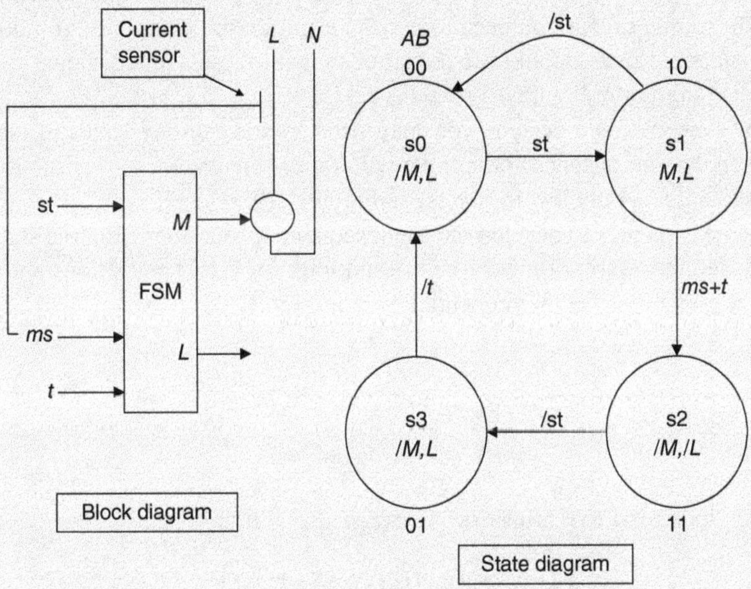


**Figure 4.18** Motor problem with fault detection and test input.

On the left of the figure is the block diagram of the system and on the right is the state diagram developed in order to meet the specification described below.

In this state diagram operating the start input st causes the motor to start by moving from state s0 to s1. The motor can be stopped by turning off the start input to move the state machine from s1 back to s0. This is the normal operation of the system.

If, once the motor has started, an over-current is detected *ms* going from 0 to 1 the state machine moves from s1 to s2 where the motor is turned off, and the LED indicator *L* is turned on (active low output). The system remains in state s2 if the start input is still on. When the start input is turned off, the state machine moves to s3. Here it will move directly back to state s0 if the test button *t* is not pressed.

**Now the effect of the *t* input**

Note that to test the system the test button *t* can be pressed with the motor running and the state machine will move to state s2. If start is still pressed, the state machine will wait for it to be released before moving to s3. Here it will wait until the user releases the test button *t* before returning to state s0, assuming st is not pressed.

Note that the event cells *A* and *B* follow a unit distance coding.

The most difficult part of the design is, of course, developing the state diagram. However, this is done much in the same way you developed synchronous state diagrams, except you need to use the event-driven equations. You do need to consider the fact that all state transitions must follow a unit distance code. In the state diagram in Figure 4.17, this is easy.

However, in some designs you may need to add dummy states in order to make up the unit distance coding. We will look at this later.

Note that *L* is only low in state s2 in Figure 4.19.

The next stage is to develop the event sequential equations from the state diagram. Here we start with event cell *A* and look for 0 to 1 transitions and write down the set terms for the sequential equation for *A*.

$$0>1 \quad 1>0$$

$$A = \Sigma \text{sa} + A\Sigma/\text{ra}.$$

$$A = \text{s0} \cdot \text{st there is only one.}$$

Then look for 1 to 0 transitions between states. In this case there are two:

$$A = \text{s0} \cdot \text{st} + A \cdot /(\mathbf{s1} \cdot /\mathbf{st} + \mathbf{s2} \cdot /\mathbf{st}).$$

Now replace the s0, s1, and s2 with SSVs:

$$A = /A/B \cdot \text{st} + A \cdot /(A \cdot /B \cdot /\text{st} + AB \cdot /\text{st}).$$

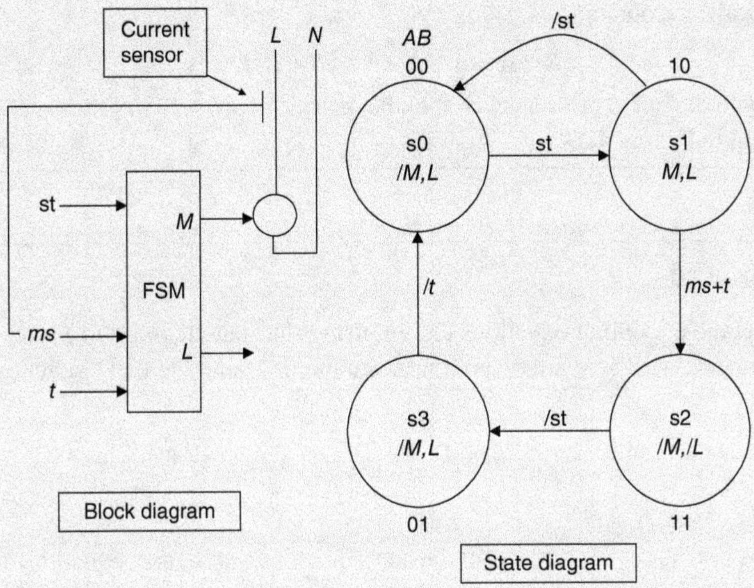


**Figure 4.19** Motor with fault detection.

In this particular example you should be able to see the logical adjacency in the two turn-off terms.

$$A = /A/B \cdot \text{st} + A \cdot /\left(A \cdot /\text{st}\left[/B + B\right]\right) = /A/B \cdot \text{st} + A \cdot /\left(A/\text{st}\right).$$

$$A = /B \cdot \text{st} + A \cdot /\left(/\text{st}\right).$$

The final form in this particular event cell is therefore $A = /B \cdot \text{st} + A \cdot \text{st}$.

**Now try to deduce the event cell equation for *B*.**

When you have done this, turn to **Frame 4.38**.

### *Frame 4.38*

From the state diagram illustrated in Frame 4.37:

$$B = \Sigma\text{sb} + B \cdot \Sigma/\text{Rb} = \text{s1}\left(ms + t\right) + B \cdot /\left(\text{s3} \cdot /t\right) = A\left(ms + t\right) + B \cdot /\left(/A \cdot /t\right).$$

$B = A \cdot ms + A \cdot t + B \cdot \left(A + t\right)$. After using De-Morgan's rule.

Finally, we obtain:

$$\boldsymbol{B = A \cdot ms + A \cdot t + /AB + B \cdot t.}$$

If you did not get this answer, look back over Figure 4.19 in Frame 4.37.

Turn to **Frame 4.39**.

### *Frame 4.39*

Next come the output equations (again, using the state diagram of Figure 4.19 in Frame 4.37). There are two outputs: the motor $M$ and the LED indicator $L$.

$$M = \text{s1} = A \cdot /B.$$
$$L(\text{active low}) = /\text{s2} = /(AB).$$

This is a NAND gate.

The next task is to convert the event equations and output equations into a schematic circuit diagram.

Turn to **Frame 4.40**.

### *Frame 4.40*

The equations are:

$$A = /B \cdot \text{st} + A \cdot (\text{st}).$$
$$B = A \cdot ms + A \cdot t + /AB + B \cdot t.$$
$$M = A \cdot /B.$$
$$L = /A + /B. \text{ Just using De-Morgan's rule.}$$
$$L = /(AB) \text{ as well.}$$

In the equation for event cell $A$ there is an AND gate $/B$·st, another AND gate $A$·st, and an OR gate.

In the equation for event cell $B$, there are four AND gates $A \cdot ms$, $A \cdot t$, $AB$, and $B \cdot t$ and an OR gate. Note that in the event cell for $A$ the second AND gate has one input that is fed from the cell output (feedback).

Also note that in the event cell $B$ the third and fourth terms have a feedback input $B$ from the cell output. The outputs are made up of a single AND gate form and a direct connection from the output of cell $B$ for $L$.

The final circuit is shown in Figure 4.20.

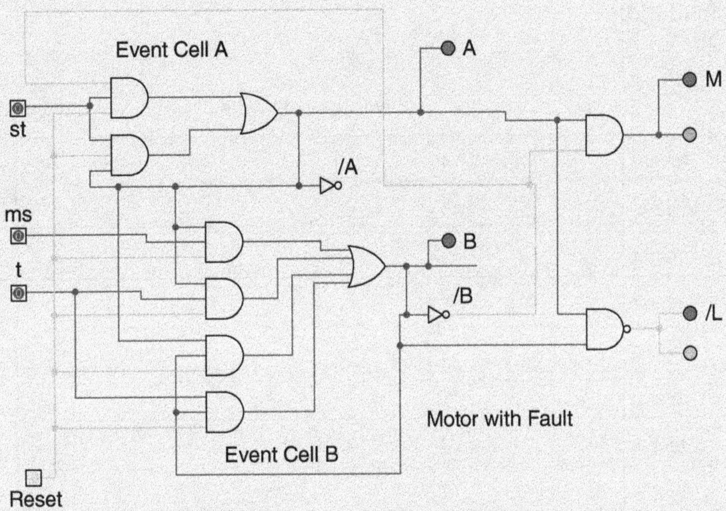


Note: The NAND gate to output /L you need to click on /L LED and answer NO to the question 'Active on High'

**Figure 4.20** Logisim circuit of motor with fault detection and test input.

Note the addition of the reset input to the AND gates of both event cells. The modified equations are:

$$A = (/B \cdot \mathrm{st} + A \cdot st) \cdot \mathbf{Reset}.$$
$$B = (A \cdot ms + A \cdot t + AB + B \cdot t) \cdot \mathbf{Reset}.$$

If the reset is logic 1 then these equations are enabled. If the reset is logic 0 then these equations are disabled and the event cell outputs are both zero.

In practice the reset input will be pulsed low to initialize the state of the FSM (into state s0).

For this reason the first state must have all of its SSVs at zero.

**You can use Logisim to simulate Figure 4.20 as you did using Figure 4.6 and Figure 4.17. It should follow the Verilog HDL simulation pattern in Figure 4.21.**

Turn to **Frame 4.41**.

### *Frame 4.41*

To illustrate how this design would work in practice, Figure 4.21 shows the design simulation.

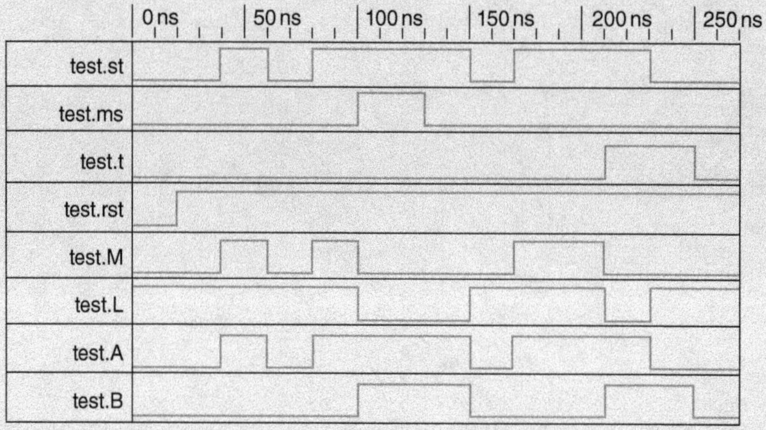


**Figure 4.21** Verilog HDL simulation of motor with fault detection and test input.

The test signals illustrate the inputs st, *ms*, *t*, and reset. The outputs are *M* and *L*. Also shown in the simulation are the SSVs *A* and *B*.

The simulation was produced by first converting the equations into Verilog HDL. Then a **test bench file** was created to test the circuit for correct operation. See Appendix A2 for a tutorial on how to use the Verilog HDL simulator SynaptiCAD.

SynaptiCAD can be obtained from www.syncad.com.

You now need to work out the test bench module sequence for this design. To do this you need to use the state diagram of Figure 4.18 or 4.19 to work this out.

Looking at the test sequence in this frame:

1. Looking at Figure 4.18, for example, you need to see if the FSM stays in s0. You can keep this as st = 0. The FSM stays in s0. Then with reset = 0 applied st = 1 remains in s0. Now st = 0 and reset = 1 should remain in s0.
2. Now st = 1 the FSM should move to s1 and stay in s1. The FSM cannot move to s0 since st is still 1, and it cannot move to s2 because *ms* and *t* are both zero.
3. Now with st = 1 make *ms* = 1. Move to s2.
4. Now, if st = 0, FSM should move to s3, then on to s0 if $t = 0$ and stop.

Since this is an event-driven FSM, the transition through s3 will be very fast, and in a simulation you may not notice the transition through s3 to s0 when $t = 0$.

If, however, you are using the 'structural flow Verilog' (see later in this chapter), where logic gates are used to describe the Verilog sequence and delays (#) have been used for each logic gate, you may see s2 as it passes from s1 to s0 (depending on the gate propagation delay).

Again, wait until the structural low is looked at in this chapter, and used by the Verilog HDL program to describe the Verilog HDL code.

To do this the simulation moves from s1 to s2, and then waits for st to go low to move to s3, then, since *t* is low, it moves to s0, thus testing the system in the absence of an over-current input.

Note how the state diagram has helped to develop the simulation sequence. This is a very common way to work out the test patterns needed. They consist of both positive tests, activating outside world inputs and negative test where outside world inputs are **not active**.

The Verilog module code and test bench can be seen later.

Note that the Verilog HDL code uses st, *ms*, and *t* inputs.

Turn to **Frame 4.42**.

## Frame 4.42

Here is the Verilog HDL module for the two event cells and outputs.

**Note that Verilog systems are considered in the Chapters** 6, 7, **and** 8 **of Minns and Elliott** (2008).

**Also, in this Verilog HDL program the structure flow method is used.**

```
module fsm(rst,st,ms,t,M,L,A,B);
 output M,L,A,B;
 input rst,st,sp,ms,t;
/// The code is described in terms of Logic Gates.
/// Below you can see the assign statement using a single
/// assign.
/// This makes use of each line using a terminating and
/// ending in a terminating semi-colon';'

assign   // Structure Flow Method.
   A=(~B&st | A&st)&rst,
   B=(A&ms | A&t | A&B | B.t)&rst,
   M=A&~B,
   L=~A | ~B;  //Could be L=/(AB) NAND
endmodule
////////////////////////////////////
```

To read these you need to be aware that in Verilog

**AND (·) is &.**

**OR (+) is | (usually the key next to the '| \' key on a MAC).**

**Invert (/) is ~ (the tilde symbol).**

The line **module** fsm(rst,st,*ms*,*t*,M,*L*,*A*,*B*); defines the name of the module (FSM in this case). Note the semicolon at the end of this line.

Output signals are: **output** M,*L*,*A*,*B*;

Input signals are: **input** rst,st,*ms*,*t*;

The definitions of the equations are placed in an assign block using the structure flow method:

```
assign
        A=(~B&st | A&st)&rst,
        B=(A&ms | A&t | A&B | Bs.t)&rst,
        M=A&~B,
        L=~A | ~B;
endmodule
```

Note each line ends with a comma (,) except the last line, which ends with a semicolon (;).

The final line **endmodule defines the end of the module.**

Turn to **Frame 4.43**.

### *Frame 4.43*

The Verilog code in Frame 4.42 describes the circuit. The Verilog compiler tools will take this code and convert it into a simulation file to test the design. Other tools (like Xilinx for example) will also produce a file that can be used to program a PLD, CPLD, or FPGA chip.

However, it is necessary to test the design.

To do this a **test bench file** is needed.

The test bench file is just another module that can be used to generate signal sequences at the inputs. It is used to simulate the design. The test bench file acts like a generator to supply input signals (in this case st, *ms*, *t*, and reset rst) to the Verilog module in the correct sequence, as these are inputs to the system.

The test bench file for the example of the motor controller is given in Frame 4.44.

The test bench module can be obtained from the state diagram as explained in Chapter 3 and in Frame 4.41.

Turn to **Frame 4.44**.

***Frame 4.44 Looking at the State Diagram Figure 4.19***

```
module test;
reg st,ms,t,rst;
//each signal is defined as a register which can store the
//current state of the signal.

fsm uut(rst,st,ms,t,M,L,A,B);
//define an instance of the verilog module now called a
Unit Under Test.
initial
 begin
   $dumpfile("motflt.vcd ");    // to get a printout of
waveforms.
   $dumpvars;                  // store the variables
    rst=0;                     // set signal rst to logic 0
    st=0;                      // set st to 0.
    ms=0;                      // set ms to 0
    t=0;                       // set t to 0 as well.
// This initialized all the input signals.
// Note it is important to ensure signals change in
// proper sequence. Also to ensure ms and sp are
// mutually exclusive.
// Just follow the state diagram Figure 4.19.
//------------------ remove reset
    #20 rst=1;                 // wait 20 time units then set rst
to logic 1.
//------------------ move to s1
    #20 st=1;                  // a further 20 time units later
set st to logic 0.
//------------------ stay in s1
//------------------ move to s0
    #20 st=0;                  // 20 time units later set st to
logic 0.
//------------------ move to s1 again
   #20 st=1;                   // and so on.
// Check these with the timing simulation in Frame 4.41 to
see the effect of the signal
// changes.
//------------------- move to s2
   #20 ms=1;
//-------------------
   #30 ms=0;
```

```
//------------------- move to s3
   #20 st=0;
//------------------- move back to s0
   #20 st=0;
//------------------- move to s1
   #20 st=1;
//------------------- stay in s1
//------------------- move to s2
   #20 t=1;
//------------------- move to s3
   #20 st=0;
//------------------- move to s0
//------------------- end of tests.
$stop(60); // stop the simulation.
 end          // end of test bench module sequence.
endmodule
```

And that is it.

Turn to **Frame 4.45**.

### *Frame 4.45*

In Frame 4.44 the **# precedes a delay value**. So, #20 means wait 20 time units before sending the following signal.

Try following the test bench module sequence and relating it to the timing simulation shown in Frame 4.41.

**Note that by using the # followed by a delay value the simulation can be made to follow prescribed propagation delay paths.**

**Note that the tutorial section in Chapter** 3 **shows a system you can study.**

Turn to **Frame 4.46**.

### *Frame 4.46 Using Relays in FSM Design*

Sometimes, there is a need to develop a design based on relays. Relays can be either electromechanical or static.

In electromechanical relays there is a relay coil through which a current can be passed. This results in a magnetic field that can be used to attract a metal armature mechanically linked to relay contacts. There are smaller relays than this such as 'reed' relays. It is also possible to make use of static semiconductor relays available these days.

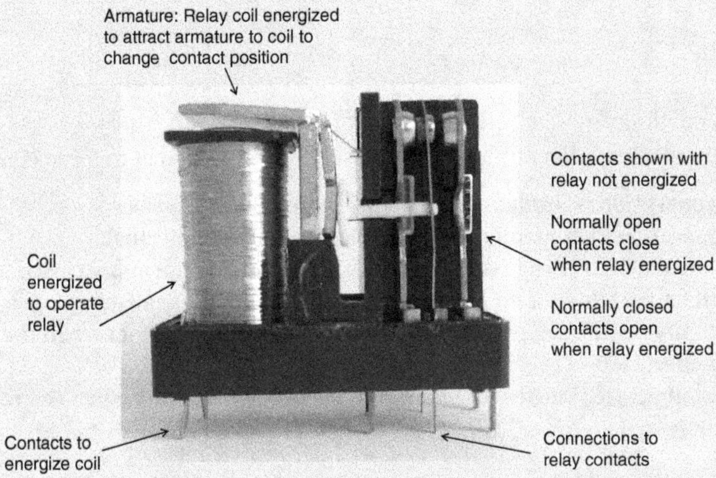


**Figure 4.22** A typical electromechanical relay.

The relay contacts can be of three types:

- **Normally open contacts – closing when the relay coil is energized.**
- **Normally closed contacts – opening when the relay coil is energized.**
- **Change-over contacts – having a normally open and a normally closed contact with the centre contact moving when the relay coil is energized. See Figure 4.22, which shows a relay with two changeover contacts, one on each side of the relay.**

The motor with fault detection looked at in the previous frames can be implemented using relays. Also you could use static semiconductor relays, which would be faster.

The main advantage of using relays is that there is no need for a low voltage power source since the relays can be obtained to work from whatever power supply is available. This could be an AC mains with simple rectification, or a battery voltage (50 V or 12 V, perhaps). The speed of operation may be slow, but in the application being considered this is not a problem.

Now back to the motor with a fault detection example using relays.

In the following design the equations are:

$$A = (/B \cdot st + A \cdot st) \cdot rst.$$
$$B = (A \cdot ms + A \cdot t + AB + B \cdot t) \cdot rst.$$
$$M = A \cdot /B \cdot$$
$$L = /A + /B.$$

These equations are in a suitable form for a relay design.

The implementation of these equations using relays is that in these equations:

- **AND operation is implemented with relay contacts in series.**
- **OR operation is implemented with relay contacts in parallel.**
- **A contact switch with invert /*A* means the contact is normally closed.**
- **A contact switch with no invert *A* means the contact is normally open.**
- **'Normally open' means the contacts are not making contact when the relay is not energized.**
- **'Normally closed' means the contacts are making contact when the relay is not energized.**

With this information it is easy to implement any Boolean equation with relay logic.

Using a textual diagram for the relay circuits, as shown below, consider the following relay arrangements. *A* = st·/*B* would look like

Gnd-----*st*------/*B*-----(coil A)------V+

**In this line drawing st is an open contact switch and /*B* is a normally closed contact. Coil A is the relay coil that will become energized when st is closed. The V+ is the positive DC voltage supply to energize the coil.**

Another example, *A* = *x*·/*B* + *y*·*A* would look like:

Gnd-----|-----*x*-----/*B*-----|-----(coil *A*)-----*V*+
Gnd-----|-----*y*------*A*-----|

Where the **|** implies connection to the above line i.e. a parallel connection. The gnd is simply the 0V (ground) point in the circuit.

Note that on the keyboard the **|** \ key is the one to press with the shift operation,

**In the equation *A* = (/*B*·st + *A*·st) what would the relay circuit look like?**

Turn to **Frame 4.47**.

***Frame 4.47***

**Remember that /*B* and st are in series while the parallel connection of *A* and st are the OR term | in parallel.**

The answer is:

```
Gnd-----|-----/B-----st-----|-----(coil A)-----V+
        |-----A------st-----|
```

In this diagram /*B* is normally closed, so when st closes the *A* relay coil will energize and the *A* normally open contact will close to hold the relay coil energized for as long as st is closed.

When st opens, the relay will de-energize.

**Now try to draw the relay circuit for the *B* relay using the equation:**

$$B = \left(A{\cdot}ms + A{\cdot}t + AB + B{\cdot}t\right).$$

When you have done this, turn to **Frame 4.48**.

***Frame 4.48***

```
Gnd-----A-----|-----ms ----|-----(coil B)-----V+
              |-----t -------|
              |-----B ------|
Gnd___B_________t _____|
```

To include the reset, simply place a series normally closed switch into each circuit as illustrated in the next frame.

Turn to Frame **4.49**.

***Frame 4.49***

There are two relays, so each needs a normally closed reset switch in series with the relay coils of both *A* and *B* relays.

```
Gnd---|-----/B-----st-----|-----(coil A)-----/reset-----V+
      |____A____st____|
```

```
Gnd-----A-----|-----ms-----|-----(coil B)-----/reset-----V+
              |____t_____|
              |____B_____|
Gnd____B________t______|
```

The /reset switch is normally closed. Operating the switch will open the contacts and hence de-energize the relays thus resetting the circuits.

The diode connected across the relay coil is not shown.

```
 −      +
-----|>|---- The cathode of the diode is to positive supply.
```

This means that any **back** EMF will flow through the diode and prevent a build-up of voltage across the coil ('Back EMF').

Turn to **Frame 4.50**.

### *Frame 4.50 FSM Outputs*

The outputs are simply equations from the ground through relay contacts and switches to the output devices (motor and LED).

For the output equation $M = A \cdot /B$.

**Draw the relay circuit.**

When you have done this, turn to **Frame 4.51**.

### *Frame 4.51*

This is:

Neutral -----*A*-----*/B*-----Motor-----Motor supply.

Here, assume a single phase AC motor, hence the neutral connection.

Also, for the other output equation: $L = /A + /B$.

```
Gnd-----|-----/B-----|-----LED-----resistor-----V+
        |____/A____|
```

Note that in an LED device current flows from the anode *A* to the cathode *C*.

*A* *C*
---|>|--- current flow →

The diode acts as a 'half wave rectifier'.

The resistor value for a rectified AC mains voltage of 230 V would be:

$$\text{Resistor} = 230x\sqrt{(2)}/10\ \text{mA} = 325/10\ \text{mA} = 32\ \text{k}\Omega.$$

**Do not forget about the diode connected across the relay coil. This is needed to suppress the 'back EMF' that is introduced in the relay coil when contacts open in the circuit path. Figure 4.24 shows how it looks.**

Turn to **Frame 4.52**.

## *Frame 4.52*

Figure 4.23 shows another example.

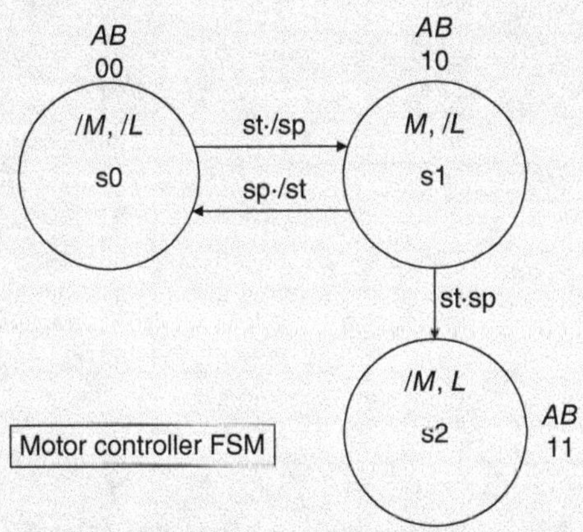


**Figure 4.23** State diagram for relay event equations.

You could try to develop the event equations and output equations from the state diagram then draw the circuit diagram for a relay-based design.

Turn to **Frame 4.53**.

*Frame 4.53*

$A =$ s0 $\cdot$ st $\cdot$ / sp $+ A \cdot /(\text{s1} \cdot / \text{st} \cdot \text{sp}) = /B \cdot \text{st} \cdot /sp + A \cdot /(/B \cdot /\text{st} \cdot \text{sp})$.

$A = (/B\text{st} \cdot /\text{sp} + AB + A/\text{sp} + A \cdot \text{st}) \cdot \text{rst}$.

$B = s1 \cdot \text{st} \cdot \text{sp} + B \cdot /0$. Note that $/0 = 1$.

$B = (A \cdot \text{st} \cdot \text{sp} + B \cdot /0) \cdot \text{rst}$.

$M =$ s1 $= A / B$.

$L =$ s2 $= AB$.

**Now try to draw the relay-based circuit from these equations. When you have completed these, turn to Frame 4.54.**

*Frame 4.54*

Owing to the complexity of this circuit, it has been drawn as a schematic diagram. Figure 4.24 is my attempt.

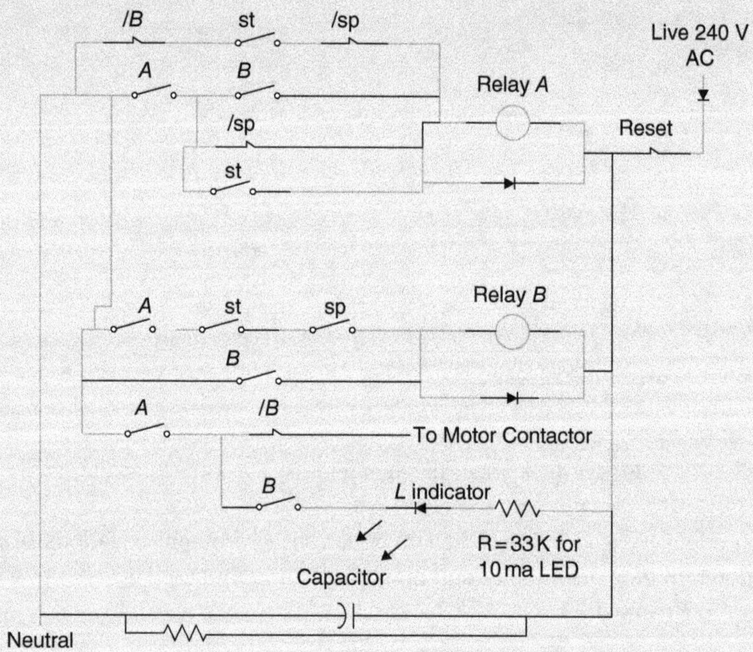


**Figure 4.24** Circuit diagram for the relays showing all components.

In this circuit diagram you can see the single diode (half-wave rectification with current flowing in the direction of the implied arrow) and smoothing capacitor to create a raw DC supply. It is assumed that the relays *A* and *B* can operate at 339 V DC.

The circles symbolize the relay coils. All contacts assume their unenergized state.

The diodes can be clearly seen across the relay coils.

**The diodes across each relay coil are to protect the relay contacts and switches when the relays de-energize. If these are not included, the contacts will produce damaging sparks due to the 'back EMF' of the relay coil.**

**Note that the diodes are reverse connected in terms of the DC voltage supply. These diodes are sometimes called 'catching diodes' and are for 'catching' the reverse 'back EMF' that can flow through the coil when the switch is opened.**

**Now draw the circuit for a design using electronic logic gates.**

Turn to **Frame 4.55**.

### *Frame 4.55*

The design in Figure 4.25 uses sum of product (SOP) topology as would be available in a PLD/CPLD device.

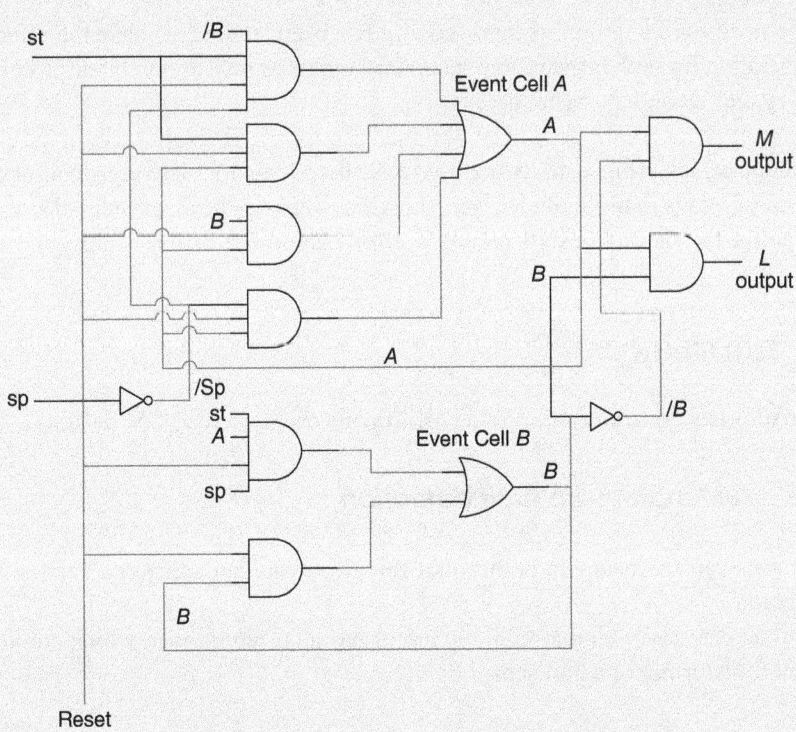


**Figure 4.25** Logic circuit for the same arrangement.

You would, of course, need to provide interface hardware between the outputs $M$ and $L$ to take account of the power levels of the motor and $L$ LED.

Turn to **Frame 4.56**.

***Frame 4.56 Future Work with FSM-based Systems***

In future it may be possible to construct nanomechanical systems. For his PhD thesis in 1991, Eric Drexler studied molecular machinery with applications to computation based on design details using nanomechanical PLDs that can be applied to FSM-based systems following reversible logic (Rice 2008). Chapter 11 of his thesis refers to these ideas. Details can be found in Google under Eric Drexler's PhD thesis (Drexler 1991).

## 4.2 SUMMARY

This chapter has looked at how asynchronous (event-driven) FSM systems can be developed by first producing a block diagram of the proposed design and then creating a state diagram for the behaviour of the FSM.

The need for using unit distance coding has been studied, as have the effects of maloperation for both primary and secondary inputs as well as combinations of both primary and secondary signal delays.

Event-driven systems need to be designed carefully taking into account the effects of maloperations. This restricts the use of such systems to relatively small designs. The use of relays in place of electronic based designs has been considered.

Finally, the use of design methods allow scope for different designs in the future.

## 4.3 TUTORIALS

These tutorials cover examples of asynchronous event-driven FSM designs.

### 4.3.1 FSM motor with fault detection

You have seen the example before, but this time you can apply the Verilog HDL simulation.

As you can see in Figure 4.26, the motor has a current sensor, which could be a current transformer or a hall sensor device.

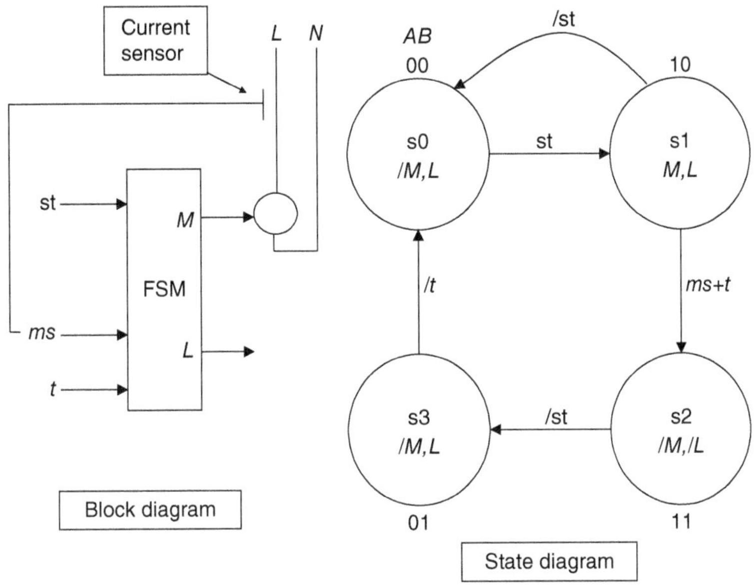


**Figure 4.26** The motor with fault and test input.

The state diagram has four states and can be used to start or stop the motor between s0 and s1. If a motor fault (over limit current) is detected, sensor input *ms* goes to logic 1 and the motor is turned off while the current limit *L* goes to logic 0 (active low output).

So the motor is now off, while the *L* indicator */L* is active low.

The system can be reset to s0 by making input st = 0.

In this sequence test input $t = 0$.

In the absence of a fault the state diagram can move from s1 to s2 when $t = 1$.

The user needs to set st = 0 to move into s2 and test input *t* must be retuned to $t = 0$ to allow the FSM to return to s0.

The state diagram can be used to produce the equations for the FSM.

These are:

$$A = /B\cdot \mathrm{st} + A\cdot \mathrm{st}.$$

$$B = A\cdot ms + A\cdot t + /A\cdot B + B\cdot t.$$

Outputs are:

$$M = \mathrm{s1} = A\cdot /B.$$

$$L = /(\mathrm{s2}) = /(AB) \text{ which is active low.}$$

You can check this against the ones in Frame 4.39.

The Verilog simulation is shown in Figure 4.27.

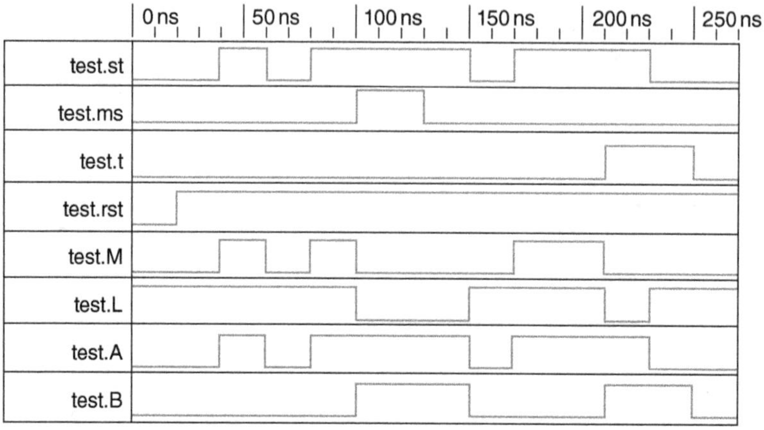


**Figure 4.27** The Verilog HDL simulations.

Listing 4.1 illustrates the Verilog HDL simulations.

The following Verilog HDL listing can be found in file **motflt.v.**

```
//////////////////////////////////////////////////////
// Motor Controller with fault current monitoring:
//
// Implemented using equations as listed in the
// fsm module assign block below.
//
// Author: PDM.
//////////////////////////////////////////////////////
module fsm(rst,st,ms,t,M,L,A,B);
  output M,L,A,B;
  input rst,st,ms,t;
// Structural Flow used..

assign
   A=(~B&st | A&st)&rst,
   B=(A&ms | A&t |A&B | B&t)&rst,
   M=A&~B,
   L=~A | ~B;
endmodule
////////////////////////////////////////

module test;
reg st,ms,t,rst;
fsm uut(rst,st,ms,t,M,L,A,B);
```

```
initial
  begin
      $dumpfile("motflt.vcd "); // to get a printout of
                                   waveforms.
      $dumpvars;
      rst=0;
      st=0;
      ms=0;
      t=0;
      // Note it is important to ensure signals change in
      // proper sequence.  Also to ensure ms and sp are
      // mutually exclusive
      //------------------- remove reset
      #20 rst=1;
      //------------------- move to s1
      #20 st=1;
      //------------------- stay in s1
      //------------------- move to s0
      #20 st=0;
      //------------------- move to s1 again
      #20 st=1;
      //-------------------- move to s2
      #20 ms=1;
      //--------------------
      #30 ms=0;
      //-------------------- move to s3
      #20 st=0;
      //-------------------- move back to s0
      #20 st=0;
      //-------------------- move to s1
      #20 st=1;
      //-------------------- stay in s1
      //-------------------- move to s2
      #20 t=1;
      //-------------------- move to s3
      #20 st=0;
      //-------------------- stay in s3
      #20 t=0;
      //-------------------- move to s0
      #20 st=0;
      //-------------------- end of tests.

  $finish(60);
  end
endmodule
```

**Listing 4.1** Motor control with fault.

### 4.3.2 The mower in four and two states

Figure 4.28 shows the block diagram in Figure 4.28a. Figure 4.28b is the FSM state diagrams.

State diagram 1 is a four-state sequence, and state diagram 2 is a two-state sequence.

In Figure 4.28, state diagram 1 shows the first system developed, while state diagram 2 illustrates the refined two-state result.

The specification for this system uses a start input st and a lever switch sf that must be pressed at the same time in order to start the mower motor. This is to prevent the user from starting the mower with the start switch only.

Note that both switches are active high in this example.

This tutorial deals with the two-state FSM only. Note that in this two-state diagram the outside world $P = 0$ in state s0, if sf is 0 as indicated.

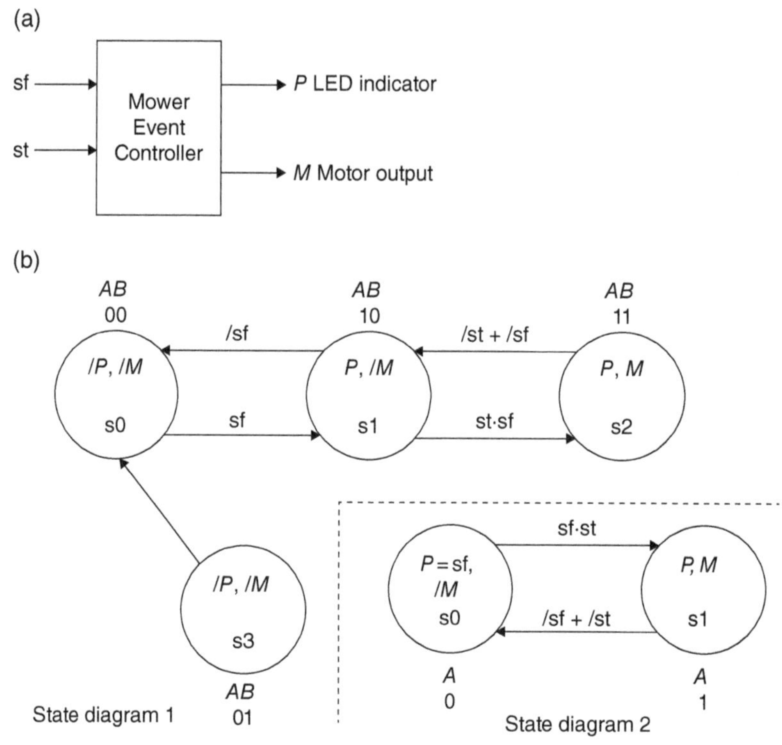


**Figure 4.28** Block diagram and state diagrams of the mower FSM.

The equations can be obtained from the state diagram (two states) as:

$$A = s0 \cdot st \cdot sf + A \cdot /\left(s1 \cdot \left[/s1 + /sf\right]\right).$$
$$A = st \cdot sf + A \cdot /\left(/st + /sf\right).$$
$$A = st \cdot sf + A \cdot //\left(st \cdot sf\right).$$
$$A = st \cdot sf + st \cdot sf = st \cdot sf.$$

Outputs:

$$\mathrm{M} = s1 = A.$$
$$P = s0 \cdot sf = sf.$$

**Make *P* = sf since the safety sensor sf should indicate *P* = 1 and indicate this with the *P* LED, as indicated in Figure 4.29**.

As you can see, the Verilog data flow is used.

The gate circuit for this is shown in Figure 4.29.

You could make use of the structural flow. Note that the reset input has been introduced into the circuit.

The Verilog simulation for this example is shown in Figure 4.30. This makes use of the two-state diagram version.

Here you can see that both st and sf need to be pressed for the mower motor to start.

Listing 4.2 shows the Verilog HDL code for the mower two state system can be found in file **mower.v**.

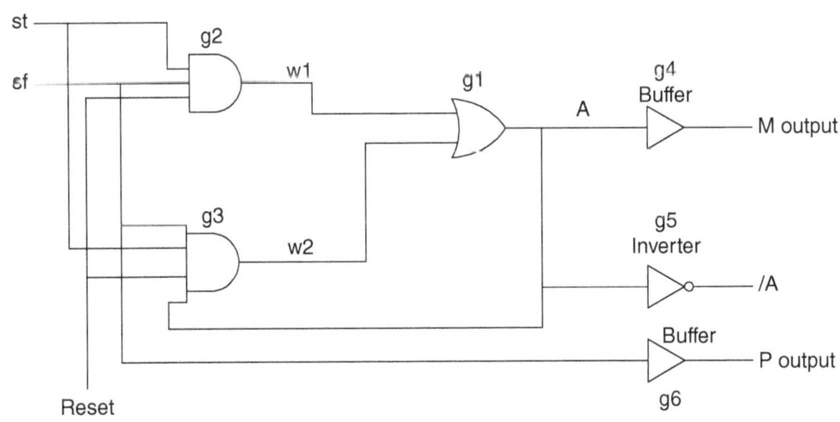


**Figure 4.29** Circuit for the mower FSM.

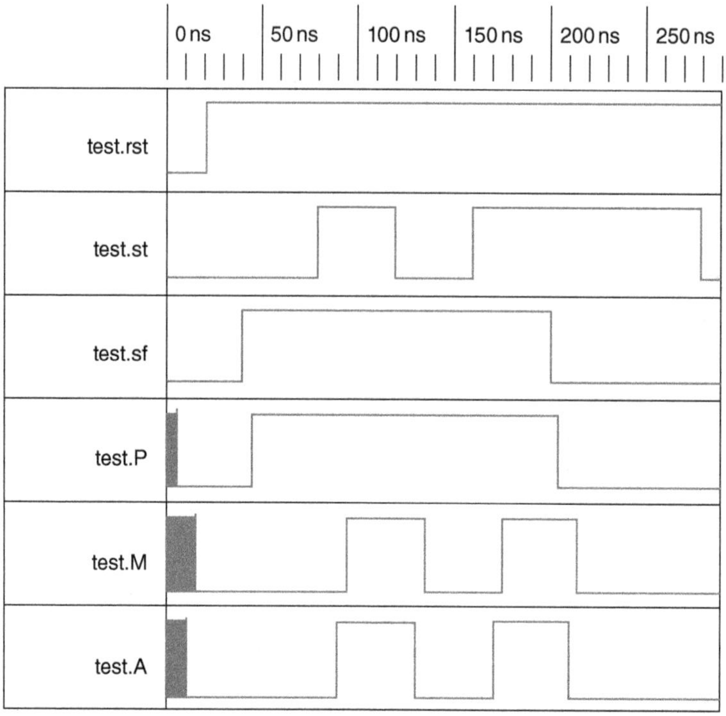


**Figure 4.30** The Verilog HDL simulation for the two-state FSM.

```
/////////////////////////////////////////////////////////////
/
// Mower FSM controller:
//
//  This is implemented from the State Diagram in
//  this section, using the 2 State Diagram arrangement.
//  It is implemented using logic gates, rather than the
//  sequential equations direct. This shows that the logic
//  gates can be given individual gate delay values to
//  meet gate tolerance rule.
//  The Reader might like to modify the circuit to make it
//  mal-operate.
//
//  Author: PDM.
module mowerfsm(st,sf,P,M,A,rst);

  input st,sf,rst;
  output P,M,A;
  wire na,nb,w1,w2;
```

```
  or #5 g1(A,w1,w2);      // The hash (#) denotes a propaga-
tion delay of 5nS.
  and #5 g2(w1,sf,st,rst);      // same here for AND gate.
  and #5 g3(w2,A,st,sf,rst);    // and here. This is an
advantage using 'Struct'; Method.
  //------------------
  buf #5 g4(M,A);
  //------------------
  not #5 g5(na,A);
  buf #5 g6(P,sf);
  //------------------
  endmodule

module test;
  reg rst, st, sf;
  mowerfsm uut(st,sf,P,M,A,rst);

  initial
    begin
      $dumpfile("mower.vcd");
      $dumpvars;
      rst=0;
      st=0;
      sf=0;
      //-----
      #20 rst=1;
      #20 sf=1;
      #20
      #20 st=1;
      #20
      #20 st 0;
      #20
      #20 st=1;
      #20
      #20 sf=0;
      #20
      #20 st=1;
      #20
      #20 st=0;
      #10 $finish;
  end
endmodule
```

**Listing 4.2** The mower two-state FSM.

Note that in this example a gate circuit has been used. Also note that propagation delays have been inserted. These are all the same value of #5 i.e. 5ns so the topology of the system is met.

## Important note

**If you make use of the structure flow method rather than the data flow method, you have control of the propagation delay path through the logic as well. However, the structure flow method is easier to follow. You do not have such a degree of control with propagation delay using the behavioural method since the behavioural method does not define low-level circuit details.**

# 5

# Clocked One Hot Method of FSM Design

## 5.1 INTRODUCTION

Up to this point the design methods for synchronous and asynchronous finite state machine (FSM) designs have made use of secondary state variables (SSVs) to identify the states in the FSM machine.

As we have seen, this can become problematic when trying to keep to a unit distance coding and can result in having to include dummy states in the state diagram. There is, however, an alternative method of design for synchronous FSM work: the one hot method.

This chapter looks at how this technique can be used to implement clocked FSM designs that do not need or use SSVs. Before beginning this work it must be stated that, at this time, there is not a method to use the one hot techniques in asynchronous (event-driven) FSM designs, although the author has done some work in this area; however, much still needs to be done.

Chapter 6 looks at this idea and compares it with the conventional event-driven method using sequential equations and SSVs.

In this chapter we look at only synchronous FSM one hot designs. Having said this, most FSM work to date is certainly using synchronous FSM designs. Indeed, most commercially available microprocessors and microcontrollers are clock driven.

This could change in the future, so look out for that. Indeed, a lot of research is going on into the use of Petri nets and asynchronous designs. See Chapter 7 for more on this.

*Digital System Design using FSMs: A Practical Learning Approach*, First Edition. Peter D. Minns.

Companion website: www.wiley.com/go/minns/digitalsystemdesign

### Frame 5.1

What is the one hot method for FSM design?

Using the one hot method in FSM design involves allocating a single flip-flop for each state in the state diagram. So if a state diagram contains five states then there are five flip-flops.

**If a state diagram contains states s0 to s13, how many flip-flops will be required?**

Turn to **Frame 5.2**.

### Frame 5.2

**If a state diagram contains states s0 to s13, how many flip-flops will be required?**

The answer is, of course, 14 flip-flops.

**What type of flip-flop is used in a one hot method?**

Well, we need to use *D* type flip-flops since these are easiest to implement in the one hot method and avoid falling into unused states.

Let us look at a simple state diagram (Figure 5.1).

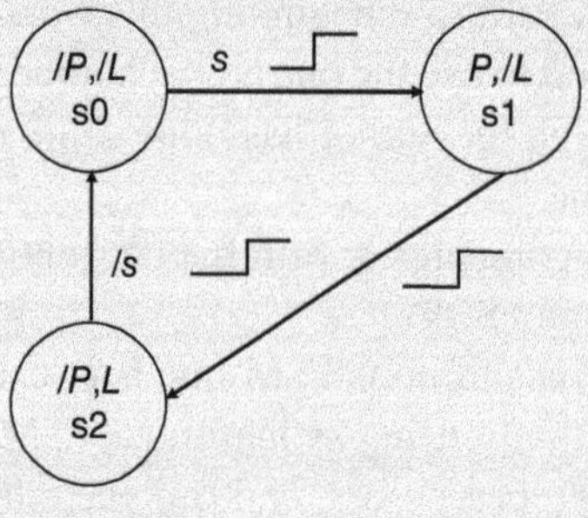


Design equations:

$s0 \cdot d = s2 \cdot /s + s0 \cdot /s$
$s1 \cdot d = s0 \cdot s$
$s2 \cdot d = s1 + s2 \cdot s$
$P = s1$
$L = s2$

**Figure 5.1** First one hot state diagram with design equations.

In this state diagram there are three states: s0, s1, and s2. This requires three flip-flops with *D* inputs s0·*d*, s1·*d*, and s2·*d*.

Each *D* type flip-flop (or DFF) has a *Q* output s0, s1, and s2 (as well as complementary outputs /s0, /s1, and /s2). Since these are clock-driven systems, the *D* type flip-flop also has a common clock connection.

**Draw three *D* type flip-flops with the correct connection labels.**

Turn to **Frame 5.3**.

### *Frame 5.3*

Three flip-flops are shown in Figure 5.2.

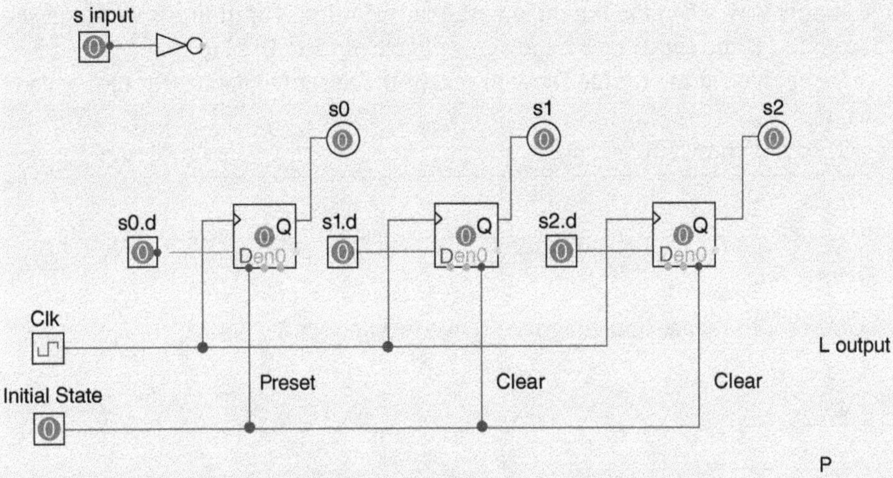


**Figure 5.2** Basic circuit for a one hot clocked design.

You can see the notation makes sense. If we are in state s0, flip-flip s0 will be set and all other flip-flops will be cleared. See Figure A2.10 for more details on *D* type flip-flops used in the Logisim Simulator.

Turn to **Frame 5.4**.

### *Frame 5.4*

At any one time there can only be one of the flip-flops set. This makes sense since the FSM can only be in one state at a time.

In this design an initialization input would be connected to each flip-flop in such a way that when the initialization signal went low (logic 0) only flip-flop s0 would be set.

**How would you connect the initialization input to the three flip-flops?**

When you have done this turn to **Frame 5.5**.

### *Frame 5.5*

The answer is that the initialization input would be connected to the preset input of flip-flop s0, and the clear input of flip-flops s1 and s2.

In this way when the initialization input went low, flip-flop s0 would preset (s0 go to 1), and flip-flops s1 and s2 would clear (s1 to 0, s2 to 0).

We now need to consider how to obtain the design equations from the state diagram.

Turn to **Frame 5.6**.

### *Frame 5.6*

. Consider the same state diagram shown in Figure 5.3

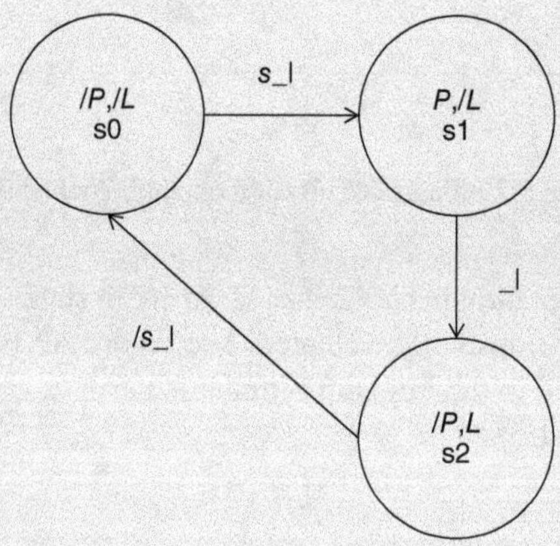


**Figure 5.3** Clocked one hot example.

This is the same as the one in Frame 5.2.

The method for obtaining the equations for the *D* input of each flip-flop is to look at the state diagram and determine:

- The conditions to enter a state.
- The conditions necessary to remain in that state.

For state s0 the condition to enter the state is that we must be already in state s2 and that the input /*s* must be true.

The condition to remain in state s0 is that we are already in state s0 and that the input condition to leave state s0 is not true.

We now have:

$$\mathrm{s0} \cdot d = \mathrm{s2} \cdot /\mathrm{s} + ?.$$

**Write down the missing term and then turn to Frame 5.7.**

***Frame 5.7***

Figure 5.4 shows the state diagram.

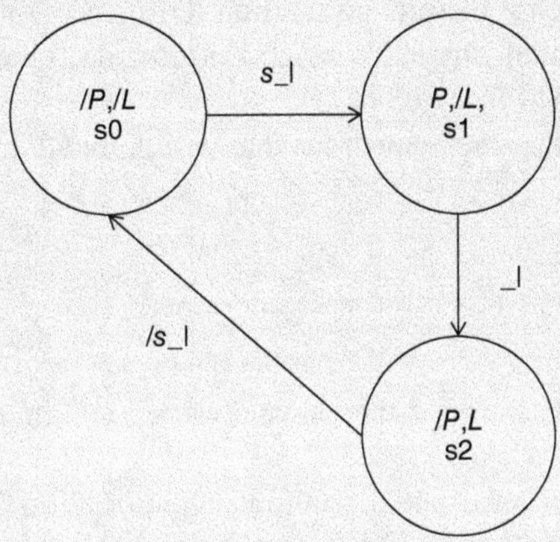


**Figure 5.4** The state diagram.

So we have:

$$s0 \cdot d = s2 \cdot /s + s0 \cdot /s.$$

The first term is s2·*/s* since to enter state s0 we must be in state s2 and the input along the transitional line between s2 and s0 is true (*/s* = 1).

The second term (called a **holding term**) is that we will remain in state s0, while the input along the transitional line between s0 and s1 is **not true**.

So the *D* input for flip-flop s0 contains a set term (s2·*/s*) and a holding term (s0·*/s*).

**Now consider the conditions to get into state s2, and how to remain there until the conditions to leave s2 occur.**

Turn to **Frame 5.8**.

***Frame 5.8***

$$s2 \cdot d = s1 + s2 \cdot s.$$

The first term (s1) is that we need to be in state s1. Note that, since there is no input along the transitional line connecting s1 to s2, there is no input signal required with this term.

The second **holding term** is s2·*s* because we are to hold in state s2, while input */s* is **not true**. That is:

$$s = 1\left(/s \text{ must be } 0\right).$$

**Finally consider the condition to enter and remain in state s1.**

Turn to Frame 5.9.

### Frame 5.9

$$s1 \cdot d = s0 \cdot s.$$

Note that in this case there is no holding term since, once the FSM reaches state s1, it will automatically leave state s1 on the next clock pulse. There is no reason to hold the FSM in state s1.

So now we have the three equations that will define the $D$ inputs for each of the flip-flops as:

$$s0 \cdot d = s2 \cdot /s + s0 \cdot /s.$$

$$s1 \cdot d = s0 \cdot s.$$

$$s2 \cdot d = s1 + s2 \cdot s.$$

The outputs are the same as they have always been, in terms of the states in the FSM.

You now need to write out the output equations for the state diagram in Figure 5.4 in Frame 5.7.

Turn to **Frame 5.10**.

### Frame 5.10

The equations are for output $P$ and $L$:

$$L = s2.$$

There are no SSVs so s2 implies that the output $L$ will be connected to the output of flip-flop s2.

$$P = s1.$$

The circuit diagram for the system is shown in Figure 5.5. This was created using Logisim with tunnels (see Appendix A2).

Study Figure 5.5 carefully. In particular, look at the connection details and note the feedback connections from the flip-flop outputs to the AND gate input for flip-flop s0 and s2.

Note: you can use Logisim to simulate Figure 5.5 to see it perform correctly. See Appendix A2 to find out how to use Logisim.

We will come back to the connection details later.

Now turn to **Frame 5.11**.

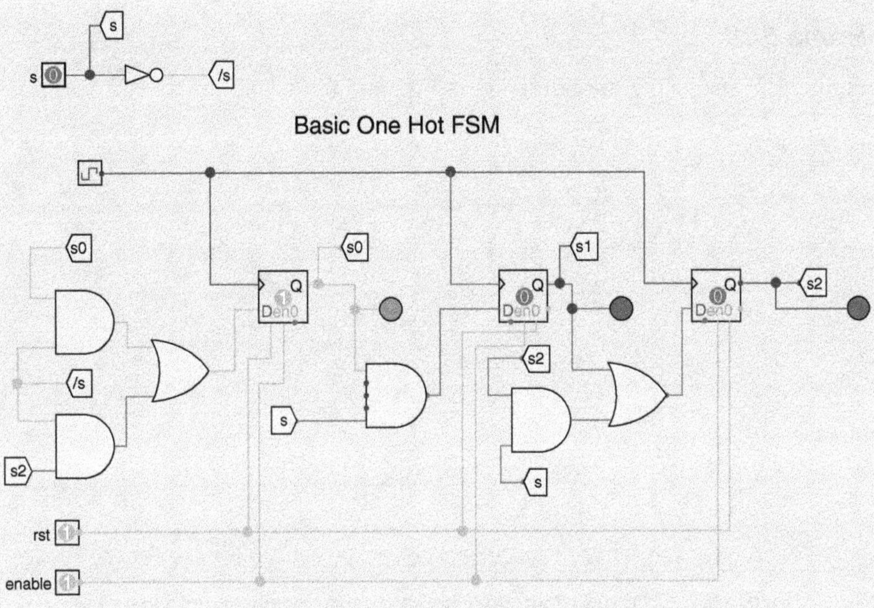


**Figure 5.5** Using Logisim with tunnels.

### *Frame 5.11*

**Produce the *D* input equations for the flip-flops and the output equations for the state diagram shown in Figure 5.6.**

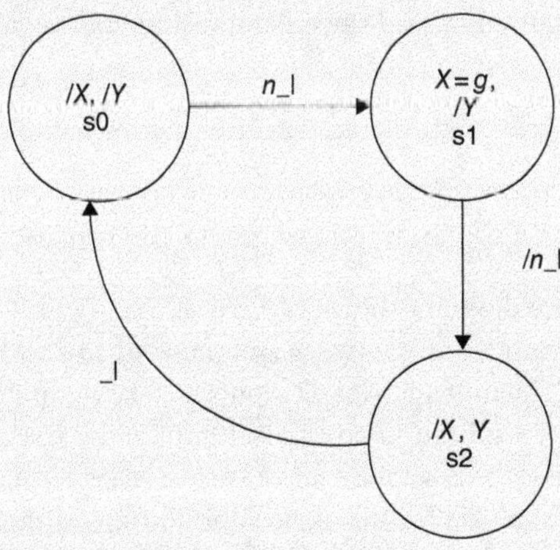


**Figure 5.6** An example for you to complete,

When you have completed this turn to **Frame 5.12**.

## Frame 5.12

Figure 5.7 repeats the state diagram in Figure 5.6.

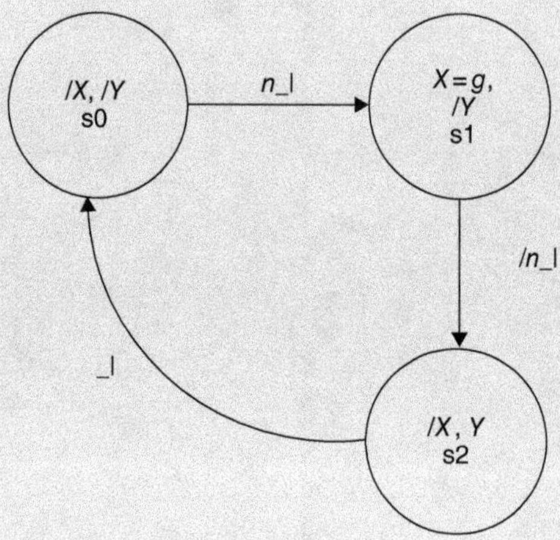


**Figure 5.7** The state diagram again to compare against your equations.

*D* type flip-flop equations are:

$$s0 \cdot d = s2 + s0 \cdot /n.$$
$$s1 \cdot d = s0 \cdot n + s1 \cdot n.$$
$$s2 \cdot d = s1 \cdot /n.$$

No hold term required in s2·*d* since the transition between s2 and s0 does not have an input.

Turn to **Frame 5.13**.

## Frame 5.13

Now look at the state diagram shown in Figure 5.8.

In this state diagram there is a two-way branch s1 to s2 when $x = 1$ and s1 to s3 when $x = 0$. The equations for the *D* inputs are shown in the figure.

For the input s0·*d* we have s2·sp as a set term and s0·/st as a hold term. For input s1·*d* we have s0·st as a set term and s3 also as a set term. But note that there is no hold term. The reason for this is because in state s1 the FSM can leave s1 for s2 if $x = 1$, or it can leave s1 for s3·*d* if $x = 0$. Either way, the state machine will leave state s1.

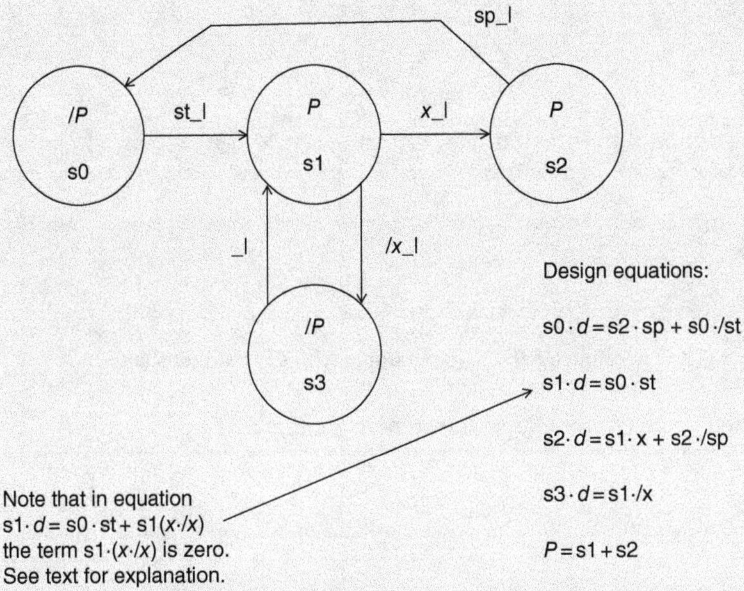


**Figure 5.8** Another example showing effect of two-way branch.

We could write the equation as:

$$\mathrm{s1} \cdot d = \mathrm{s0} \cdot st + \mathrm{s1}\left(x \cdot /x\right) + \mathrm{s3}.$$

Therefore:

$$\mathrm{s3} \cdot d = \mathrm{s1} \cdot /x.$$

Now, $\mathrm{s1}(x \cdot /x) = \mathrm{s1} \cdot 0 = 0$, so we are left with:

$$\mathrm{s1} \cdot d = \mathrm{s0} \cdot /st + \mathrm{s3}.$$

Note that when in s3 the FSM will move directly to s1 on the next clock pulse. Note that $x \cdot /x = 0$ (see Boolean algebra in Appendix A1 at the end of this book).

Turn to **Frame 5.14**.

## *Frame 5.14*

**Try to produce the D input equations and the output equations from the state diagram shown in Figure 5.9.**

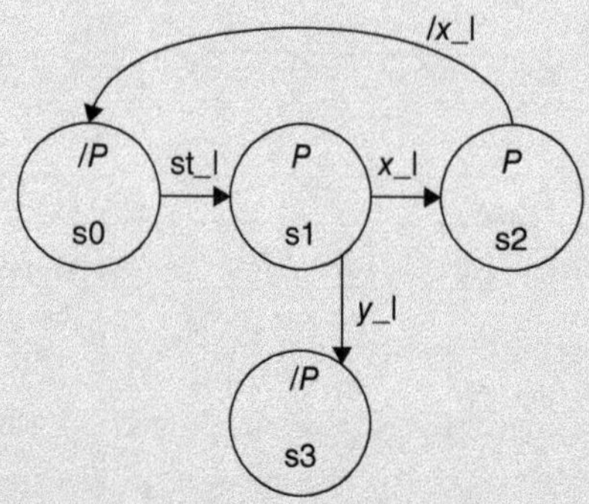


**Figure 5.9** Same example with s3 invariant state.

When you have completed this, turn to **Frame 5.15**.

### *Frame 5.15*

Figure 5.10 repeats the state diagram in Figure 5.9.

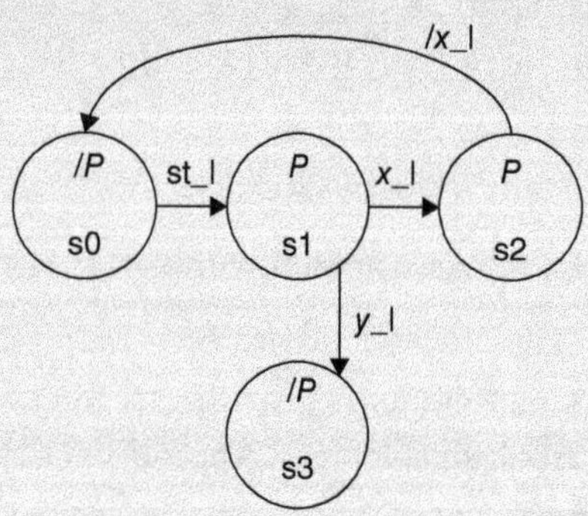


**Figure 5.10** Compare state diagram with equations.

$$\mathrm{s0} \cdot d = \mathrm{s2} \cdot /x + \mathrm{s0} \cdot /st.$$
$$\mathrm{s1} \cdot d = \mathrm{s0} \cdot st + \mathrm{s1} \cdot /x \cdot /y.$$

Note the second minterm needs to have both $x$ and $y = 0$ **to remain in s1**.

$$\mathrm{s2} \cdot d = \mathrm{s1} \cdot x + \mathrm{s2} \cdot x.$$
$$\mathrm{s3} \cdot d = \mathrm{s1} \cdot y + \mathrm{s3}.$$

Note in particular the two-way branch state s1. Here we have a two-way branch with different inputs $x$ and $y$. For the state machine to remain in state s1, both $x$ and $y$ must be zero hence the term s1(/x·/y). Also note that the term s3 in equation s1·$d$ is not needed (can you see why?).

**Note that in s1·*d* = s0·st + s1·/*x*·/*y*. You do not need s3 as there is no transition from s3 to s1.**

Also, however, when the state machine reaches state s3 **it must be held in this state**. Hence the term:

$$s3 \cdot d = s1 \cdot y + s3.$$

The only way to leave state s3 is to initialize the state machine back to state s0.

Turn to **Frame 5.16**.

### *Frame 5.16*

This next example (Figure 5.11) involves most of the ideas used in the one hot method.

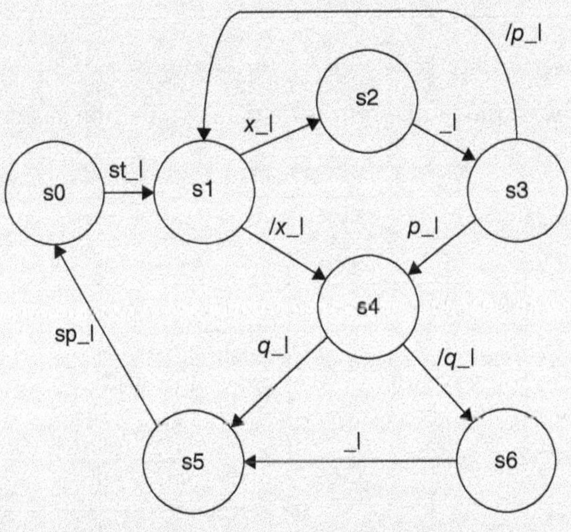


**Figure 5.11** Seven-state FSM example.

**Obtain the equations for the *D* inputs (this state diagram does not have any outputs as it is just an academic exercise).**

When you have completed this turn to **Frame 5.17**.

### Frame 5.17

Compare your answers with the solutions shown below.

$s0 \cdot d = s5 \cdot sp + s0 \cdot /st.$

$s1 \cdot d = s0 \cdot st + s3 \cdot /p.$

$s2 \cdot d = s1 \cdot x.$

$s3 \cdot d = s2$ (*do* not need s3 since it cannot hold in s3).

$s4 \cdot d = s1 \cdot /x + s3 \cdot p.$

$s5 \cdot d = s4 \cdot q + s6 + s5 \cdot /sp.$

$s6 \cdot d = s4 \cdot /q.$

If you got any wrong, check them out. You might need to go back and read over the work done in other frames to make sure you understand all the techniques used.

Turn to **Frame 5.18**.

### Frame 5.18

Figure 5.12 shows a block diagram for a data acquisition system controlled by an FSM.

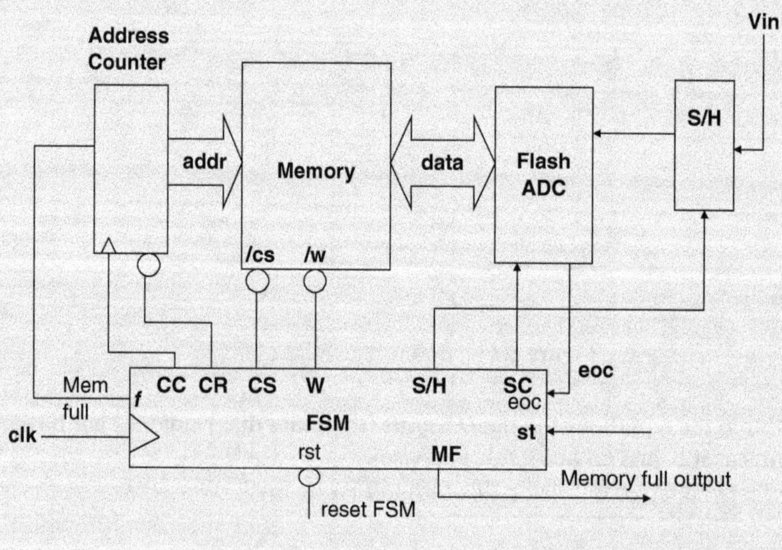


**Figure 5.12** FSM data acquisition system.

This example is much like the one you saw in Frame 2.4 (Figure 2.6). The state diagram for that example is illustrated in Frame 2.9 (Figure 2.11).

**See if you can develop the state diagram here, without looking at the solution. In this case you are to use the one hot method so you do not need to consider SSVs.**

When you have done this, turn to **Frame 5.19**.

## *Frame 5.19*

Figure 5.13 is a possible solution.

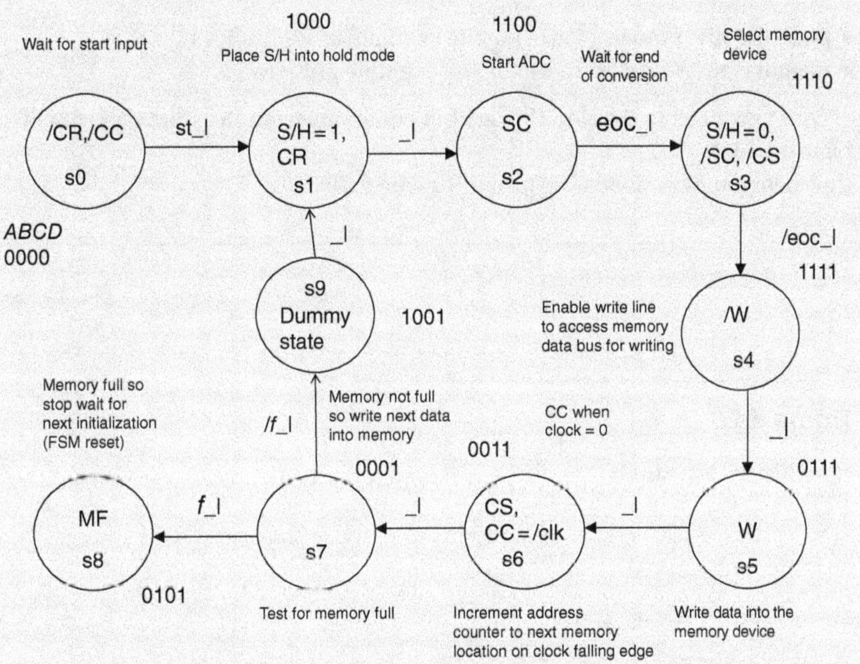


**Figure 5.13** State diagram for the data acquisition system.

If your state diagram is not the same as this, it still could be correct. However, check the sequence to see that you have not missed out any part of the sequence.

Note the need to remove the counter reset signal (CR) in state s1 to allow the address counter to be incremented. Then the sample and hold circuit needs to be placed into hold mode before making an ADC conversion in s2. Once this is done, the sample and hold circuit can be placed back into sample mode in state s3 and the chip select for the memory device asserted low. The ADC digital data then need to be written to the memory device (states s4 and s5).

After this the memory chip will be de-selected (CS = 1), then the address counter (CC) incremented in state s6 (in this example the counter is incremented on the negative edge of the FSM clock). In state s7 a test is made to find out if the end of memory has been reached, and if not the whole sequence is repeated again, storing the next ADC values into consecutive memory locations (states s7 to s1, through to s7). When the end of memory is reached, the FSM moves to state s8 where the memory full (MF) output is enabled.

**In Figure 5.13 SSVs *ABCD* have been added. These, of course, are not needed by the one hot method but can be used with *D* flip-flops, as in Chapters 1–3. The reader might like to try this out and simulate (using Logisim perhaps).**

**See the Logisim examples:**

- **fsm_das_dff_tunnels_4.circ for the example of Figure 5.13.**
- **memory_tester_dff_svv.circ for the example of Figure 5.17.**

**You now need to develop the one hot equations from the state diagram of Figure 5.13.**

When you have done this, turn to **Frame 5.20**.

### *Frame 5.20*

Here are the equations:

$$s0 \cdot d = s0 \cdot /st \;(\text{no other terms required, except access the reset}).$$

$$s1 \cdot d = s0 \cdot st + s9.$$

$$s2 \cdot d = s1 + s2 \cdot /eoc.$$

$$s3 \cdot d = s2 \cdot eoc + s3 \cdot eoc.$$

$$s4 \cdot d = s3 \cdot /eoc.$$

$$s5 \cdot d = s4.$$

$$s6 \cdot d = s5.$$

$$s7 \cdot d = s6 \;(\text{the two-way-branch has complementary inputs}).$$

$$s8 \cdot d = s7 \cdot f + s8 \;(\text{hold}).$$

The output equations are:

$$/CR = /s0.$$
$$S/H = s1 + s2.$$
$$/CS = /(s3 + s4 + s5).$$
$$/W = /s4.$$
$$CC = s6 \cdot /ck \text{ this is a Mealy output.}$$
$$MF = s8.$$

Turn to **Frame 5.21**.

### *Frame 5.21*

Finally, Figure 5.14 shows the data acquisition FSM simulation. This was obtained by using the equations in Frame 5.20 implemented in Verilog HDL.

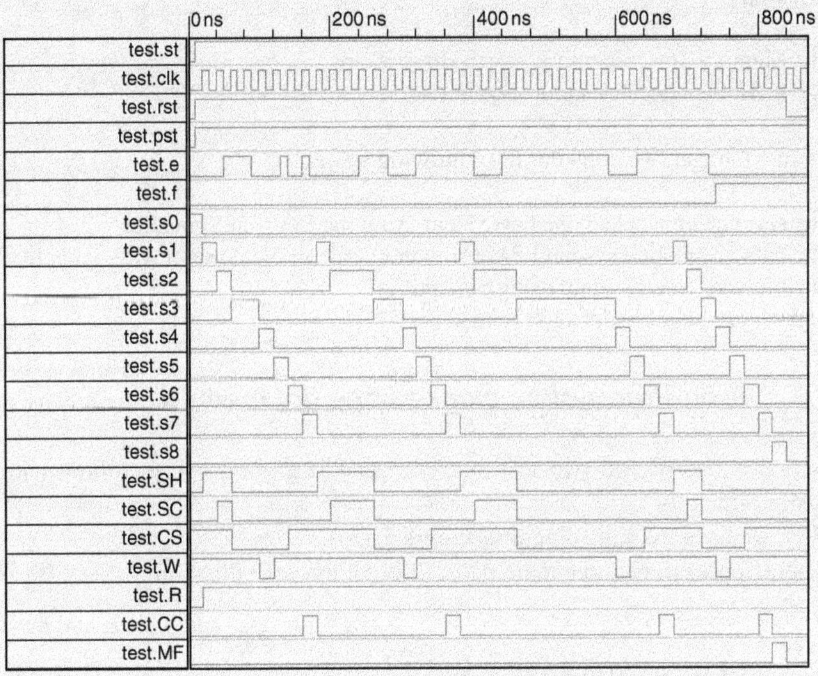


**Figure 5.14** Simulation for the data acquisition system.

**Note that the signal R in Figure 5.14 is in fact signal CR in Figures 5.12 and 5.13.**

Figure 5.14 uses the one hot method, and you can clearly see that at any one time only one state is active. Try sequencing through the state diagram and the simulation-timing diagram noting the outputs and inputs.

**One final note when using the one hot Method with *D* type flip-flops: if the FSM falls into an unused state, the FSM will automatically reset to s0 as the unused state will make any unused flip-flop clear to zero.**

**However, if *T* type flip-flops (TFFs) are used (or indeed *JK* or SR flip-flops), the FSM could fall into an unused state – so beware!**

This is a very useful point to observe when using the one hot method, which can potentially have many unused states.

## 5.2 TUTORIALS ON THE CLOCKED ONE HOT FSM METHOD

These tutorials cover the clocked one hot method. They provide complete examples of the seven-state diagram FSM. There are two other incomplete tutorials for the reader to try:

- The Memory Tester memory_tester_dff_ssv.circ
- The 8-Bit Sequence Detector seqdec.v.

In this tutorial, we consider the following systems:

- The seven-state clocked one hot system using the Logisim Simulator.
- The data acquisition system FSM using the Verilog HDL system with SynaptiCAD.
- The memory tester system using SynaptiCAD.
- Eight-bit sequence detector system.

### 5.2.1 Seven-state system clocked one hot method

This makes use of two two-way branches and is a good one to follow with the Logisim Simulator.

First we have the state diagram (Figure 5.15).

There appear to be three main paths through the state diagram.

s0 → s1 → s2 → s3 → s4 → s6 → s5 → s0. With st, $x$, and $p$ all 1, and $q = 0$.

Need to make st = 0 to return to s0.

s0 → s1 → s4 → s5 → s0.

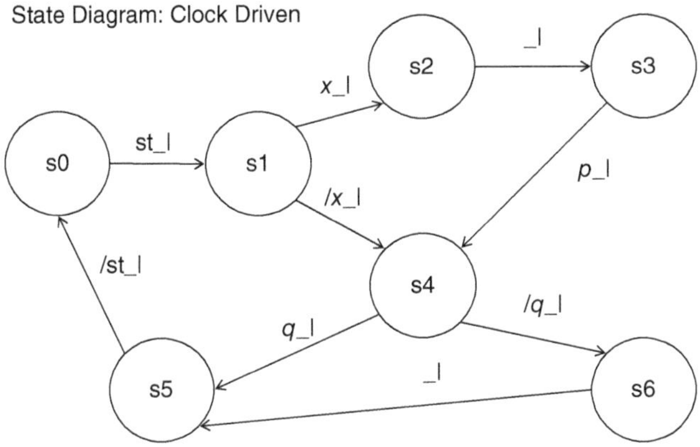


This example has seven states with two two-way branches s1 and s4.

**Figure 5.15** Seven-state one hot FSM.

With st = $q$ = 1.

$$s0 \to s1 \to s4 \to s6 \to s5 \to s0.$$

With st = 1 when in s5 st = 0.
The state diagram equations are:

$$\mathrm{s0} \cdot d = \mathrm{s5} \cdot /st + \mathrm{s0} \cdot /st.$$
$$\mathrm{s1} \cdot d = \mathrm{s0} \cdot st \; (\text{no hold due to two-way branch}).$$
$$\mathrm{s2} \cdot d = \mathrm{s1} \cdot x (\text{no input on s2 leaving term}).$$
$$\mathrm{s3} \cdot d = \mathrm{s2} + \mathrm{s3} \cdot /p.$$
$$\mathrm{s4} \cdot d = \mathrm{s1} \cdot /x + \mathrm{s3} \cdot p.$$
$$\mathrm{s5} \cdot d = \mathrm{s4} \cdot q + \mathrm{s6} + \mathrm{s5} \cdot st.$$
$$\mathrm{s6} \cdot d = \mathrm{s4} \cdot /q.$$

You should try to draw the circuit using Logisim and simulate it with a very slow clock of, say, 1 Hz so you can follow it. See Appendix A2 to find out how to use Logisim. Also see Logisim program examples like the memory tester and the eight-bit sequence detector.

Also, look at other Logisim examples to see other ideas of what you can do in Logisim. You can of course do this example using $D$ type flip-flops and state variables, as used in Chapter 3. This can also be done using the Verilog HDL code and SynaptiCAD method.

### 5.2.2 Memory tester FSM

The next example (Figure 5.16) illustrates using the clocked one hot method to implement the memory tester system.

In this system the FSM is used to control an address counter (to keep track of memory locations), a memory chip, and a digital comparator. Looking at the diagram, you may be able to decide how to set up the sequence for the state diagram.

Note the two tri-state buffers used to control the direction of data from the memory data bus lines using the read and write controls from the FSM. Note that both read and write lines are active low (hence the negating circles on each of the rd and wr inputs to the memory device).

The Digital Comparator is devised using a number of exclusive NOR gates to form the eight-bit comparator.

When both the output data lines from the memory are the same as the data compare lines (in this case 55h or 0101 0101) the $A = B$ output line from the comparator goes high at the 'FAB' input to the FSM.

Note that in Logisim there is an eight-bit comparator simulation you can look at to see how it works (**comparator8.circ**).

The state diagram for the FSM is shown in Figure 5.17.

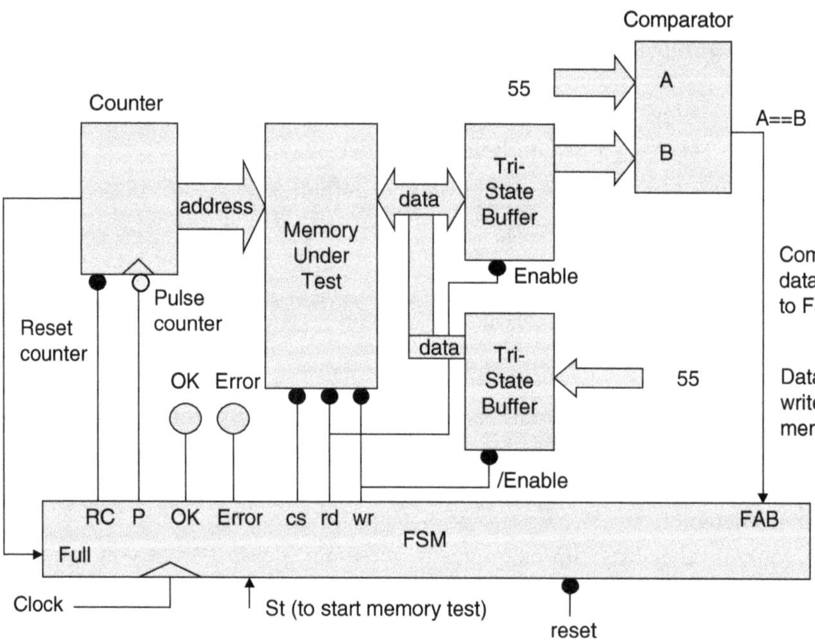


**Figure 5.16** Memory tester FSM.

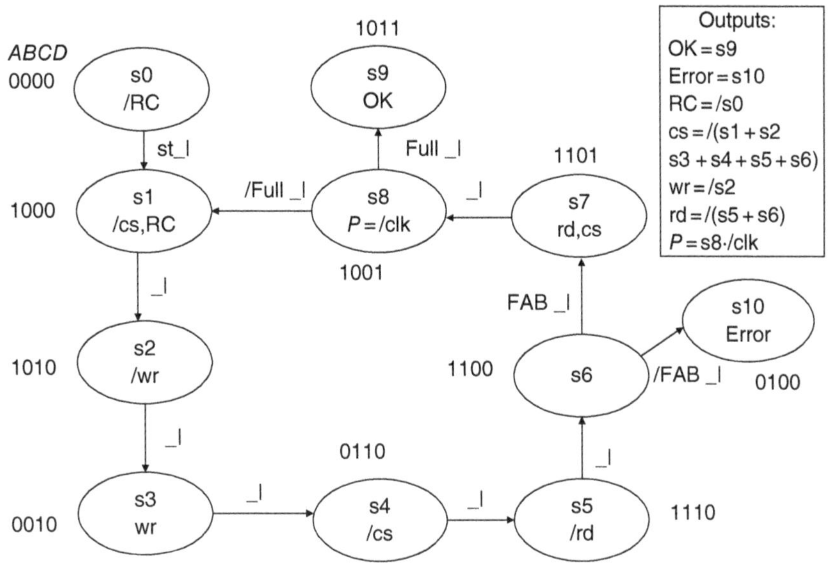


**Figure 5.17** State diagram of memory tester with outputs.

Figure 5.17 shows the state diagram for the FSM. It also shows the outputs from the FSM used to control the system.

**Note that 'FAB' is an input and that some outputs are shown in lower case. This is just to show that some people may mix case sensitivity. It is the actual variable name that matters.**

**Never use the same variable name for two different variables.**

The one hot method is used to produce the flip-flop equations. Perhaps you might like to try to produce these yourself? The one hot equations for this system are:

$s0{\cdot}d = s0{\cdot}/st + reset.$

$s1{\cdot}d = s0{\cdot}st + s8{\cdot}/\mathrm{full}.$

$s2{\cdot}d = s1.$

$s3{\cdot}d = s2.$

$s4{\cdot}d = s3.$

$s5{\cdot}d = s4.$

$s6{\cdot}d = s5.$ No hold term since two – way branch.

$s7{\cdot}d = s6{\cdot}fab.$

$s8{\cdot}d = s7.$

$s9{\cdot}d = s8.\ Full + s9.$ Uses s9 as a hold ter*m*.

$s10{\cdot}d = s6{\cdot}/FAB + s10.$ Hold term $s10$.

The outputs are shown in Figure 5.17.

If you like you could try using the SSVs in Figure 5.17 to produce a clock driven *D* type FSM (perhaps later).

The Verilog HDL SynaptiCAD code can be found in the file **memorytester.v**.

```
// Memory Tester FSM
// Flip Flop module with preset & clear.
// Author: PDM
// This code needs to be completed by the reader.
module D_FF(q,d,clk,rst,pst);
 input d,clk,rst,pst;
 output q;
 reg q;
 always @ (posedge clk or negedge rst or negedge pst)
  case ({rst,pst})
    2'b00: q<=d;
    2'b01: q<=0;
    2'b10: q<=1;
    2'b11: q<=d;
 endcase
endmodule
//-------------------------------------------------------
module fsm(rst,pst,clk,st,FAB,full,cs,rd,wr,ok,err,rc,p,s0,s1
,s2,s3,s4,s5,s6,s7,s8,s9,s10);
  input rst,pst, clk,st,FAB,full;
  output cs,rd,wr,ok,err,rc,p,s0,s1,s2,s3,s4,s5,s6,s7,s8,
s9,s10;

  wire s0d,s1d,s2d,s3d,s4d,s5d,s6d,s7d,s8d,s9d,s10d;

  D_FF s0ff(s0,s0d,clk,1'b1,pst);
  D_FF s1ff(s1,s1d,clk,rst,1'b1);
  D_FF s2ff(s2,s2d,clk,rst,1'b1);
  D_FF s3ff(s3,s3d,clk,rst,1'b1);
  D_FF s4ff(s4,s4d,clk,rst,1'b1);
  D_FF s5ff(s5,s5d,clk,rst,1'b1);
  D_FF s6ff(s6,s6d,clk,rst,1'b1);
  D_FF s7ff(s7,s7d,clk,rst,1'b1);
  D_FF s8ff(s8,s8d,clk,rst,1'b1);
  D_FF s9ff(s9,s9d,clk,rst,1'b1);
  D_FF s10ff(s10,s10d,clk,rst,1'b1);

  assign
    s0d=s0&~st,
```

```
    s1d=s0&st | s8&~full,
    s2d=s1,
    s3d=s2,
    s4d=s3,
    s5d=s4,
    s6d=s5,
    s7d=s6&FAB,
    s8d=s7,
    s9d=s8&full | s9,
    s10d=s6&~FAB | s10,

    ok=s9,
    err=s10,
    rc=~s0,
    cs=~(s1|s2|s3|s4|s5|s6),
    wr=~s2,
    rd=~(s5|s6),
    p=s8&~clk;

endmodule
// test bench for fsm testing..

module test;
  reg clk,st,rst,pst,FAB,full;

  fsm uut(rst,pst,clk,st,FAB,full,cs,rd,wr,ok,err,rc,p,s0,s1,
s2,s3,s4,s5,s6,s7,s8,s9,s10);

   initial
     begin
      $dumpfile("memtester.vcd");
      $dumpvars;
      rst=0;
      pst=0;
      clk=0;
      st=0;
      FAB=0;
      full=0;
      //-----------
      #10 rst=1;
          pst=1;
      #10 clk=~clk;
      #10 clk=~clk;
      #10 st=1;
```

```
repeat (40)
       #10 clk=~clk;
       $finish (10);
       end
   //  FSM ends up in state s10 with error.
 // repeat (40)
   initial
     begin
       #20 FAB=1;
       #5 full=0;
       #85 full=1;
       #10 rst=0;
           pst=0;
           full=0;
       #10 rst=1;
      #85 FAB=1;
      #10 FAB=0;
     end
// The above code loops 40 times around s1 to s8 to fill
memory with data.
endmodule
```

**Listing 5.1 Memory tester FSM.**

Perhaps you might like to run this file with the SynaptiCAD system to produce the Verilog simulation. Note this system does not include the comparator, memory, or tri-state devices.

You could try to add these yourself.

If you look into Logisim, there is a **memory_tester_dff_ssv.circ** simulation contained in a memory tester folder to try out. The folder also contains some slides describing the example.

### *5.2.3 Eight-bit sequence detector FSM*

This is a sequential detector to identify the eight bits of a sequence of bits used to identify an incoming set of eight bits of data.

The block diagram for this FSM is shown in Figure 5.18.

The idea is that an input octal data latch accepts an eight-bit 'C code' C0 to C7 that can be presented to the eight bits of a set of one-bit comparators (exclusive NOR) and this is presented to an FSM system (not shown here).

If each of the RC1 to RC7 bits are the same (0 or 1), to each of the serial '*D* input stream' presented to the other eight bits of the bit comparators the identity of the serial bits, the output D0 to D7 lines will be 1; otherwise, they will be 0.

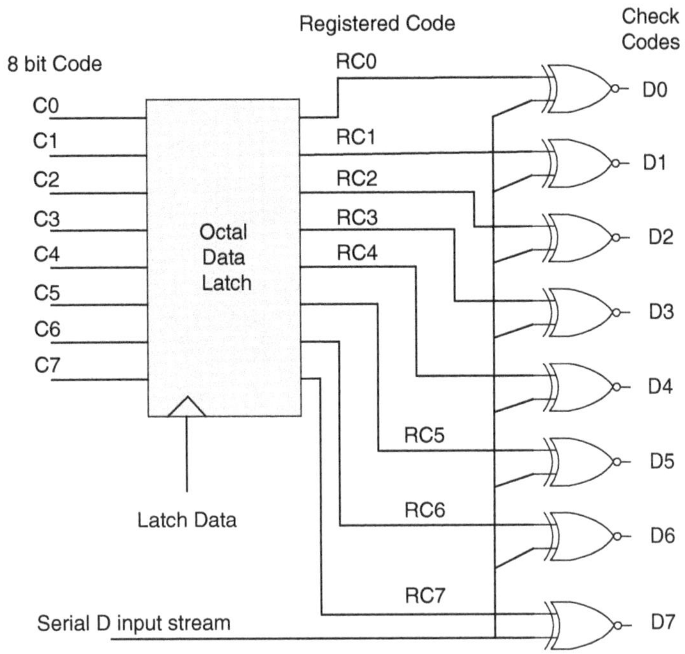


Comparator scheme for checking data stream codes

**Figure 5.18** The octal data latch and comparators.

In Figure 5.19, the FSM is added to the block diagram.

So from the state diagram of Figure 5.20, it can be seen that the FSM needs to check each of the RC0 to RC7 bits against each of the two-bit exclusive NOR comparators and set the D0 to D7 bits to either a logic 1 bit the same as the RC bit, or 0 bit if not the same as the RC bit.

Figure 5.20 shows the FSM sequence.

Initialize input starts the FSM in s0.

When enable En starts the FSM s1 loads the octal data latch (*Ld* = 1) with the C0 to C7 (see Figure 5.19) data to be detected. From s2 to s16 performs the detection process.

If the FSM reaches s9 it uses En to decide if another eight-bit C0 to C7 code is to be detected, or if En = 0 returns to s0.

In this state diagram we start by looking for the En enable signal to start the sequence. Then the data latch is pulsed to load the RC0 to RC7 bit pattern to look for in the FSM. We assume that the data inputs to the FSM are D0 to D7 as indicated in Figure 5.20. Then each bit of the incoming sequence is read into the FSM and tested against the bits RC0 to RC7 to find a match.

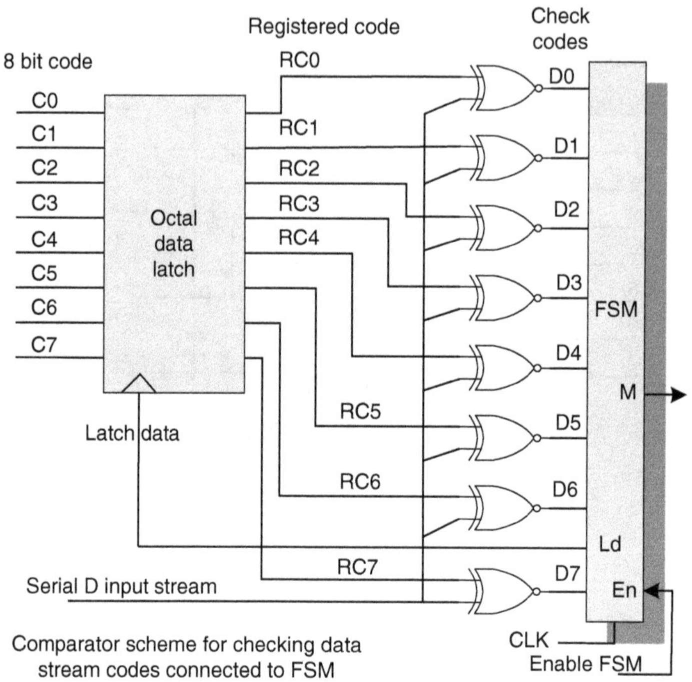


**Figure 5.19** Adding the FSM to the block diagram.

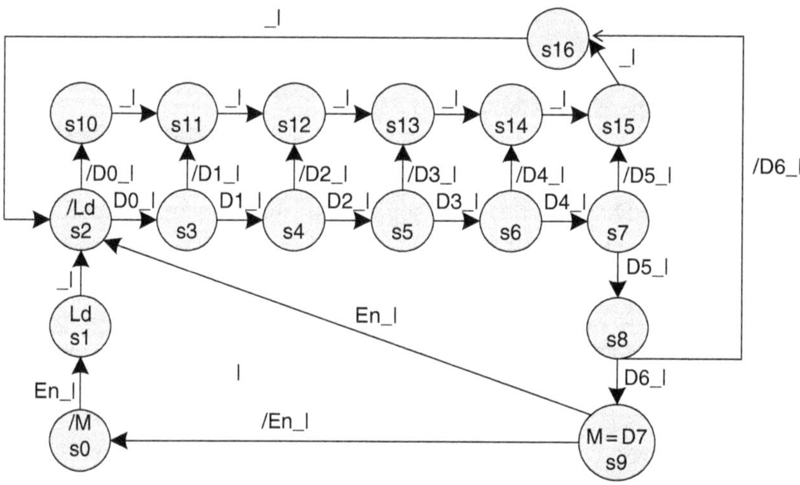


**Figure 5.20** The state diagram of the FSM.

**If the D0 to D7 bits (outputs of the comparators) are the same as the RC0 to RC7 bits, the D0 to D7 in the state diagram are set to logic 1; otherwise, they are left at logic 0.**

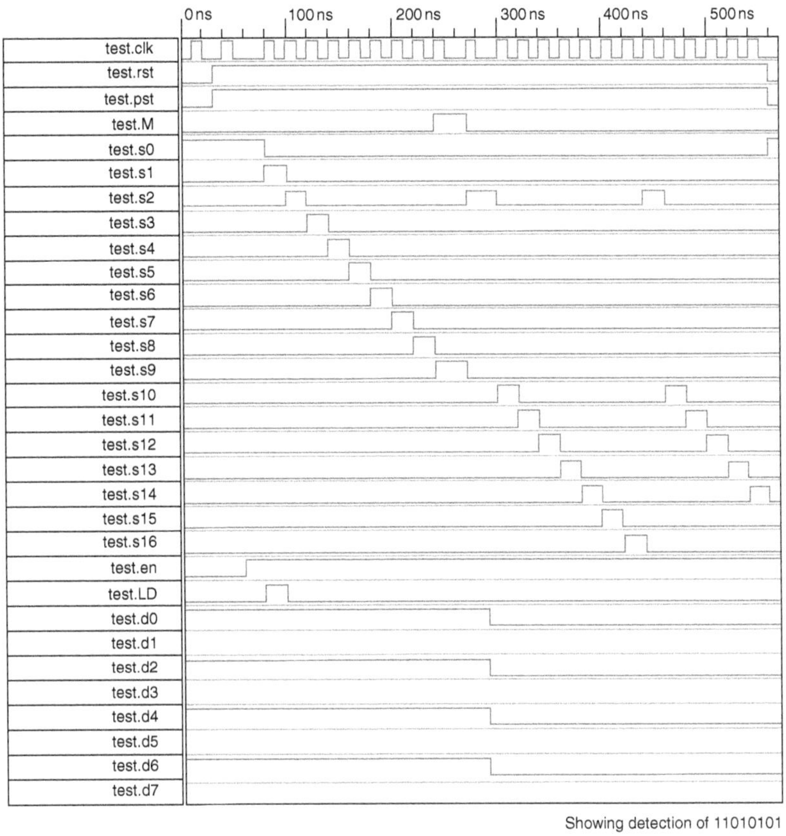


**Figure 5.21** A Verilog HDL simulation of the FSM system.

From the sequence shown in the state diagram of Figure 5.20, it should be apparent that, if all the D0 to D7 bits are 1, then a match has been found.

Note that the state diagram sequence must follow the eight-bit sequence. All D0 to D7 bits at the input to the FSM will be logic 1 and a match has been found. If any of the D0 to D7 bits is logic 0, this means a match has not been found.

You might like to try to create a Verilog HDL file for this system using the SynaptiCAD software, or perhaps the Logisim Simulator in order to try this system out (Figure 5.21).

Remember that Appendix A2 can be used to produce the required simulations using either SynaptiCAD or Logisim or both.

# 6

# Further Event-Driven FSM Design

## 6.1 INTRODUCTION

In Chapter 4 we covered the design of asynchronous (event-driven) finite state machine (FSM) using a method of design of digital FSM systems originally looked at by Zissos (1979), and considered in this book. In this design it is found that the method is very reliable as long as unit distance coding is used, and that the primary gate delays Tp are less than the secondary gate delays Ts. The topology of the circuits is always arranged (and indeed the design method always produces a design).

In Chapter 5 we covered the one hot method used with a clock-driven FSM. This allowed us to be able to design FSM systems without using secondary state variables (SSVs) and hence avoid the need for unit distance coding of these SSV.

Many people have made use of the idea of using the one hot method with the asynchronous method using the Muller C gate and Request/Acknowledge handshake signals to avoid the effects of glitch generations that can cause the system to maloperate (Myers 2001; Yakovlev and Koelmans 1998).

This chapter begins by looking at the event-driven method used in Chapter 4.

Later it discusses ideas for implementing event-driven one hot designs without using the Muller C gate and Request/Acknowledge handshakes. An example using just two states is considered and some of the problems in one hot design are discussed and how the method is implemented.

The chapter then looks at the idea of using handshakes to control three independent systems: an FSM, an analogue-to-digital converter (ADC), and a microcontroller.

At the end of Chapter 7 the reader is introduced to the Muller C gate and Request/Acknowledge handshakes to see how they work (Yakovlev and Koelmans 1998).

Now move to **Frame 6.1**.

*Digital System Design using FSMs: A Practical Learning Approach*, First Edition. Peter D. Minns.

Companion website: www.wiley.com/go/minns/digitalsystemdesign

## *Frame 6.1*

Consider the following Verilog HDL (Figure 6.1).

**Verilog HDL Description**

```
//declaring the list of variables
module example1(P,s0,s1,w1,w2,w3,w4,x,rst);
//declare list of outputs
output P,s0,s1,w1,w2,w3,w4;
//declaring list of inputs
input x,rst;
wire nx,nrst;
not g1(nrst,rst);
not g2(nx,x);
//combinational logic for s0
and g3(w1,s0,nx);
and g4(w2,s1,nx);
or g5(s0,w1,w2,nrst);
//combinational logic for s1
and g6(w3,s0,x,rst);
and g7(w4,s1,x,rst);
or  g8(s1,w3,w4);
//output P
buf g9(P,s1);
endmodule
```

**Test Bench**

```
module test;

reg x,rst;

wire P,s0,s1,w1,w2,w3,w4;
//reset the module example1
Example0 fsm(P,s0,s1,w1,w2,w3,w4,x,rst);

initial
begin
//initializing the variables
x=0;
rst=0;
#10 rst=1;
#30 x=1;
#30 x=0;
#100 $finish
end
endmodule
```

**Figure 6.1** Test bench of logic gates FSM.

Figure 6.1 shows a typical Verilog HDL listing. In this you can see the listing of logic gates forming a typical structure of an FSM.

One thing you can do in Verilog is to introduce delays for each logic gate using the # symbol followed by a delay value. So #5 would represent a gate delay of 5 ns or $5 \times 10^{-9}$ of a second.

Turn to **Frame 6.2**.

## *Frame 6.2*

Here is some Verilog HDL code with some propagation delays. In this example, the key Verilog words are in **bold**.

```
module example(P,s0,s1,w1,w2,w3,w4,x,rst);
     output P,s0,s1,w1,w2,w3,w4;
     input x,rst;
     wire nx,rst;
```

```
    not #5 g1(nrst,rst);
    not #5 g2(nx,x);
    and #5 g3(w1,s0,nx);
    and #5 g4(w2,s1,nrst);
    or #5 g5(s0,w1,w2,nrst);
    and #5 g6(w3,s0,x,rst);
    and #5 g7(w4,s1,x,rst);
    or #5 g8(s1,w3,w4);
    buf #5 g9(P,s1);
endmodule
```

The propagation delays are all the same (5 ns) but could be different to reflect the expected delays of the individual gates used in the design. It is also possible to simulate the circuit using a logic simulator **Logisim.**

Logisim is a gate level logic simulator that was developed by Carl Burch in 2011. It was developed using the programming language JAVA, and as such the program Logisim can be executed on any computer using an operating system such as Windows, MAC OS X, or Linux. You just have to select the platform you are working on.

**Remember, Appendix A2 provides some details on how to use the Logisim simulator in order to produce a Logisim simulation.**

Turn to **Frame 6.3**.

### *Frame 6.3*

A list of different Verilog HDL platforms available for the Windows PC[1] are:

1. VeriLogger by SynaptiCAD (http://www.synapticad.com/syn_down.htm).
2. Modelsim-PE Mentor Graphics (http://www.model.com/resources/student_edition-download.asp).
3. Xilinx ISE Xilinx (http://www.xilinx.com/ise/logicdesign_prod/webpack.htm).
4. Active-HDL Aldec Incorporated (http://www.aldec.com/education/student/).

In point 4 you can access the Active-HDL student edition for free.

Turn to **Frame 6.4**.

[1]If you are using a MAC computer you can buy Parallels and Windows 10 to work on your MAC and hence use this software seamlessly.

### Frame 6.4

Figure 6.2 gives an example of an event-driven FSM using sequential equations and SSVs. Note: the method uses *A* and *B* SSVs. See Chapter 4 for the method.

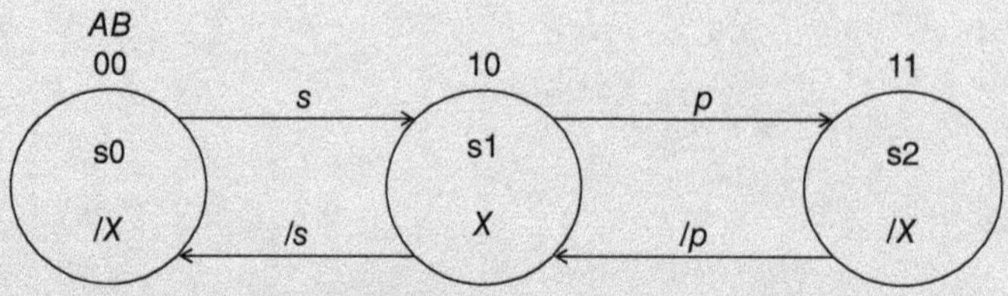


| | |
|---|---|
| Equations are: | $A = \text{Sum}(Sa) + A \cdot /\text{Sum}(Ra)$ |
| So: | $A = /B \cdot s + A \cdot B + A \cdot s$ |
| Likewise: | $B = A \cdot p + /AB + B \cdot p$ |

**Figure 6.2** Example using *s* and *p* inputs.

The state diagram shows a system with three states (s0, s1, and s2) in which inputs *s* and *p* control the state of output *X*.

You can see the two equations *A* and *B* for the event-driven FSM in Figure 6.2. The figure shows the equations for the state diagram using SSVs with the variable s0 = 00, s1 = 10, and s2 = 11.

**Draw the circuit.**

When you have done this, turn to **Frame 6.5**.

### Frame 6.5

Figure 6.3 shows the Logisim circuit diagram for the event-driven one hot design using the SSVs and sequential equations in Chapter 4.

If you have made any mistakes, check them out and make sure you understand where you made the mistake. Look back over the frames in Chapter 4 for the event-driven example method.

Note this version works well even with term $A \cdot p$.

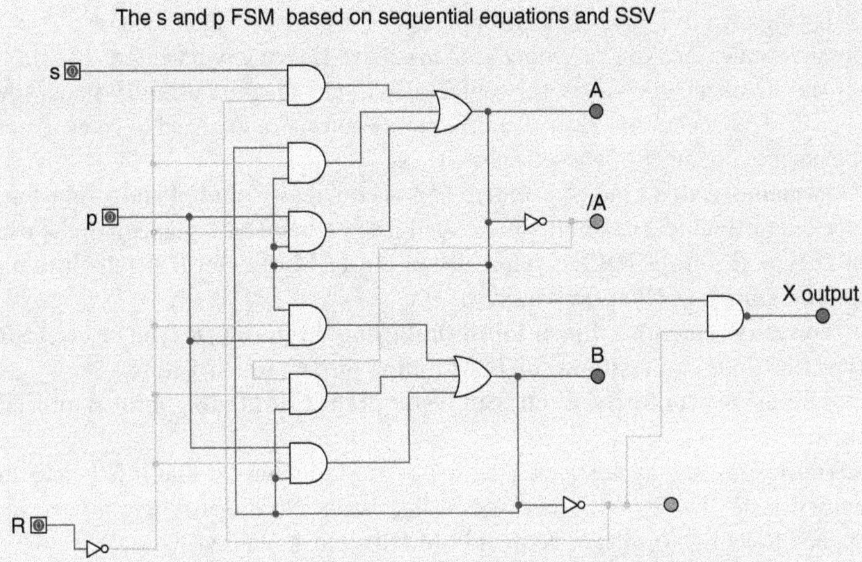


**Figure 6.3** Logisim circuit for the *s* and *p* input FSM.

### *Problems caused by propagation delays*

Forms of maloperation can be a problem with large one hot event-driven designs. It is normally not present in the event-driven method described above, due to the topological design used, i.e. following the delay rules in Frame 4.36. Note, however, the need for SSVs in the design, and this implies the use of unit distance codes. Larger designs can lead to maloperation. See Yakovlev and Koelmans (1998).

When designing large event-driven designs the user can make use of design tools to aid in the design (search on the Web). It is also possible to use Muller C gates and Request/Acknowledge handshakes (Yakovlev and Koelmans 1998).

Please turn to **Frame 6.6**.

## *Frame 6.6*

### *Unused states*

It is also possible for an FSM design to fall into an unused state due to the fact that the FSM design can have a number of unused states in its design. This can happen in both clock- and event-driven FSM systems. Also an **event-driven one hot design** can lead to a great many unused states in a given design.

If you have five states in a given design, for example, then there are $2^3 = 7$ possible states and you have only used up to five (leaving two unused states).

This number can increase very quickly. In a state diagram using 19 states you need $2^5 = 32$ states, and $32 - 19 = 13$ unused states. So you need to be aware of the potential number of unused states.

**Remember, this kind of problem can occur in any kind of state diagram, clocked or unclocked. But if you are using a clocked FSM, you can make use of *D* type flip-flops (DFFs). This allows the FSM to reset if it falls into an unused state (see Chapter 3 and 5).**

**You may remember I mentioned that using *T* type or *JK* type or even SR type flip-flops can result in the FSM falling into an unused state.**

**Also using event-driven cells can result in the FSM falling into an unused state.**

However, you may not want a particular FSM to return to s0 if it falls into an unused state. This is not addressed in this book. Note that research is being applied to look into what to do if an FSM falls into an unused state.

Event-driven FSMs do not have this protection.

**If you are using Verilog HDL you can make use of the idea of adding Verilog code to your system that causes the FSM to fall into state s0 if it encounters an unused state and it is using behavioural Verilog code (see Appendix A4). This is important since the behavioural method can select any type of flip-flop to use.**

The code for this is:

- Define the states used and those not used.
- Define event cells used in the design.
- `2'bxxxx` defines a Verilog HDL four-bit binary number.

```
Parameter s0=2' b000, s1=2' b001, s2=3' b011, s3=2' b010,
s4=2' b100, s5=2' b101, s6=2' b110, s7=2' b111;
// note define state sequence for FSM with s4 to s7 unused
states.
always @ (posedge clk or negedge rst)
if (~rst)
state <=s0;
else
case (state) // This defines what each state does.
s0: if (s) state<=s1; else state <=s0;
s1: state <=s2;
s2: if (~x) state <=s3; else state <=s1;
s3: if (s) state <=s0; else state <=s3;
```

```
// That ends the State FSM sequence. Now follows picking
up 'unused' states.
s4: state <=s0; // note these all direct FSM to s0.
s5: state <=s0;
s6: state <=s0;
s7: state <=s0;
endcase
```

Note that event-driven FSMs do not use the behavioural Verilog method. They use the Logic gate entry method (struct method). You would need to develop a program to work out the required unused state to include in such code as shown in larger designs.

**While in practice the behavioural method could be used to produce the resulting Verilog HDL code, it is not under the fine control of the designer, so beware.**

**You can now begin to see why most FSM are clocked. Also *D* type flip-flops are to be considered best practice.**

Please turn to **Frame 6.7**.

### *Frame 6.7*

Now look at a typical event-driven FSM system using the event method. And note in the figure that there is only one input change along the transitional lines.

In the example in Figure 6.4, use is made of only single input variables along transitional lines.

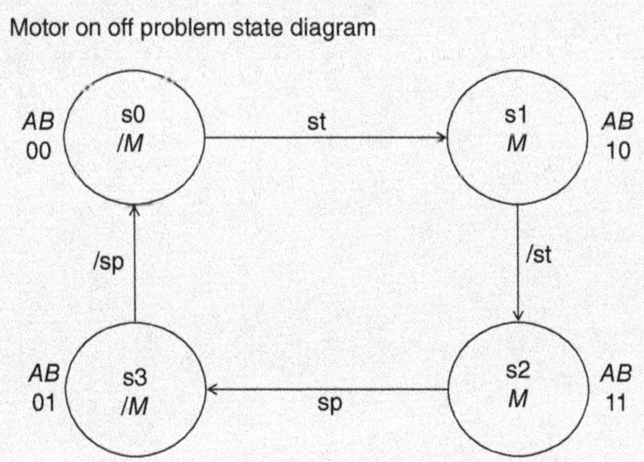


**Figure 6.4** Using one input in each transition.

The FSM waits for st = 1 to move from s0 to s1 to turn the motor on. Then when st = 0 it proceeds to s2 keeping the motor on. When sp = 1 the FSM moves to s3 turning the motor off and then waits for sp = 0 to return the FSM to s0.

*A* = Sum(Sa) + *A*·/Sum(Ra) –(1)
*A* = s0·st+ *A*·/(s2·sp) = /*B*·st+ *A*·/(*B*·sp)

<u>*A* = /*B*·st + *A*·/*B* + *A*·/sp</u>

*B* = Sum(Sb) + *B*·(Rb)
*B* = s1·/st+ *B*·/(s3·/sp) = *A*·/st+ *B*·(/*A*·/sp)
*B* = *A*·/st+ *B*·(*A*+sp)

<u>*B* = *A*·/st + *AB* + *B*sp</u>

**Figure 6.5** Sequential equations for Figure 6.4.

The sequential equations are shown in Figure 6.5.

**It should be apparent that M = *A*/*B* + AB = *A*.**

**Now draw the circuit diagram for the FSM. Then turn to Frame 6.8 to check the answer.**

Remember, you can always look at Chapter 4 to see how this is done. This example works well. Follow the sequence shown in the Figure 6.4.

Note Figure 6.6 was drawn using the Logisim Simulator. You could try to use Logisim to simulate the FSM design. See Appendix A2 for help on how to use this.

Turn to **Frame 6.8**.

### *Frame 6.8*

Figure 6.6 shows the circuit for the system mentioned in the previous frame.

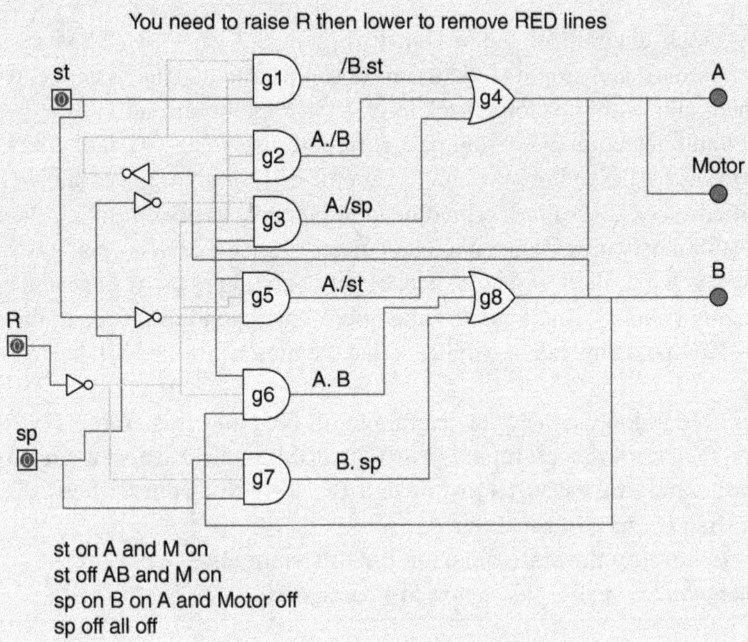


**Figure 6.6** The circuit of the system.

Now consider the following problem in **Frame 6.9**.

### *Frame 6.9 Motor Problem with Fault*

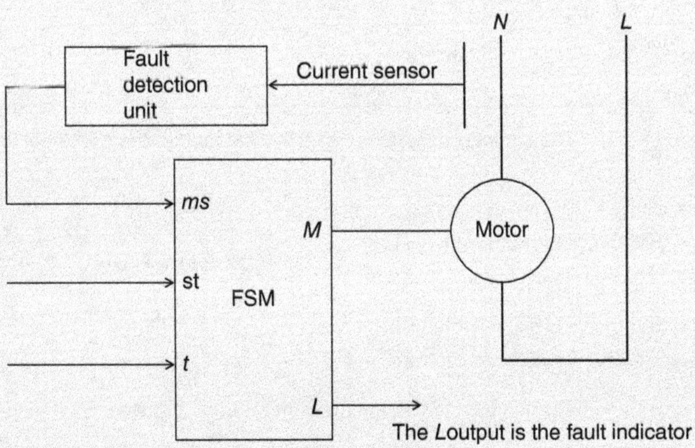


**Figure 6.7** Motor problem with fault.

If a fault current is detected (by a current detector) the motor will be stopped and the current detector indicator $L$ will indicate a fault (Figure 6.7).

The test input $t$ can be used to test the system in place of the fault current detector.

The block diagram of the system in Figure 6.7 shows an FSM system to control a motor and stop it if the motor develops a fault. The system can detect possible faults with the motor, such as excessive motor current.

The fault detection unit could be a current transformer, or a hall effect transducer to detect the motor current. The motor drive system is not shown in this simplified block diagram. It would need to be capable of controlling the motor and would most certainly need to use power transistor drivers (perhaps such as insulated gate bipolar transistors) capable of operating at the voltage and current of the motor supply. But these considerations are beyond the scope of this book.

The $L$ output is intended to indicate that the motor is turned off due to a faulty motor.

You need to look at the state diagram to find out how this example was done.

Note the use of the test input $t$ to test the FSM in the absence of a real fault.

**Hint: it may be easier to just design the state diagram without the *t* test input, then try to add this later.**

**Try to develop the state diagram for this yourself.**

When you are ready, please turn to **Frame 6.10**.

## *Frame 6.10*

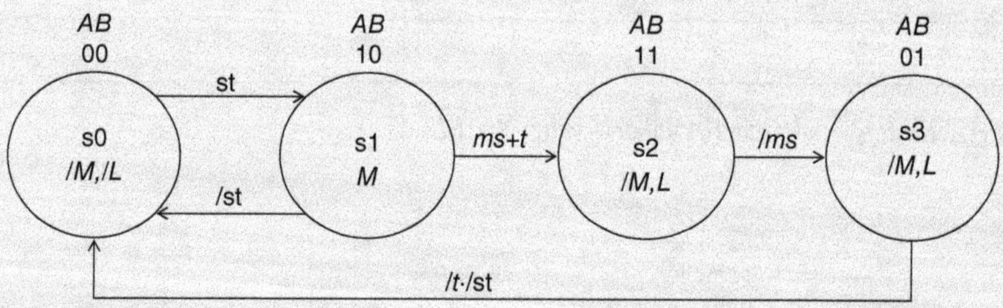


State Diagram The $t$ input is used to act as a system fault test to check out the system

Equations:

$A = \text{Sum(Sa)} + A\cdot/(\text{Sum[Ra]})$ -- (1)

$\underline{A = /B\cdot st + AB\cdot \underline{ms} + A/B\cdot st + A\cdot st\cdot \underline{ms}}$

$B = \text{Sum(Sb)} + B\cdot/(\text{Sum[Rb]})$ – (2)

$\underline{B = A\cdot \underline{ms} + A\cdot \underline{t} + A\cdot \underline{B} + B\cdot \underline{t} + B\cdot st}$

Outputs:

$M = s1 = A\cdot/B$

$L = s2 + s3 = AB + /AB = B$

**Figure 6.8** Motor with fault and test input.

In Figure 6.8, the *A* term and *B* term work well and the FSM can be sequenced through all states s0 to s3 and back to s0. In this example, the input *t* is used to present a possible fault condition (not a real fault) in order to test the system.

If you look at the state diagram in Figure 6.8 and imagine removing the *t* input (leaving the *ms* input), you will also need to remove the */t* input along the s3 to s0 transitional line. That is in Figure 6.8 the term '/*t*·/st' becomes '/st'.

You might like to complete the circuit schematic for the logic gates needed for this system and turn to **Frame 6.11**.

### *Frame 6.11*

The schematic circuit using Logisim is shown in Figure 6.9.

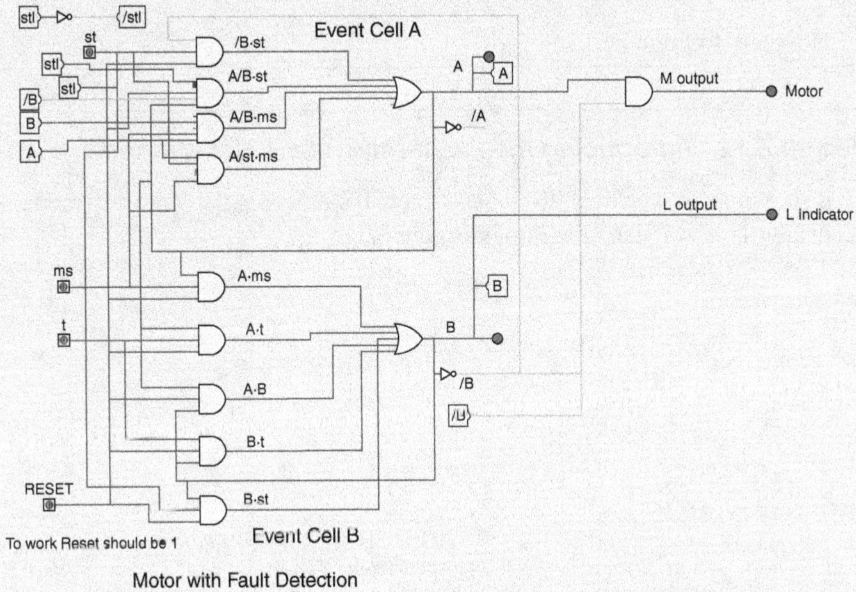


**Figure 6.9** Logisim circuit details.

Look at the solution carefully.

Figure 6.10 shows the simulation using Verilog HDL. This Logisim program is called **fig6.9_motor_with_fault_1.circ**.

The sequence for testing this program is indicated in the Logisim circuit showing the step-by-step sequence.

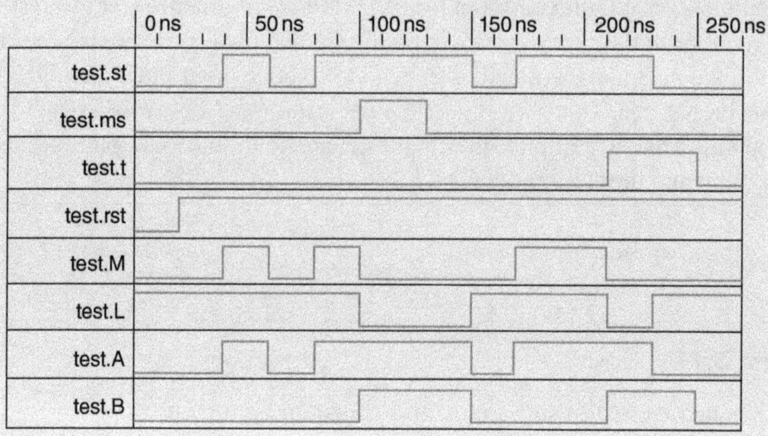


**Figure 6.10** Simulation showing its details.

Now turn to Frame 6.12.

### *Frame 6.12 Introducing the Event-Driven One Hot Method*

It is possible to perform an event-driven FSM using the one hot method. Consider the FSM state diagram in Figure 6.11.

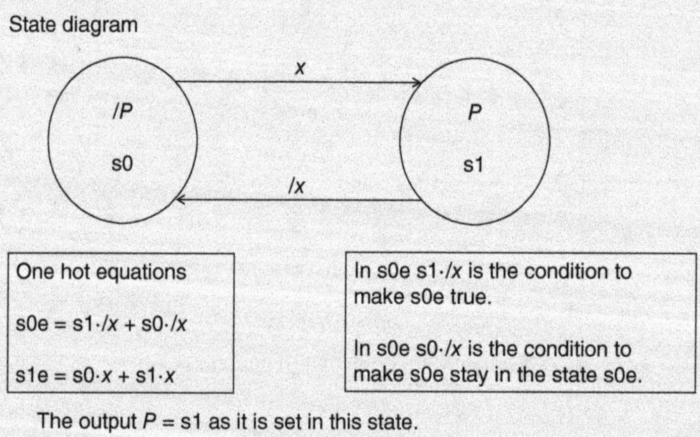


**Figure 6.11** State diagram for event-driven one hot example note equations.

In this example there are two states s0 and s1. When in s0 the output $P = 0$. If input $x = 1$ the FSM will move to s1 where output $P = 1$.

When in s1, if input $x = 0$ the FSM will move to s0 again making $P = 0$.

The one hot equations are as illustrated in Figure 6.11. Here s0e is the output of event cell s0 and you can predict its form from the equation for s0e:

$$s0e = s1 \cdot /x + s0 \cdot /x.$$

The term '$s1 \cdot /x$' is used to get into the event cell s0e (see state diagram), while the second term '$s0 \cdot /x$' is used to keep the event cell in s0e. That is, the equation consists of a term to move the event cell into s0e and the second term is used to keep the event cell in s0e.

The same actions are used to obtain the equation for event cell s1e.

In order to ensure that the FSM always starts from cell s0e, you need to ensure that, when the reset input is rst = 1, event cell s0e is set to 1. Adding an input rst to the equation for s0e can do this:

$$s0e = s1 \cdot /x + s0 \cdot /x + rst.$$

This ensures that when the reset input rst = 1 the event cell s0e will be set. It is important that the input rst is then made equal to zero.

$$\text{rst} = 0.$$

This allows the FSM system to work. The next stage is to draw the circuit for the one hot event of the FSM. This is drawn in Logisim so it can be simulated.

Note you could replace the *x* input with a pulsing input (pulse *x* using a clock input) and set it for 1 Hz so as to see it in operation. To use you must release the rst from its logic 1 to logic 0 when *x* is at logic 0 so as to see the FSM move from s0 to s1 and back again.

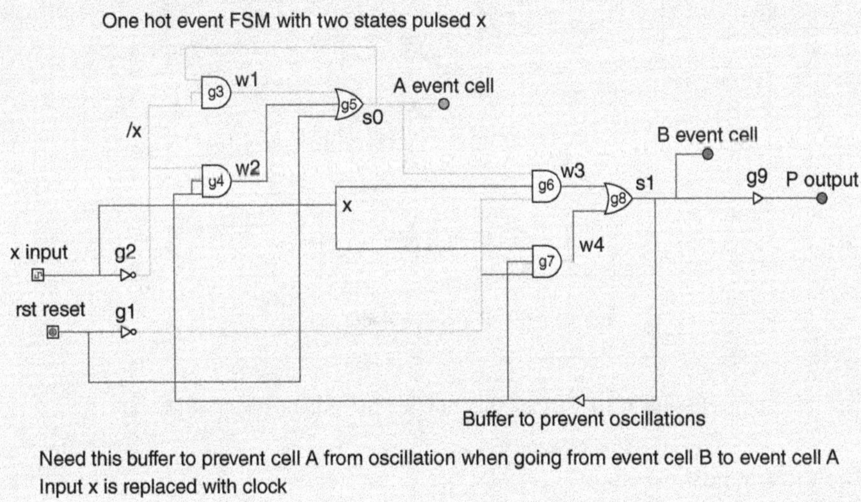


**Figure 6.12** Two-state one hot event circuit using Logisim.

In Figure 6.12, you can see that a buffer needs to be inserted in the feedback hold between gates g8 and g7 to provide event cell *B* with a longer turn-off time to prevent event cell s0 from going into oscillations when the FSM tries to move from s1 to s0. This does not seem to happen in the Verilog simulations, owing to the closeness of the gate delays.

In this example the input *x* is to be used as a pulse so that the FSM can move between s0 and s1.

Note that when using Logisim with *x* connected to a clock use rst = 1 to set *A* but make rst = 0 when *x* = 0 to see it work.

See **fig 6.12 eventonehot 2 state_with *x* pulsed.circ** in the Logisim simulations.

**Having tested a number of event-driven one hot designs, it is important that the hold time Th of event cell s*n* must be longer than the set time Ts of the event cell following s*n* + 1. This gives the following event cell time to set before the prior event cell turns off.**

**That is, Ts < Th to prevent failure of the FSM.**

Turn to **Frame 6.13**.

## *Frame 6.13 The Three-Handshake FSM System*

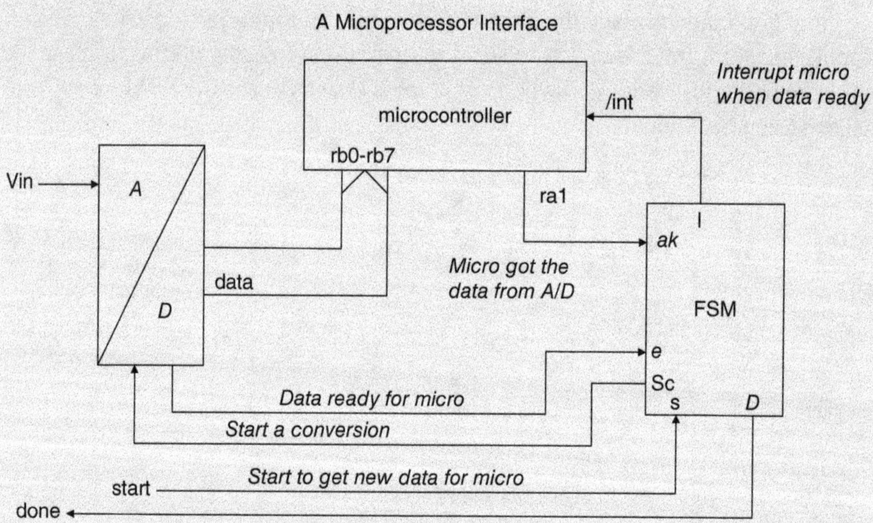


**Figure 6.13** Block diagram for three-handshake FSM.

The three handshake signals (Figure 6.13) are:

- The start and done pair to the FSM. They could be from an external device or from the microcontroller (not shown).

- The start conversion (Sc) and data ready (e) pair used by the FSM to start and look for the end of the ADC conversion from the ADC.
- The pair /int and ra1 between the FSM and the microcontroller.

The system works by using the first handshake pair to send the start line high to the FSM to ask it to perform an ADC. The FSM does this by sending the Sc signal high and waiting for the ADC to perform a conversion. When the ADC has completed this task it sends its e line high to indicate to the FSM that the converted data are now available at the digital inputs rb0 to rb7 of the microcontroller.

At this point the microcontroller is interrupted by the FSM by sending its int line low (it is an active low input).

The microcontroller receives the interrupt signal and causes the microcontroller to stop whatever it is doing and move to its interrupt service routine where its software goes to retrieve the data waiting at the output of the ADC and reads them into the port rb0 to rb7 and hence now has these data available to store in its own memory.

The microcontroller can now raise its ra1 output line to signal to the FSM that it now has the ADC data and the FSM can now raise the done *d* line to signal that the transaction has completed.

So you can see how the three handshake signals can synchronously control all three devices in the system. This is a good example of the use of handshake signals to control the three devices in the system.

Perhaps at this point you might like to consider trying to design the state diagram of the FSM system (Figure 6.14).

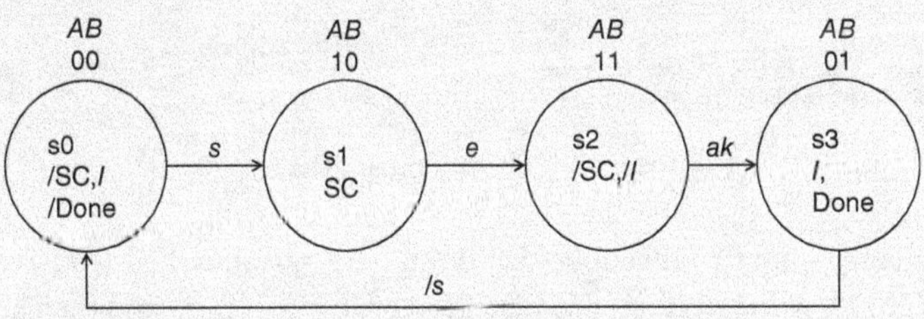


State diagram

Event-driven with *A* and *B* variable:

*A* = Sum(Sa + *A*·/Sum(Ra) --- (1)
*A* = s0·*s* + *A*/(s2·*ak*) = /*B*·*s* + + *A*·/(*B*·*ak*)
<u>*A* = /*B*·*s* + *A*/*B* + *A*·/*ak*</u>

*B* = Sum(Sb) + *B*·/Sum(Rb) --- (2)
*B* = s1·*e* + *B*·/(s3·/*s*) = *A*·*e* + *B*(/*A*·/*s*)
<u>*B* = *A*·*e* + *A*·*B* + *B*·*s*</u>

One hot method:

s0e = s3·/*s* + s0·/*s*

s1e = s0·*s* + s1·/*e*

s2e = s1·*e* + s2·/*ak*

s3e = s2·*ak* + s3·*s*

Outputs: SC = s1, /I = /s2, Done = s3

**Figure 6.14** The three-handshake state diagram.

Finally, the circuit is drawn (and can be simulated with Logisim).

If you are concerned about the time for the ADC to return *e* back to logic 0 then there are a few ways to monitor the *e* input.

One is to modify the state diagram by changing the s3 to s0 transition input from /*s* only to /*s*·/*e*. This will only change the *B* event cell equation to include the term '*B*·*e*' (*A* event equation remains unchanged).

Another way could be to allow the microcontroller to monitor the *e* signal before issuing the *ak* signal high.

The delay in *e* returning low is quite small so you may not need this modification.

Still, the circuit diagram in Figure 6.15 shows the modification in place.

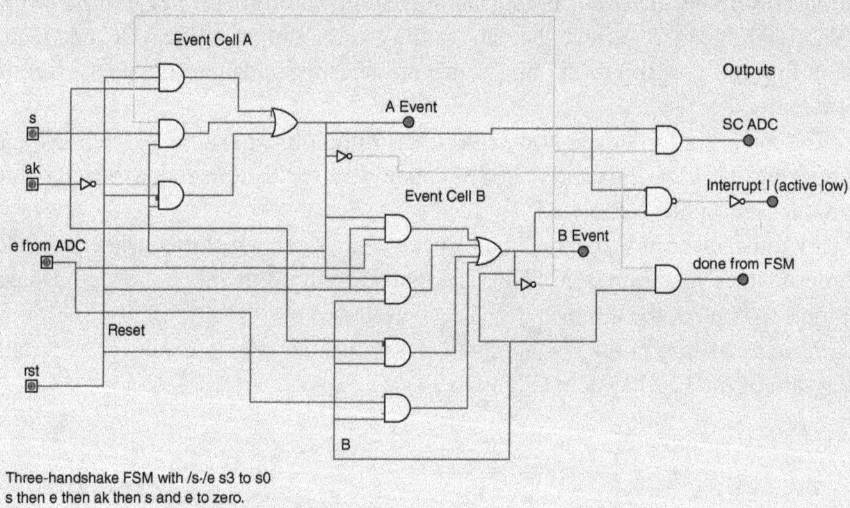


**Figure 6.15** The three-handshake circuit.

The reader should now consider doing the one hot FSM design with the SynaptiCAD Verilog system, produce the Verilog program, and then simulate the design. Note input signals in circuit diagram are shown in capitals (it is just how it was drawn).

The circuit of Figure 6.15 is very reliable and can be found as fig6.15 **threehs-FSM.circ.**

## 6.2 CONCLUSIONS

In this design the event-driven one hot method avoids using the SSVs and hence the need to use unit distance coding. It is, however, be susceptible to errors. The SynaptiCAD Verilog HDL software is better, but the Logisim simulation must be used carefully to consider possible gate delay problems.

On the other hand, the event-driven method using the sequential equation approach is more reliable and, provided that the primary delays Tp < the secondary delays Ts, the design is going to work – as long as unit distant coding is used.

In most cases people are still using the clocked-driven method for FSM design.

Using unit distant coding can lead to very reliable designs. The clock helps to keep the delays manageable since the clock guarantees time between clock pulses to help keep the system glitch-free.

Using event-driven FSM with unit distance coding and the sequential equations can lead to very reliable designs for small FSM systems.

You can usually prevent the FSM falling into unused states by:

- Keeping the design so that it uses all possible states (i.e. 4 states, 8 states etc.).
- In clock driven FSM the DFF is the best option to help avoid unused state problems.

Finally, for this chapter, it is possible to introduce the idea of using C gates with Request/Acknowledge handshakes into the design of asynchronous FSM (Yakovlev and Koelmans 1998).

As discussed in this chapter, there is a need to design event-driven systems carefully, and to use synchronous techniques to help in the design of these systems.

This is left to Chapter 7, after looking at Petri nets. In that chapter, Yakovlev and Koelmans (1998) introduces ideas on why this is done, and also how it can be done, and describes the use of the C gate and Request/Acknowledge handshakes to produce reliable asynchronous one hot systems.

# 7 Petri Net FSM Design

## 7.1 INTRODUCTION

Up to this point we have been using state diagrams to implement finite state machines (FSMs). In a state diagram, the FSM can reside in only one state. There can be only one state 'active' at any given time.

However, there are a lot of situations where there is a need to do things in parallel in Petri nets. This involves having more than one state 'active' at a time. The state diagram cannot be used in these cases – well, not very easily anyway. (You could have more than one state diagram in a system, with handshakes between them, but this is not the same as having Petri nets.)

The Petri net is a useful tool that can be used to describe systems where a system can reside in more than one state. The Petri net allows us to be able to describe systems that exhibit both sequential and parallel behaviour and has been used for many years to model systems that contain both sequential and parallel conditions. A typical example would be flow control in a manufacturing system where some parts can be assembled in parallel, while others need to be in sequence due to the nature of the assembly process.

It is now possible to create a Petri net then translate this into an electronic circuit that exhibits both sequential and parallel control.

This chapter shows how this can be achieved. Once we have an electronic circuit for our Petri net, we can use Verilog hardware descriptive language (HDL) to implement it in either a complex programmable logic device (CPLD) or a field programmable gate array (FPGA) device.

There is a website at: http://www.informatik.uni-hamburg-de/TGI/PetriNets/ that deals with all things Petri you might like to take a look at it some time. In

*Digital System Design using FSMs: A Practical Learning Approach*, First Edition. Peter D. Minns.

Companion website: www.wiley.com/go/minns/digitalsystemdesign

addition, there is a workflow Petri net designer (WoPeD) simulator available from www.woped.org that you can install and use to simulate some Petri nets.

There are a number of Petri net simulators around. Try: http://www.informatik.uni-hamburg-de/TGI/PetriNets/tools/java/Braunl/. Note that it does not support enabling/disabling arcs: You need to run this from the Web page. There are others.

The following frames show how Petri nets can be used to develop and synthesize systems, using ideas from Fernandes et al. (1997) with copyright permission.

Turn to **Frame 7.1**.

### *Frame 7.1*

Figure 7.1 shows a state diagram and a Petri net alongside each other.

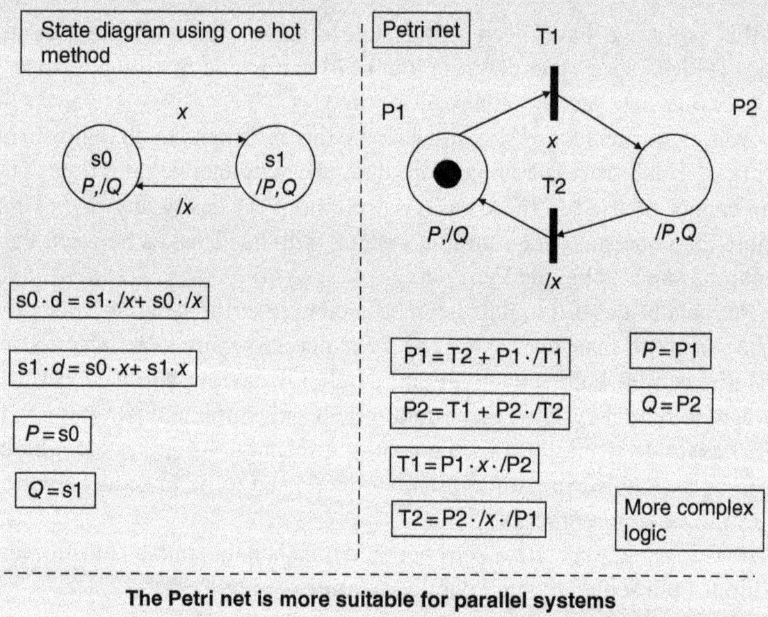


**Figure 7.1** One hot state diagram alongside Petri net.

They both do the same thing, so it is easy to see the differences.

In a state diagram there are states and the transitional lines to show how to get from one state to the next state.

In a Petri net there are **placeholders** P1 and P2, in this example they are connected by **arcs** (the arrows) to other **placeholders** via **transitions.** A transition is a trigger point that causes transitions from one placeholder to the next to take place. Since Petri nets can support parallelism, there are **tokens** (the black dots inside the placeholder) to keep track of more than one placeholder being active.

If a **placeholder** has a **token,** that placeholder is '**active**'.
If the **placeholder** does not have a **token,** that placeholder is '**not active**'.
**Which of the placeholders is active in the above Petri net P1 or P2?**
When you have decided, turn to **Frame 7.2.**

### *Frame 7.2*

The answer is, of course, P1, as this is the active placeholder because it holds the token.

In the figure of Frame 7.1, there are two outputs alongside the placeholders *P* and *Q*. Since P1 is active, the outputs will be $P = 1$, $Q = 0$. P2 is not active so the output condition $P = 0$, $Q = 1$ will not be true.

Looking further at the figure in Frame 7.1, there is an input *x* alongside transition T1, and the input */x* alongside transition T2.

If the token is in P1 (as shown in the figure) then when $x = 1$ transition T1 will fire and the token will move from placeholder P1 to P2 in the direction of the arcs.

**If the token is now in placeholder P2, under what conditions will the token transition T2 fire, and what will happen?**

When you have decided, turn to **Frame 7.3.**

### *Frame 7.3*

The answer is that to fire T2, $x = 0$. The result will be that the token will move from P2 back to P1.

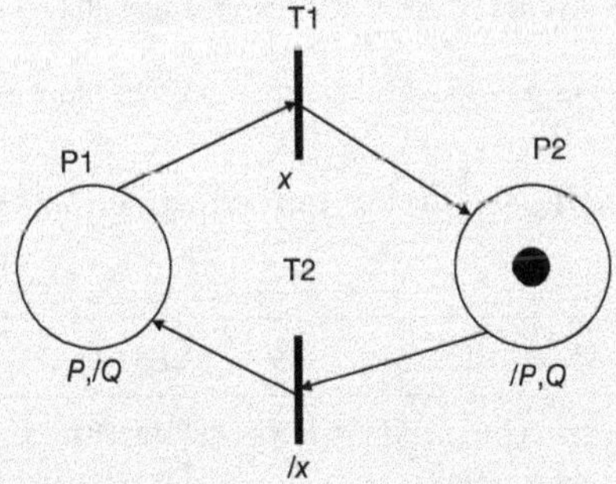


**Figure 7.2** Petri net with two placeholders.

In Figure 7.2, the token is in P2, so P2 is active.
**What are the outputs *P* and *Q*?**
When you have decided, turn to **Frame 7.4.**

## Frame 7.4

The outputs will be $P = 0$, $Q = 1$ as defined under placeholder P2.

Now turn to **Frame 7.5**.

## Frame 7.5

Look at the Figure 7.3.

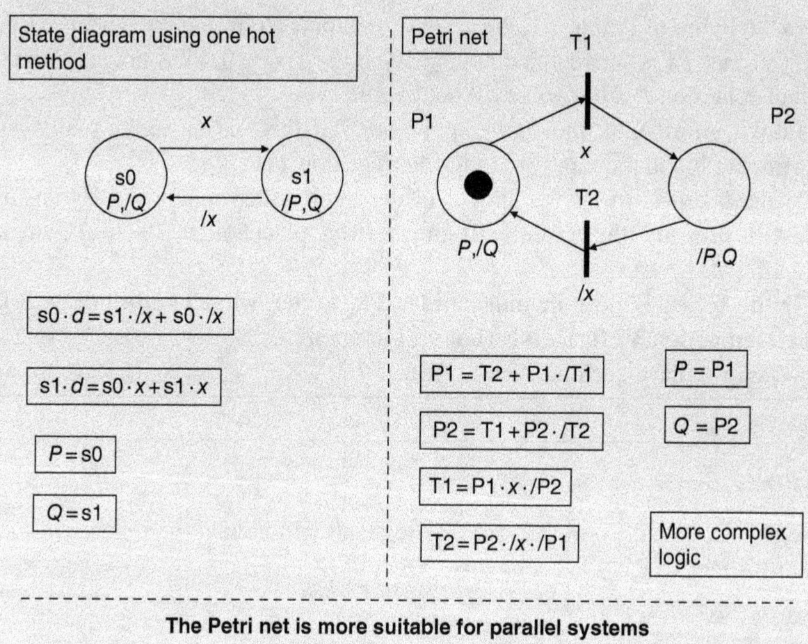


**Figure 7.3** The one hot state machine and two placeholder Petri nets.

You can now see that the layout for both state diagrams and Petri nets are very similar.

| | State diagram | Petri net |
|---|---|---|
| **Inputs** | along transitional lines | at transitions |
| **Outputs** | inside state | alongside placeholders |

The reason outputs are alongside placeholders is that the placeholder needs to be kept clear for the token.

**So why the token?**

When you have considered this, turn to **Frame 7.6**.

### Frame 7.6

The token is needed because in a Petri net more than one placeholder can be active at any one time. Contrast this with a state diagram where there can only be one state active at any one time.

The token allows us to be able to see all the tokens active in a Petri net diagram at any one time.

You will notice that in the Figure 7.3 of Frame 7.5 there are equations shown for both the state diagram and the Petri net.

**If you look closely at the equations for the state diagram, does it remind you of anything?**

When you have thought about this, turn to **Frame 7.7**.

### Frame 7.7

In this figure, the equations for the state diagram are one hot.

It also turns out the equations for the Petri net are very similar to the sequential equations we use in Chapter 4 on event (asynchronous) FSMs. As you will see later, this has implications on the way we can use Petri nets.

In Figure 7.4 on the left Pn and Pm both have tokens so both placeholders are active.

**This means that Pn and Pm are running in parallel. This will continue until the input *L* becomes logic 1 (active) at which point transition tx will fire.**

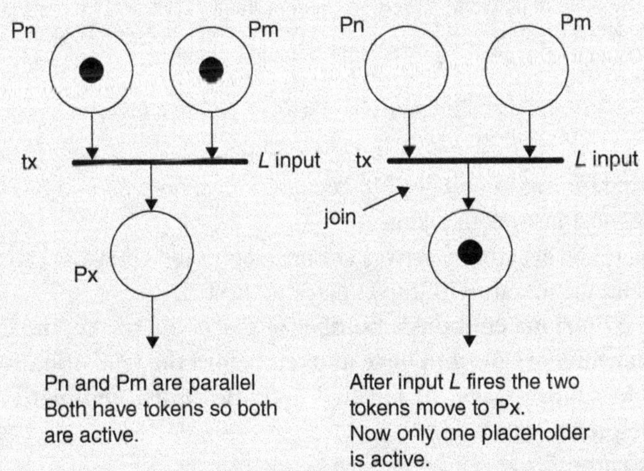


**Figure 7.4** The join mechanism.

When tx fires (as a result of *L* becoming logic 1) the tokens in Pn and Pm will move to placeholder Px.

Now there is only one placeholder active and the Petri net has moved from a parallel state to a sequential one.

**This sort of behaviour is called a 'join' because a parallel behaviour has joined to become a sequential one.**

Turn to **Frame 7.8**.

### *Frame 7.8*

On the left of Figure 7.5, Pn has a token but Pm does not (perhaps because it has not yet arrived).

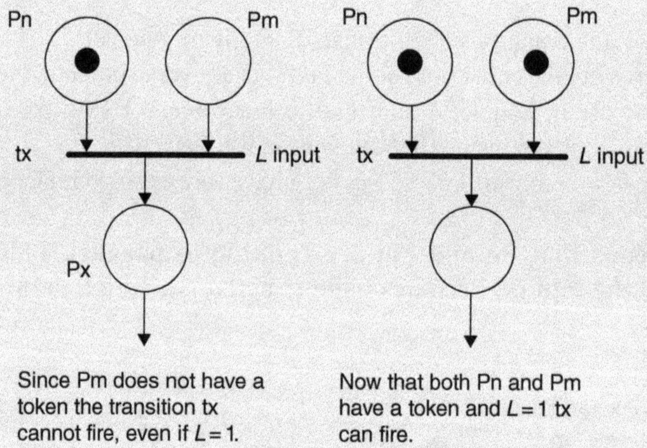


**Figure 7.5** Both Pn and Pm must have a token before a join can be completed.

If $L = 1$, the transition will not fire because to fire both Pn and Pm feeding the transition tx must have both tokens.

When the token eventually arrives at Pm and provided that $L = 1$ the transition tx will fire and the tokens will join at placeholder Px.

**So when a Petri net contains a number of placeholders feeding into a transition all placeholders need to have a token before the transition will fire.**

**This makes sense since all parallel activities must complete before the following sequence can begin.**

Turn to **Frame 7.9**.

### Frame 7.9

In Figure 7.6, you can clearly see the sequence of token movement.

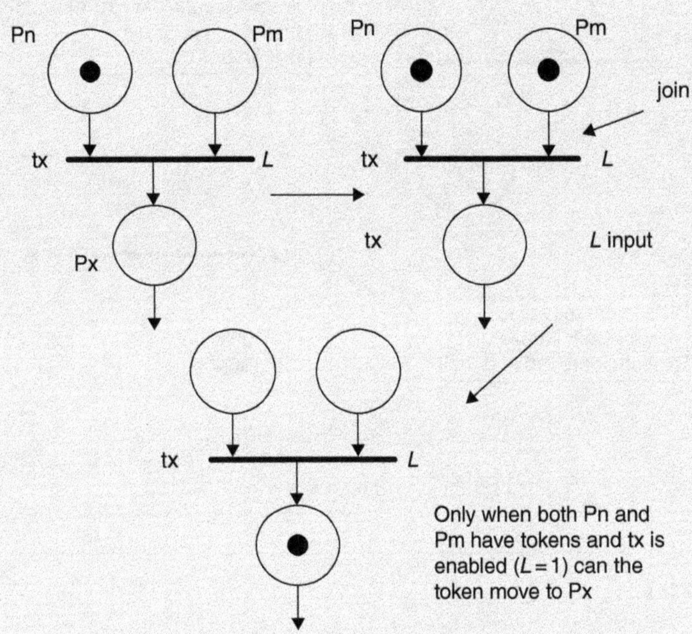


**Figure 7.6** The result of a join.

Note that when going from two or more parallel placeholders to a single placeholder (parallel to sequential) is called a 'join'.

Turn to **Frame 7.10**.

### Frame 7.10

In Figure 7.7 you can see the other situation where a sequential process becomes a parallel process. This is called a 'fork'.

In this case when the transition fires the token in Pn 'forks' to parallel processors Pm and Pk.

Turn to **Frame 7.11**.

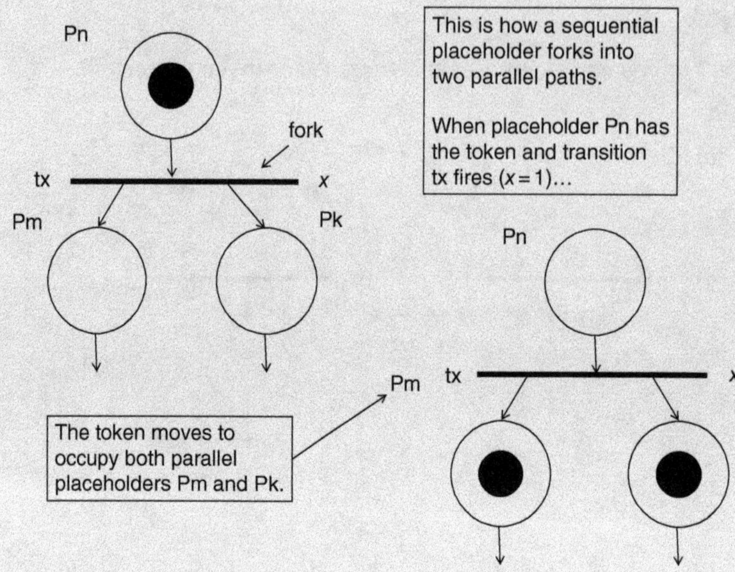


**Figure 7.7** The result of a fork.

### *Frame 7.11*

When using Petri nets, you need to consider certain conditions that must not be allowed to happen (Figure 7.8).

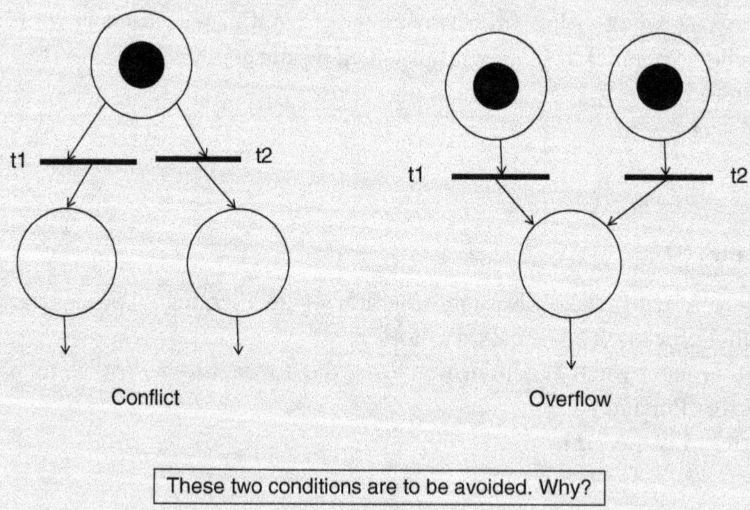


**Figure 7.8** The effect of conflict and overflow.
*Source:* Fernandes, J. M. et al., 1997 / Institution of Engineering and Technology (IET)

The left of Figure 7.8 is one such condition. It is called the 'conflict condition' and occurs when one placeholder tries to move its token into two or more placeholders via separate transitions.

The reason this must be avoided is that if transition t1 fires before t2 (or t2 fires before t1) then the token could move to only one of the two placeholders. This would then prevent the other placeholder from becoming active.

The condition on the right in Figure 7.8 shows what is called an 'overflow condition'.

In the overflow condition if t1 fires before t2 the token flows via the transition t1 to the lower placeholder.

**This then blocks t2 from moving its token because a token can only move to an empty placeholder.**

If the placeholder the token is trying to move to already contains a token then the **transition is blocked.**

**Can you think how these two conditions could be avoided?**

When you have thought about this, turn to **Frame 7.12**.

### *Frame 7.12*

Figure 7.9 shows how the conditions of conflict and overflow can be avoided.

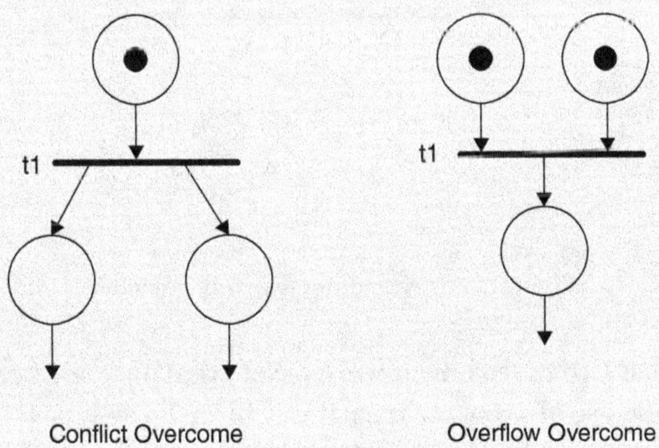


**Figure 7.9** How to overcome conflict and overflow.

It is now obvious that we need to avoid having more than one transition, because by using a single transition we easily overcome both problems.

You will see lots of examples of Petri nets and be able to see what is desirable and what is not.

Now there are occasions when we want to use the 'conflict' arrangement.

**Can you think of a possible situation?**

After thinking about this, turn to **Frame 7.13**.

### *Frame 7.13*

The answer is, of course, when we want to make decisions (two-way branches).

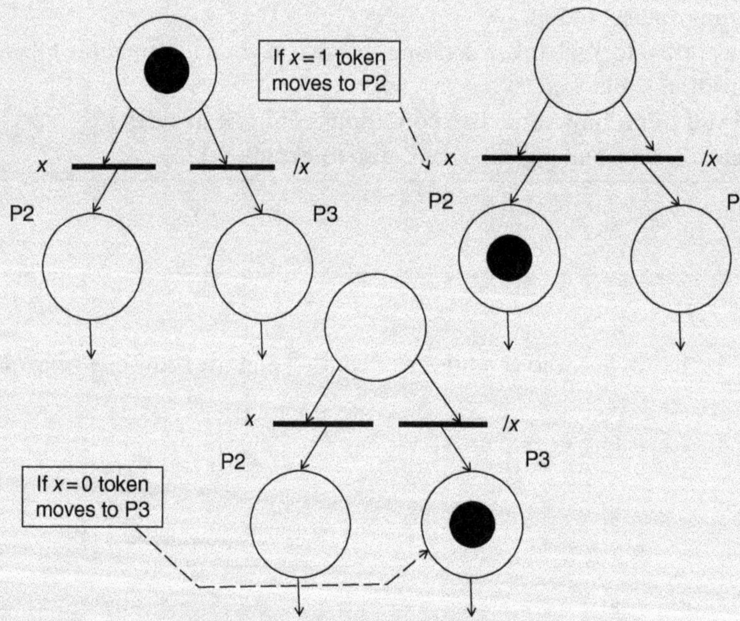


**Figure 7.10** Handling two-way branches.

**Study Figure 7.10 to see how it works. Note that only one branch is taken. Also note the use of separate transitions in each case.**

If you cannot see the answer, look back over Frames 7.11–7.13.

Turn to **Frame 7.14**.

### Frame 7.14

Now consider Figure 7.11.

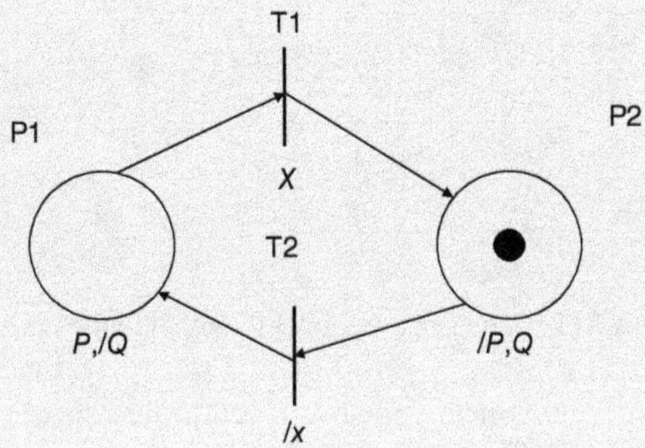


**Figure 7.11** A two-placeholder Petri net.

Three equations can be obtained from the Petri net that can be used to synthesize the circuit for the Petri net.

- Placeholder equations (P1 and P2 in this example).
- Transition equations (T1 and T2).
- Output equations (*P* and *Q* in this example).

Placeholder equation for P1:

$$\mathrm{P1} = \mathrm{T2} + \mathrm{P1} \cdot / \, \mathrm{T1}.$$

This defines how to move the token to placeholder P1 · T2 must fire or we must be already in P1 and T1 has not fired.

**Write down the equation for P2.**

Turn to Frame **7.15, which repeats the figure in Figure 7.12**.

### Frame 7.15

The answer is P2 = T1 + P2 · /T2 since to get into P2 T1 needs to fire or we will need to be in P2 and T2 has not fired.

Now consider the transition equation:

$$\mathrm{T1} = \mathrm{P1} \cdot x \cdot / \, \mathrm{P2}$$

which means that P1 must be active **and** input $x = 1$ **and** P2 must not be active.

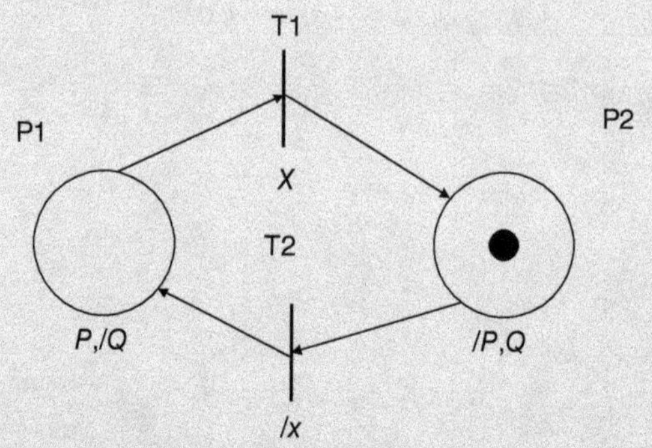


**Figure 7.12** The same Petri net to compare the equation.

Note: we have still not considered how to synthesize the circuit – more on that later.

**Now write down the equation for transition T2.**

Turn to **Frame 7.16**.

## *Frame 7.16*

The answer is:

$$T2 = P2 \cdot /x \cdot /P1.$$

Because placeholder P2 needs to be active **and** $x = 0$ **and** P1 not active.

The outputs are as always dependent on the placeholders so:

$$P = P1.$$

since only in P1 is $P = 1$. Also:

$$Q = P2 \text{ for } Q = 1.$$

Turn to **Frame 7.17**.

## *Frame 7.17*

Consider the Petri net diagram in Figure 7.13.

This is the pump motor problem we dealt with in other chapters. Note this is a sequential problem, but we use a Petri net to solve it.

Sequential Petri net pump – Spin Motor Problem

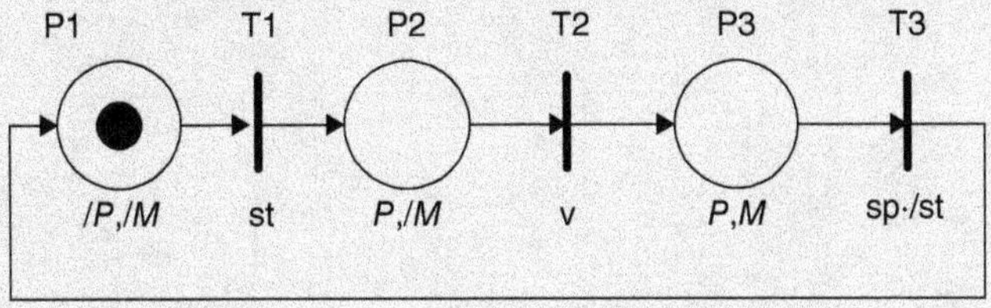


Produce the Petri net equations for this controller.

T1 =
T2 =
T3 =

P1 =
P2 =
P3 =

**Figure 7.13** Pump motor example using sequential Petri net.

**Write down the transition equations and the placeholder equations for the Petri net diagram.**

When you have done this, turn to **Frame 7.18**.

### *Frame 7.18*

The transition equations are:

$$T1 = P1 \cdot st \cdot /P2 \qquad T2 = P2 \cdot v \cdot /P3T3 = P3 \cdot sp \cdot /st \cdot /P1.$$

The placeholder equations are:

$$P1 = T3 + P1 \cdot /T1 \qquad P2 = T1 + P2 \cdot /T2 \qquad P3 = T2 + P3 \cdot /T3.$$

The output equations are:

$$P = P2 + P3 \qquad M = P3.$$

Look carefully at the equations and the circuit. You should be able to see how these come together. Note that in Figure 7.14 outputs *P* and *M* are not shown.

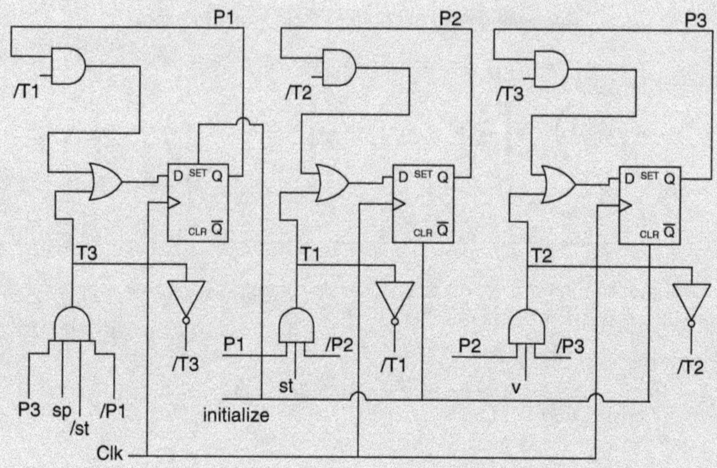


**Figure 7.14** Circuit for the pump motor Petri net.

For example, the transition equations T1, T2, and T3 are AND circuits, while the placeholder equations are AND/OR circuits with feedback from the flip-flop output.

This is not always the case (though, see later).

Turn to **Frame 7.19**.

## *Frame 7.19*

**Try to obtain the transition equations, the placeholder equations and the output equations for the Petri net shown in Figure 7.15.**

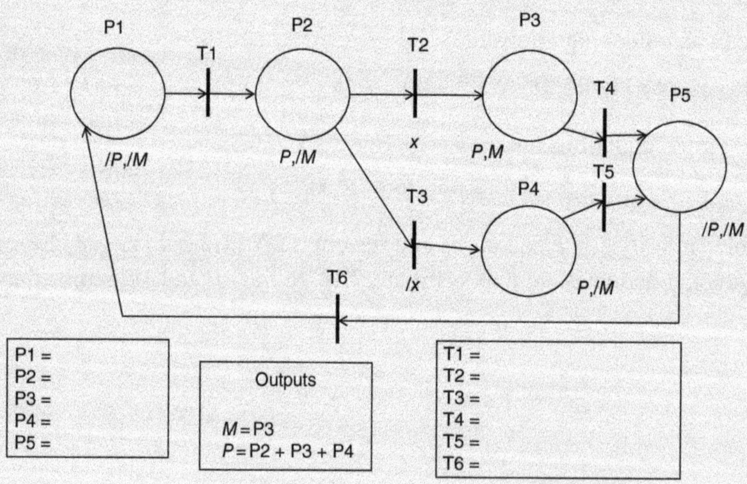


**Figure 7.15** Work out equations from the Petri net.

**Try to work out what the placeholder terms P1 to P5 are in Figure 7.15. Also try to work out the transient terms T1–T6 in Figure 7.15.**

When you have done these, turn to **Frame 7.20**.

## *Frame 7.20*

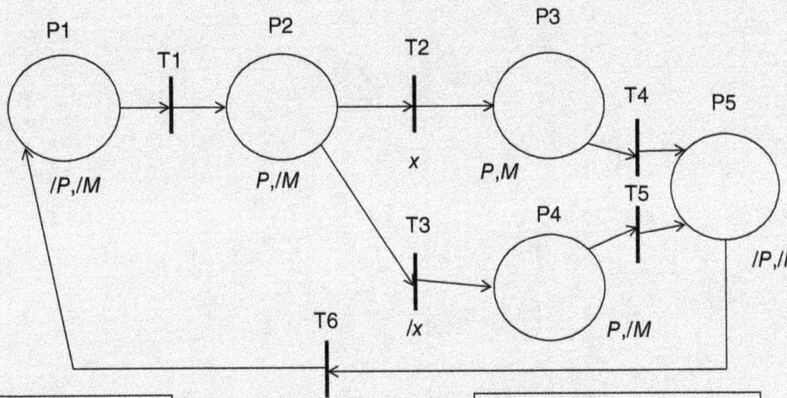


| |
|---|
| P1 = T6 + P1 · /T1 |
| P2 = T1 + P2 · /T2 · /T3 |
| P3 = T2 + P3 · /T4 |
| P4 = T3 + P4 · T5 |
| P5 = T4 + T5 + P5 · /T6 |

| Outputs |
|---|
| M = P3 |
| P = P2 + P3 + P4 |

| |
|---|
| T1 = P1 · /P2 |
| T2 = P2 · /P3 · x |
| T3 = P2 · /P4 · /x |
| T4 = P3 · /P5 |
| T5 = P4 · /P5 |
| T6 = P5 · /P1 |

**Figure 7.16** Completed equations for the Petri net.

As you can see from Figure 7.16, these follow the rules we have introduced in earlier frames.

Note that there are separate paths for P3 to P5 and P4 to P5 with their own T4 and T5 transitions so that the T2 and T3 paths from P2 form a decision.

Turn to **Frame 7.21**.

## *Frame 7.21*

Now look at Figure 7.17.

This shows the Petri net together with its equations and circuit diagram.

Study this carefully, in particular the P1 connection from the flip-flop output to the AND gate, and the P2 connection to the AND gate. Here you should be able to see the feedback paths.

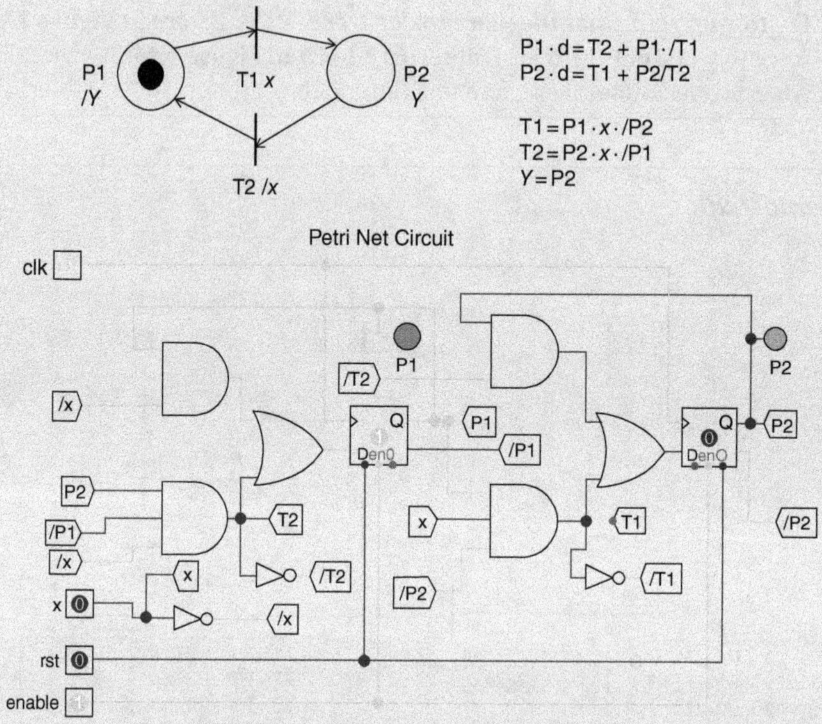


**Figure 7.17** The Petri net, equations, and circuit.

Also notice the transition logic made up of AND gate logic with inverting buffers for /T1·/T2.

The Petri net circuit is initialized so that the P1 flip-flop is set, while the P2 flip-flop is cleared. This is in line with the need for P1 to contain a token.

Compare these with the circuit arrangements in Chapter 4 on asynchronous FSM design. You should be able to see that they have similar circuit topology.

Now Turn to **Frame 7.22**.

## *Frame 7.22*

Look again at the circuit diagram of the Petri net (Figure 7.18).

**Note placeholder P1 must be active (with token) and P2 not active (without token) at reset.**

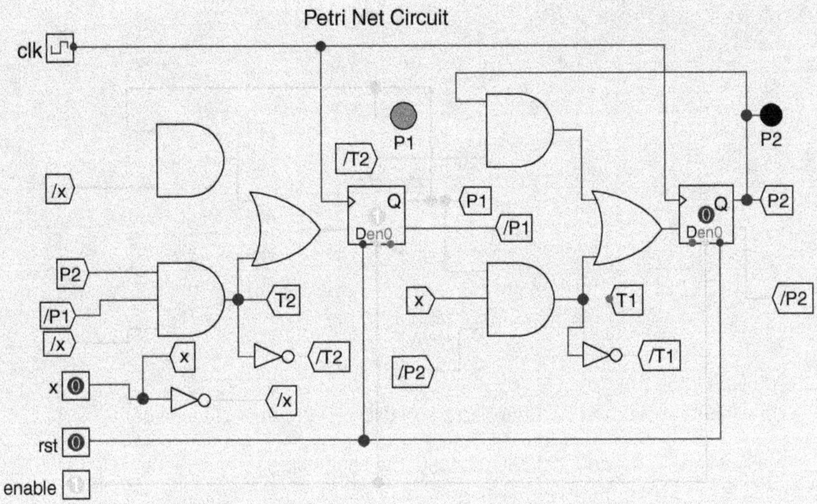


**Figure 7.18** The Logisim circuit diagram.

This example is called **fig 7.17 full cycle.circ** and is located in the Wiley Logisim web folder. Note figure 7.17 and fig 7.18 contain the same Logisim circuit.

Now turn to **Frame 7.23**.

## *Frame 7.23*

**Look at Figure 7.19. It shows the basic topology. The *D* type flip-flop (DFF) makes this a clocked (synchronous) Petri net.**

Pn = tin + Pn/tout ------------------------------------------------ (1)

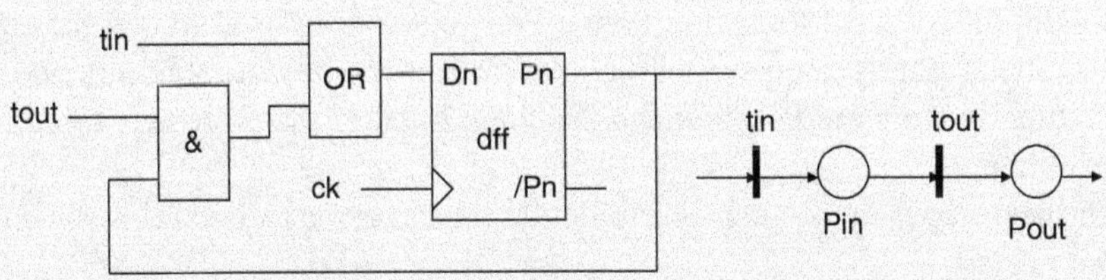


The tin term in Equation (1) is:

Tin = input placeholder AND input enable AND NOT (output placeholder)

Equation (1) describes the structure of the synthesized circuit of the Petri net.

**Figure 7.19** A clock-driven Petri net.

Now look at Figure 7.20.

Pn = tin + Pn/tout ............................................ (1)

The tin term in Equation (1) is:

Tin = input placeholder AND input enable AND NOT (output placeholder)

Equation (1) describes the structure of the synthesised circuit of the Petri net.

**Figure 7.20** An event-driven Petri net.

Figure 7.20 is the same as Figure 7.19 but without the *D* type flip-flops.

You might like to compare Figure 7.20 with the event cells used in Chapter 4.

In this case the feedback path allows the circuit to act as a memory element and you should be able to see the resemblance to that of the event-driven circuits in Chapter 4. **This shows that any given Petri net can be used in either synchronous or asynchronous form.**

Turn to **Frame 7.24**.

## *Frame 7.24*

At this point we have looked at how a Petri net can be used to control electronics using purely sequential systems. We can see how to obtain the equations directly from the Petri net and use them to build the electronic circuit to realize them.

The next step is to look at how parallel Petri nets can be realized electronically in more detail.

Make sure that you are comfortable with the methods of obtaining the equations from the Petri net, by going back over the previous frames.

**This is an important point to ensure that you are happy with the method so far. You will be using this method whenever you use the Petri net system.**

Turn to **Frame 7.25**.

### *Frame 7.25*

Consider the part of a Petri net diagram in Figure 7.21.

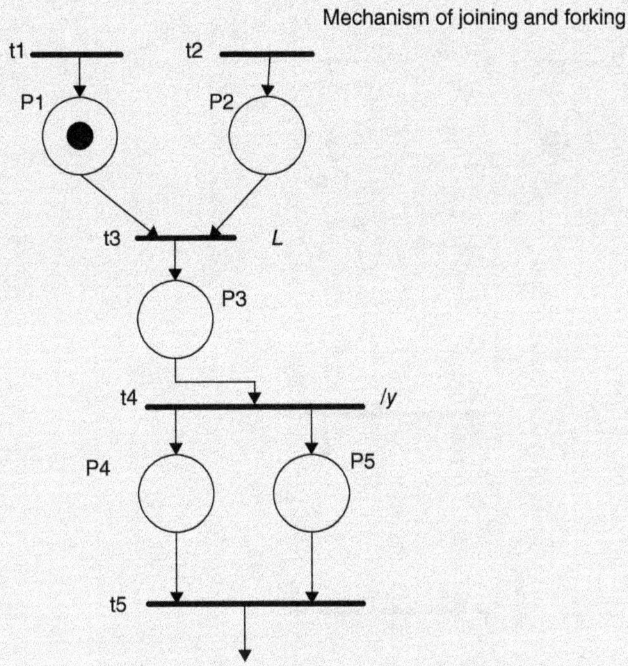


**Figure 7.21** Forks and joins in a Petri net.

P1 and P2 are parallel branches joining at t3 to form a sequential section P3 that then forks at t4 to become parallel branches P4 and P5. This then joins again at t5. The details after this are not shown in Figure 7.21.

The Petri net segment is, of course, incomplete, but we can use it to look at the equations from P1 to P5 and transitions t3 and t4.

**Rule 1: Consider the join mechanism of P1, P2, t3, and P3:**

$$t3 = P1 \cdot P2 \cdot L \cdot / P3.$$

Here t3 can only fire when both P1 and P2 have the token. Thus, the two parallel branches P1 and P2 must be active before t3 can fire, and placeholder P3 is not active, having room to accept a token.

Turn to **Frame 7.26**.

### *Frame 7.26*

Now consider P3, t4, P4, and P5 in Figure 7.22.

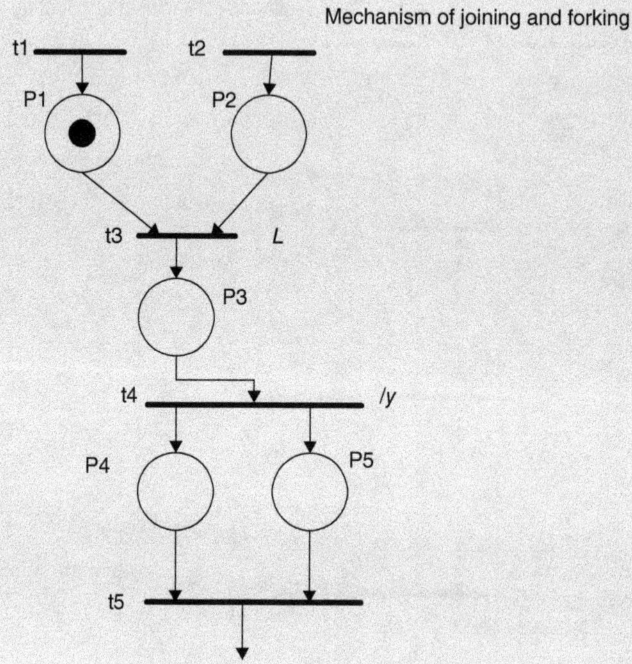


**Figure 7.22** The forking mechanism.

$$t4 = P3 \cdot / y \cdot / P4 \cdot / P5.$$

**Rule 2: Here we see that for a fork to occur t4 must fire:**

**P3 must have the token AND input *y* = 0 AND both P4 AND P5 must not have tokens.**

In this case the parallel branch P4 and P5 must be free (not have any tokens) for the fork to occur.

Turn to **Frame 7.27**.

## *Frame 7.27*

Now consider the example shown in Figure 7.23.

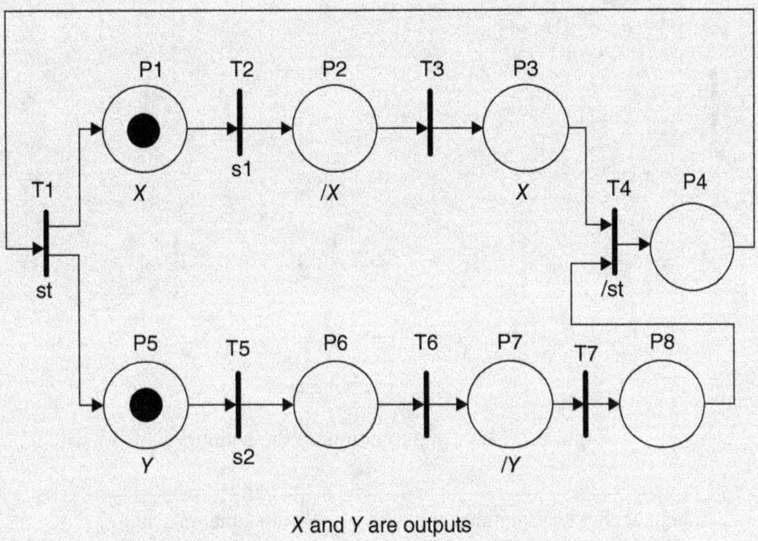


**Figure 7.23** Example of parallel to sequence.

**Produce the equations for placeholders, transitions, and outputs for this Petri net.**

When you have done this, turn to **Frame 7.28**.

## *Frame 7.28*

From the Petri net diagram we see that there is a fork T1, P1, P5, and a join P3, P8, T4, and P4. And use Figure 7.24 to compare the equations.

The Petri net has two parallel paths:

- One consisting of P1 to P3.
- The other consisting of P5 to P8.

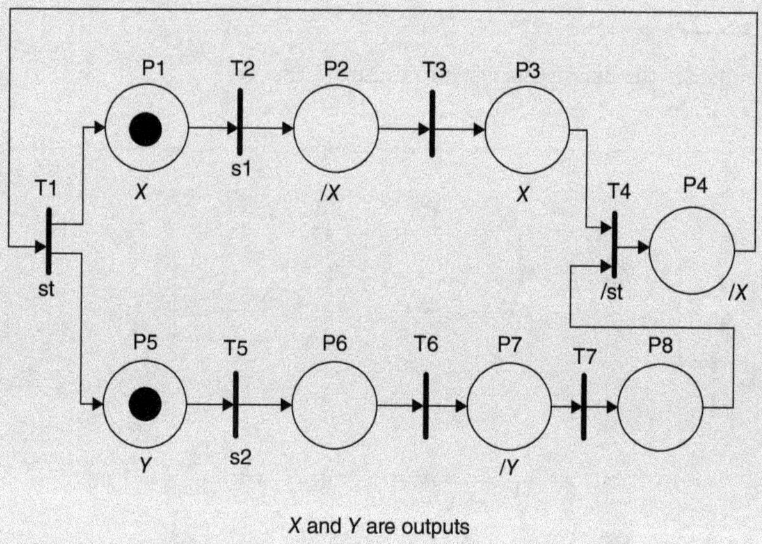


**Figure 7.24** Used to compare the equations.

| The placeholder equations are: | The transition equations are: |
|---|---|
| $P1 = T1 + P1 \cdot /T2$ | $T1 = P4 \cdot st \cdot /P1 \cdot /P5$ – fork (1) |
| $P2 = T2 + P2 \cdot /T3$ | $T2 = P1 \cdot s1 \cdot /P2$ |
| $P3 = T3 + P3 \cdot /T4$ | $T3 = P2 \cdot /P3$ |
| $P4 = T4 + P4 \cdot /T1$ | $T4 = P3 \cdot P8 \cdot /st \cdot /P4$ – join (2) |
| $P5 = T1 + P5 \cdot /T5$ | $T5 = P5 \cdot s2 \cdot /P6$ |
| $P6 = T5 + P6 \cdot /T6$ | $T6 = P6 \cdot /P7$ |
| $P7 = T6 + P7 \cdot /T7$ | $T7 = P7 \cdot /P8$ |
| $P8 = T7 + P8 \cdot /T4$ | |

Outputs:

$$X = P1 + P3; Y = P5 + P6.$$

In particular note the fork and join transitions in Equations (1) and (2), also note that in this Petri net there are two placeholders P1 and P5 that both have tokens in the initial state of the Petri net.

**If placeholders P3, P4, and P8 have the token what happens next?**

When you have considered this, turn to **Frame 7.29**.

***Frame 7.29***

If placeholders P3, P4, and P8 have the token then T4 still cannot fire since P4 has the token so the Petri net cannot proceed.

Turn to **Frame 7.30**.

***Frame 7.30***

Now consider the Petri net in Figure 7.25.

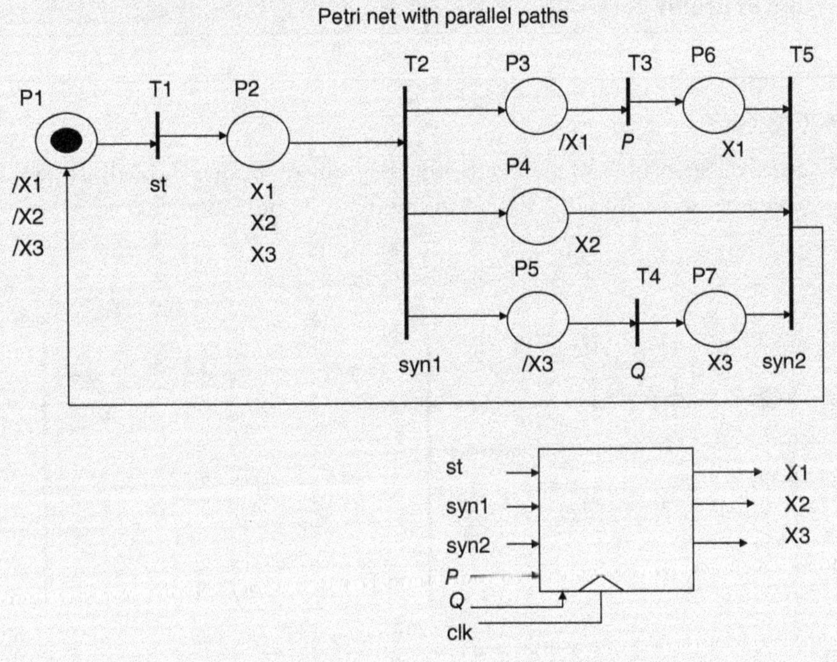


**Figure 7.25** A running example of the action of the Petri net.

**Write down the equations for the placeholders, the transitions, and the outputs.**

When you have done this, turn to **Frame 7.31**.

## Frame 7.31

The Petri net equations are:

| | |
|---|---|
| P1 = T5 + P1· /T1 | T1 = P1 · st · /P2 |
| P2 = T1 + P2 · /T2 | T2 = P2 · syn1 · /P3 · /P4 · /P5 – fork |
| P3 = T2 + P3 · /T3 | T3 = P3 · *p* · /P6 |
| P4 = T2 + P4 · /T5 | T4 = P5 · *Q* · /P7 |
| P5 = T2 + P5 · /T4 | T5 = P6 · P4 · P7 · syn2 · /P1 – join |
| P6 = T3 + P6 · /T5 | Outputs: |
| P7 = T4 + P7 · /T5 | X1 = P2 + P6 |
| X2 = P2 + P4 | X3 = P2 + P7 |

Turn to **Frame 7.32**.

## Frame 7.32

A series of snapshots of the Petri net diagram of Frame 7.32 illustrates its possible operation. Initially, P1 is waiting for st input (Figure 7.26).

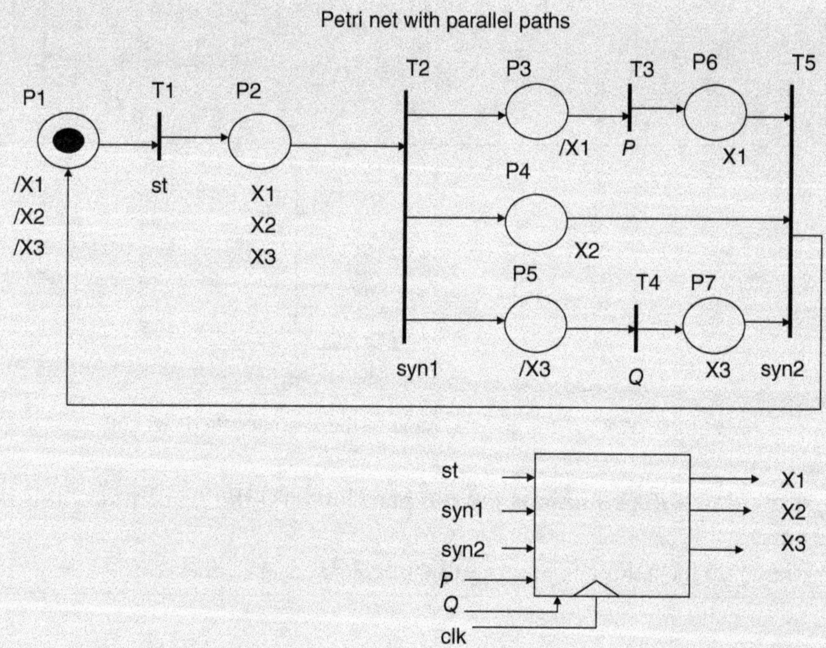


**Figure 7.26** Starting from placeholder P1.

After the st input token moves to P2 and outputs X1, X2, and X3 become asserted (Figure 7.27).

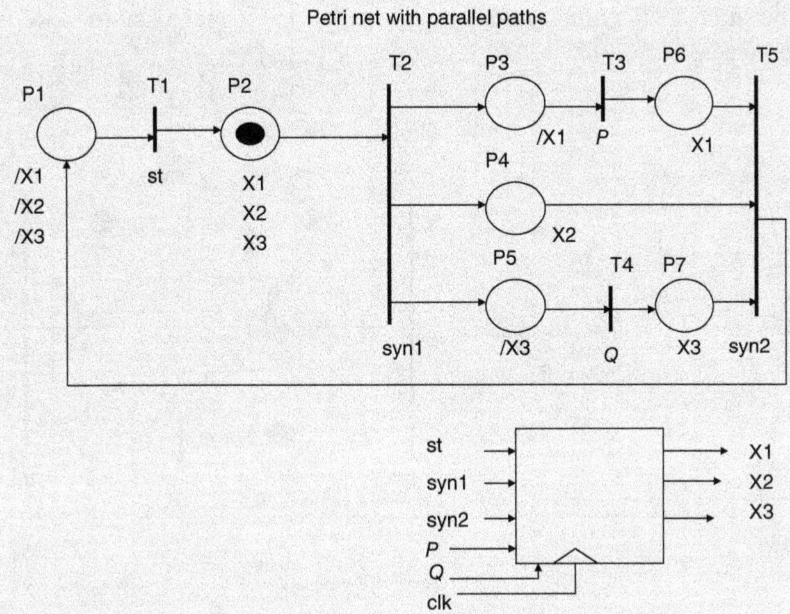


**Figure 7.27** Now waiting for transition from P2 via T2.

When syn1 goes to logic 1, token moves to P3, P4, and P5 (fork) to start the three parallel actions (Figure 7.28).

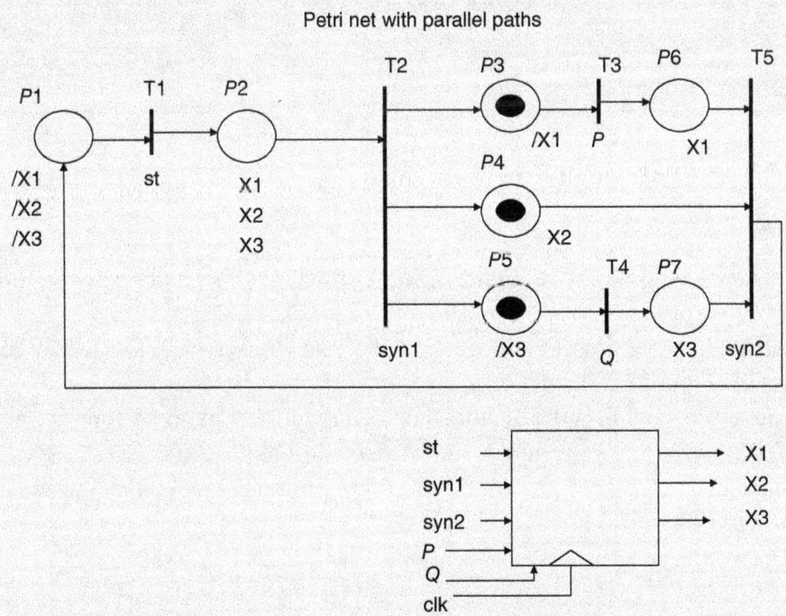


**Figure 7.28** Waiting for the inputs *P* and *Q*.

Transition T3 fires first (in this case with $P = 1$) causing token to move from P3 to P6 (Figure 7.29).

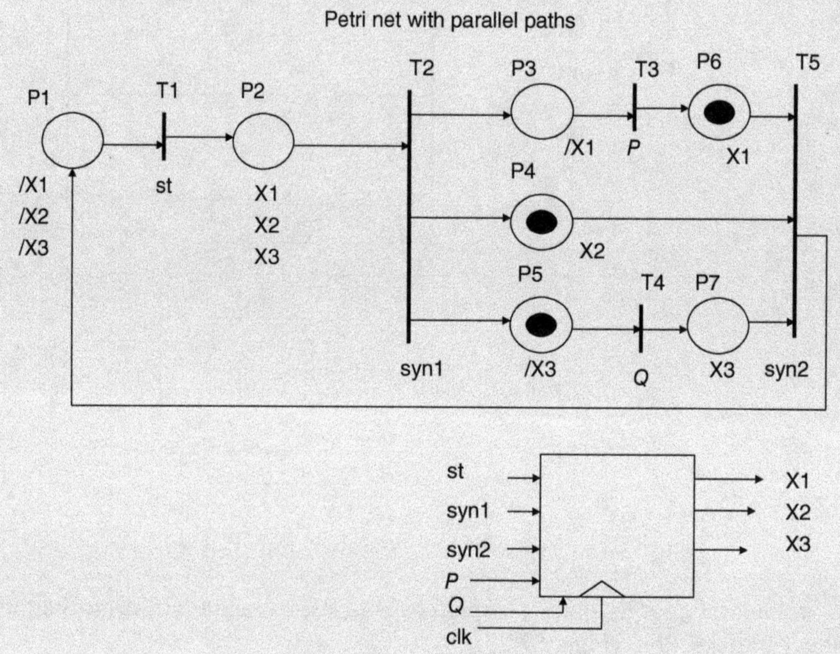


**Figure 7.29** After input *P*.

Turn to **Frame 7.33**.

### *Frame 7.33*

The sequence continues in Figure 7.30. Also when *Q* becomes 1 token moves into P7.

Eventually syn2 fires T5 and since P4, P6, and P7 have the token we are ready to fire T5 (Figure 7.31).

Now when syn2 fires T5 the token is seen to move back to P1 (join).

In this way the Petri net is seen to control both the fork and the join mechanisms.

Turn to **Frame 7.34**.

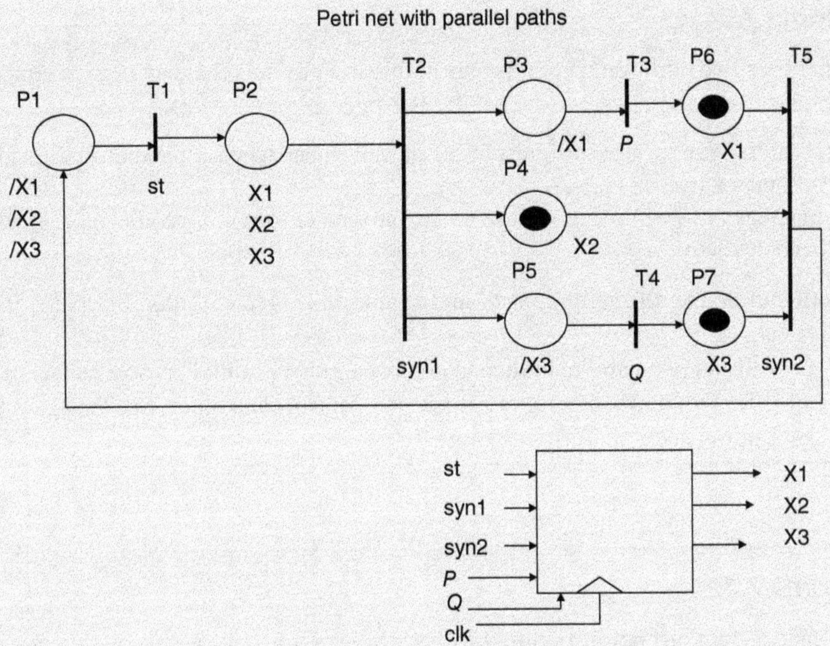


**Figure 7.30** After input *Q*.

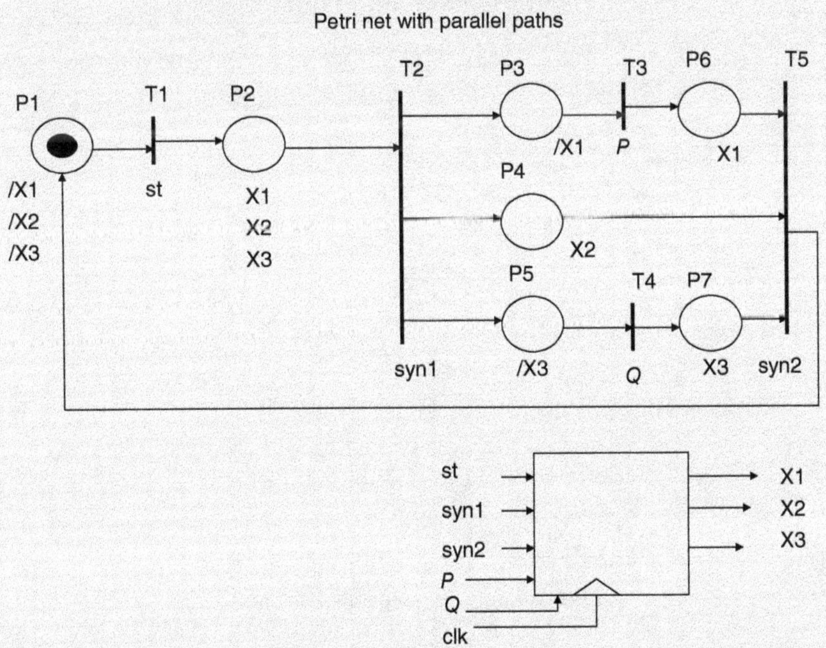


**Figure 7.31** After input syn2.

### *Frame 7.34*

There are two additional arc types available in Petri nets that can help to control synchronization between two parallel Petri net graphs.

- Disabling arcs – used to prevent token movement when a parallel placeholder becomes active.
- Enabling arcs – used to allow token movement when a parallel placeholder becomes active.

Both can be used to help control and synchronize separate parallel paths in a Petri net.

This allows one parallel branch to control another parallel branch. In fact, it is possible for two or more parallel branches to control each other! See the example in the Tutorial section. Turn to **Frame 7.35**.

### *Frame 7.35*

Consider the Petri net in Figure 7.32.

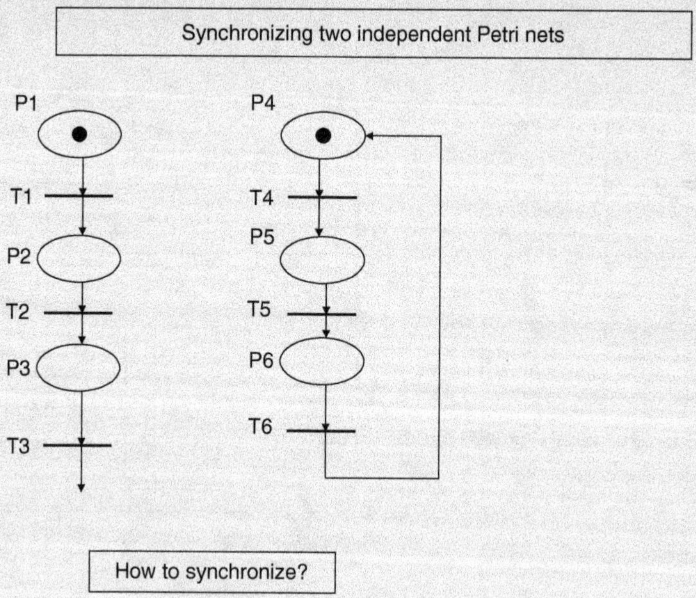


**Figure 7.32** Two separate Petri nets.

In the Petri net the two parallel paths are independent. We can make use of enabling/disabling arcs to gain control from one path over the other.

This allows one Petri net to act as a synchronizing point for other parts of the Petri net. Consider the next Petri net in Figure 7.33.

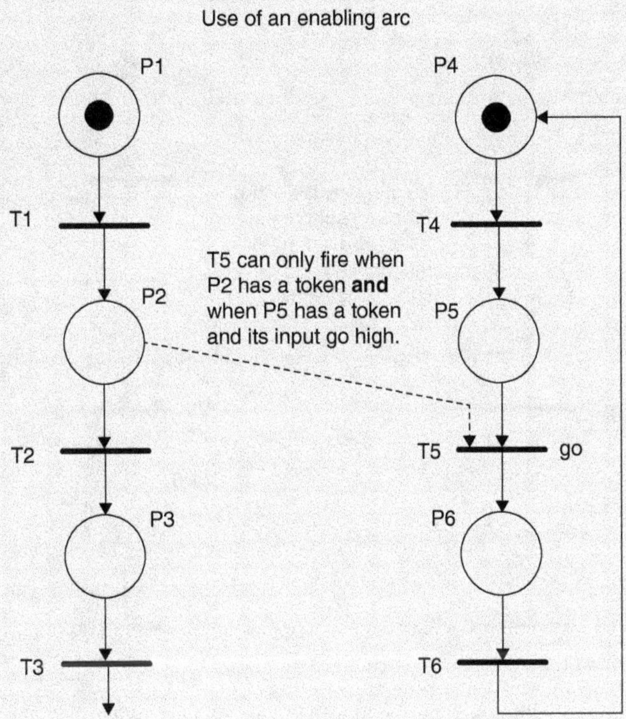


**Figure 7.33** An enabling arc affecting transient T5.

If the token P4 moves to P5 before the token P1 moves to P2 then even if input go is true T5 cannot fire. It must wait for the token to arrive at P2 before T5 can fire. The enabling arc from P2 to T5 acts as a synchronizing point to allow the two parallel paths to synchronize.

Turn to **Frame 7.36**.

### *Frame 7.36*

Now consider the Petri net shown in Figure 7.34.

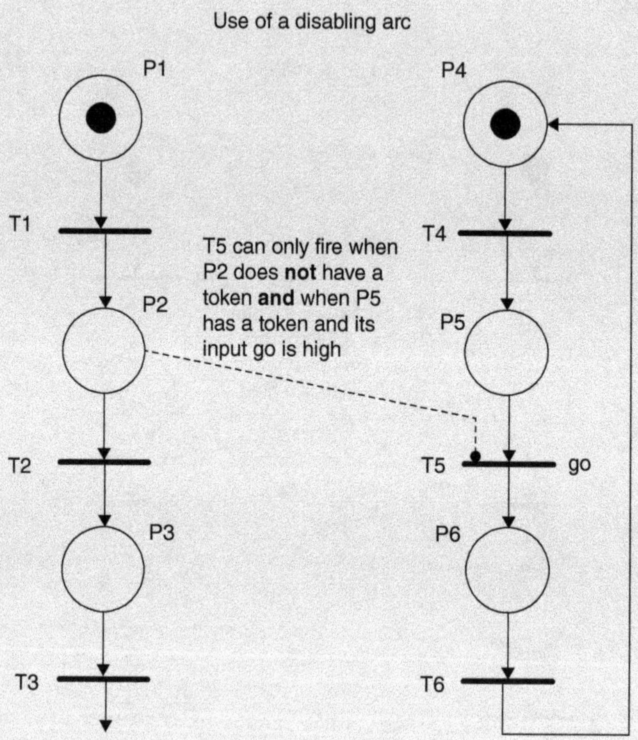


**Figure 7.34** A disabling arc to transient T5.

In this case the arc from P2 to T5 is a disabling arc.

If T1 fires the token from P1 into P2 before the token from P4 arrives at P5 then as long as the token is in P2 the transition T5 cannot fire, even if the token is in P5 and go is true.

Using enabling arcs and disabling arcs can simplify the control of complex Petri nets.

Turn to **Frame 7.37**.

### *Frame 7.37*

Look at Figure 7.35. Study it carefully.

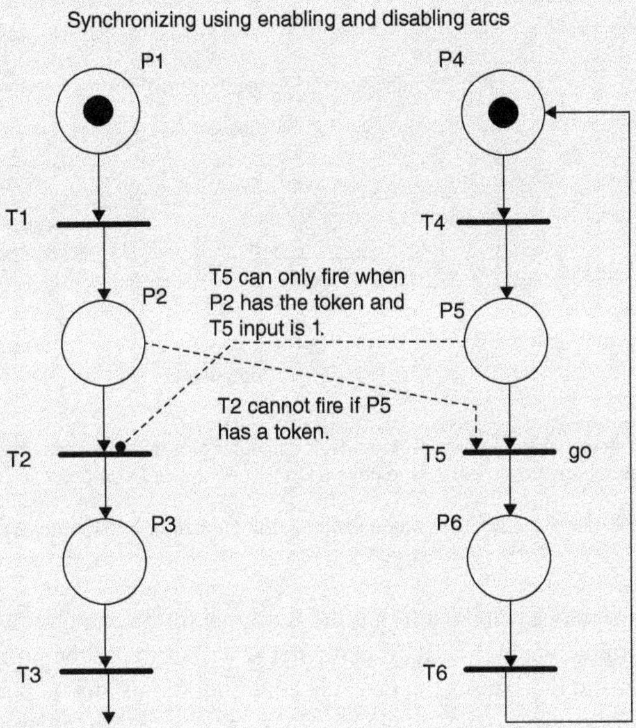


**Figure 7.35** The effect of both enabling and disabling arcs.

Here we see an example with both an enabling arc and a disabling arc. **What happens if the token arrives at P2 first and T2 fires?** When you have considered this, turn to **Frame 7.38**.

### *Frame 7.38*

If the token arrives at P2 first and T2 fires then when the token arrives at P5 T5 cannot fire when go is true, because the enabling arc is no longer true.

In a real Petri net it would make sense for the P1–P3 parallel path to form a loop so that at some future time the token would come back to P2 to allow the transition T5 to fire. In this way the left-hand parallel path would act as a regular synchronizing point.

Now look at a more practical example. Turn to **Frame 7.39**.

### *Frame 7.39*

In the system shown in Figure 7.36 two computers A and B share access to a printer via tri-state buffers.

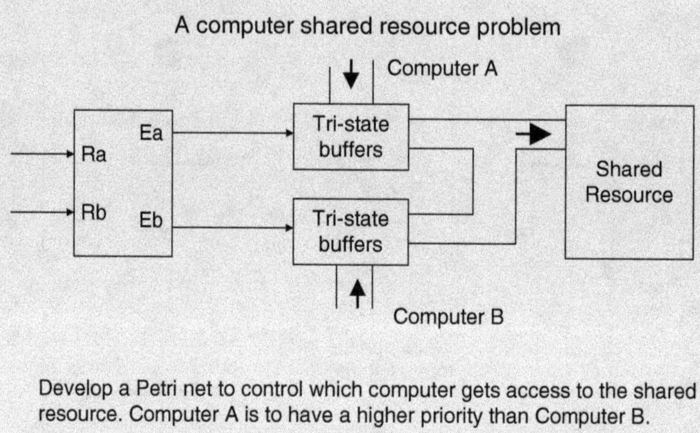


**Figure 7.36** How a Petri net can be used to share a printer between two computers.

The control block (on the left) has the Ra signal from computer A, and a control signal from computer B Rb. If computer A wishes to use the printer it raises its Ra signal and this causes the Petri net controller to raise its output Ea to turn on the tri-state device so that computer A has direct access to the printer.

If, on the other hand, computer B raises its Rb signal line then Eb turns on the tri-state device so that computer B has direct access to the printer.

Computer A is to have priority access to the printer such that even if computer B already has access to the printer it will release its control if computer A tries to access the Printer.

Under these conditions after computer A has finished using the printer, computer B will be allowed to re-access the printer.

**Try to think how you could create a Petri net to do this using enabling and/or disabling arcs.**

Turn to **Frame 7.40**.

***Frame 7.40***

A possible Petri net solution is shown in Figure 7.37.

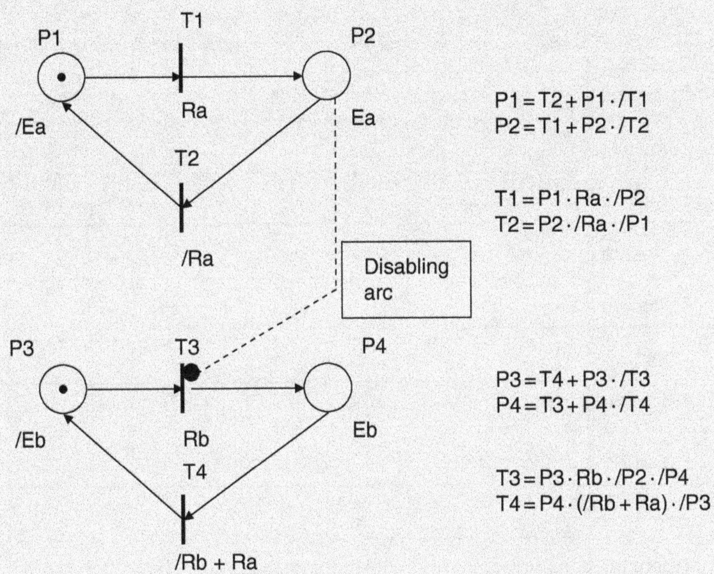


**Figure 7.37** The Petri net solution for the shared resource.

If Ra is raised, the token in P1 will move to P2 and block the possible transition T3 until Ra is released. This uses the disabling arc from P2 that disables T3.

If, on the other hand, Rb were raised so that the token in P3 moved to P4, Ra going to 1 would cause T4 to fire and the token in P4 would be forced to move back to P3. At the same time the token in P1 would move to P2 to allow computer A to access the printer.

Note that on releasing Ra, the token in P2 would move back to P1 and, since Rb would still be raised, the token in P3 would move to P4 to allow computer B access to the printer.

This is in effect a form of interrupt that becomes apparent due to the disabling arc.

Quite a good idea, yes? **Note that you could construct an 'interrupt priority scheme' using this idea.**

Here we have used a disabling arc from P2 to T3 to give computer A a higher priority over computer B.

Turn to **Frame 7.41**.

*Frame 7.41*

The Petri net can be implemented using Verilog HDL.

The simulation of this Petri net is shown in Figure 7.38.

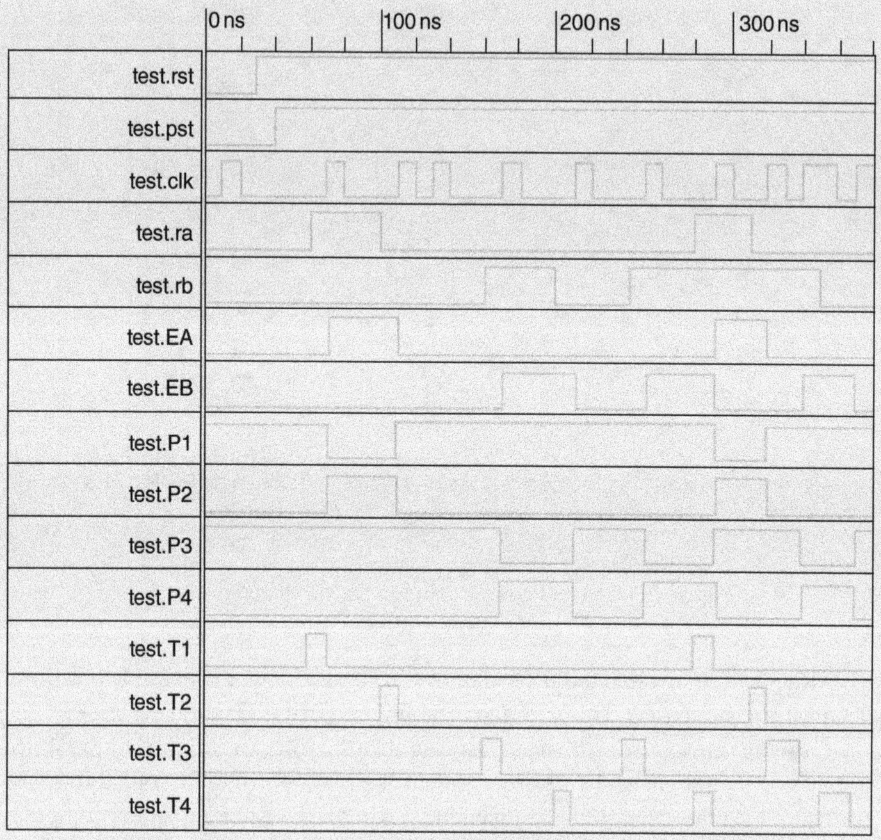


**Figure 7.38** The Verilog HDL simulation for the shared resource.

In this diagram you can clearly see the placeholders P1 and P3 both active at the beginning of the simulation. When Ra becomes true because Ea becomes asserted the token moves from P1 to P2. Note that in Figure 7.38 some variables are slightly different but use the same letters – this is OK.

When Ra is dis-asserted, Ea goes low and P1 gets the token, while P2 loses the token.

Try looking at the rest of the sequence in conjunction with the Petri net.

Turn to **Frame 7.42**.

### Frame 7.42

The Petri net of Frame 7.40 (Figure 7.37) is to be implemented using a computer program that extracts the equations from the Petri net diagram. First the computer-generated equations are implemented in Verilog HDL. Note that here we use '~' for NOT, '&' for AND, and '|' for OR.

$$P1 = T2 \mid P1 \,\&\, {\sim} T1$$
$$P2 = T1 \mid P2 \,\&\, {\sim} T2$$
$$P3 = T4 \mid P3 \,\&\, {\sim} T3$$
$$P4 = T3 \mid P4 \,\&\, {\sim} T4$$
$$T1 = P1 \,\&\, Ra \,\&\, {\sim} P2$$
$$T2 = P2 \,\&\, {\sim} Ra \,\&\, {\sim} P1$$
$$T3 = P3 \,\&\, Rb \,\&\, {\sim} P4 \,\&\, {\sim} P2$$
$$T4 = P4 \,\&\, \left(Ra \mid {\sim} Rb\right) \,\&\, {\sim} P3$$
$${\sim} Ea = P1$$
$$Ea = P2$$
$${\sim} Eb = P3$$
$$Eb = P4$$

Now move to **Frame 7.43**.

### Frame 7.43

The computer program will use the equations to obtain the Verilog HDL file:

```
//PN-HDL Petri net Workbench
module Petri (clock,reset,~Ea,Ea,~Eb,Eb,Ra,Rb);
// Input/Output Declaration
output ~Ea,Ea,~Eb,Eb;
input clock,reset,Ra,Rb;
reg  P1,P2,P3,P4,T1,T2,T3,T4;   // Need to keep these
stored for //use.
// Define Output Variable Equation
assign ~Ea=P1;
assign Ea=P2;
assign ~Eb=P3;
assign Eb=P4;
```

```
always @(posedge clock or posedge reset)
begin
if(reset)
 begin
     assign P1=1;          //placeholder P1 set.
     assign P2=0;
     assign P3=1;          //placeholder P2 set.
     assign P4=0;
 end
else
 begin
     assign T1=Ra &P1 &~P2;
     assign T2=~Ra&P2 &~P1;
     assign T3=Rb &P3 &~P4 &~P2;
     assign T4=P4 & (Ra|~Rb)&~P3;
     assign P1=T2 |(P1 &~T1);
     assign P2=T1 |(P2 &~T2);
     assign P3=T4 |(P3 &~T3);
     assign P4=T3 |(P4 &~T4);
 end
end
endmodule
```

You should look carefully at the above Verilog code and compare it with the Petri net diagram and equations. From this you should be able to see how to generate Verilog program code for almost any Petri net.

This example is available in the tutorial section later. This uses a rather different Verilog HDL arrangement so it is worth looking at.

Note that the names of some variables may be changed, i.e. Ra, Rb, Ea, and Eb.

Turn to **Frame 7.44**.

### *Frame 7.44*

In this frame the screen of a Petri net program is illustrated (Figure 7.39).

This program was used to automate the generation of Petri net diagrams so that this program could obtain the equations and Verilog HDL.

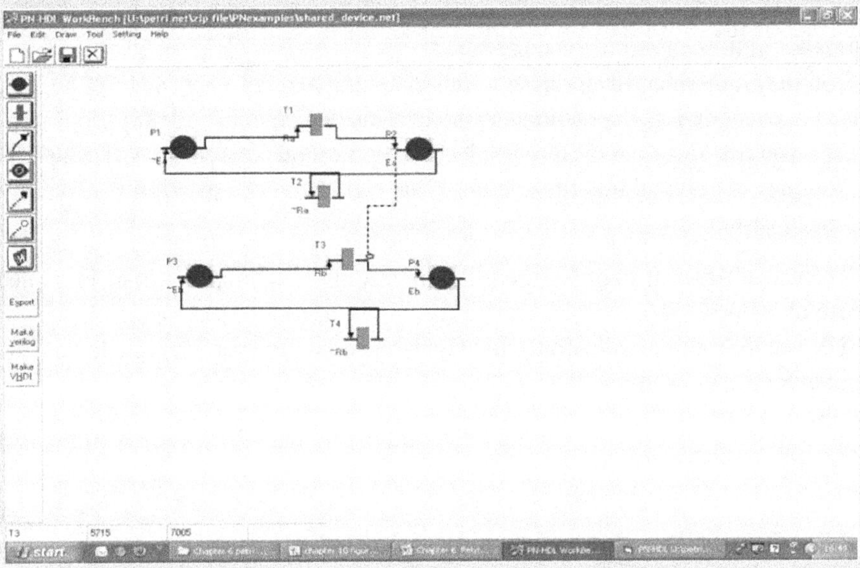


**Figure 7.39** Sample shot of a Windows program to design Petri nets.
*Source:* Petri Nets.

The diagram drawing part of the program needs further development to improve its appearance but it certainly produces the correct equations and Verilog program code.

It is possible to use this program to produce the equations, the Verilog code, and the VHDL code for the Petri net and incorporate them all into physical systems.

Unfortunately, the Petri net program no longer works due to Microsoft Visual Basic no longer being available to run on Microsoft Windows beyond Windows XP.

The original Basic program could be translated into Python, or even C++ (both languages can be produced using GUI style). There is also a number of GUI C++ compilers available for the Apple MAC that could use XCODE.

If you have an old PC computer with Windows XL, you may be able to get the original code to run.

See Section 7.2.3 for information on the use of event-driven Petri net systems after reading the Tutorial examples in Section 7.2.

## 7.2 TUTORIALS USING PETRI NET FSM

In this tutorial, the Petri net is considered.

You may remember from earlier in this chapter that a Petri net can be either clock driven or event driven, depending on whether you decide to include flip-flops or not. In this case flip-flops are used to produce a synchronous (clock-driven) design.

You may also recall that the one hot FSM can also make use of this idea, as we indicated earlier.

1. The use of the control of a shared resource example is looked at first. This makes use of a clock.
2. The second example illustrates the eight-bit serial receiver example looked at in Chapter 3, but using the Petri net.

Both examples make use of Verilog HDL and the SynaptiCAD system and you can try them out.

### 7.2.1 Controlled shared resource Petri nets

This example has been looked at in this chapter already but here it is considered as a build-up to a simulation.

The block diagram for this is illustrated in Figure 7.40.

The idea is to provide access to shared resource, in this case a printer.

Two computers (A and B) are joined, via two access control inputs, to the Petri net (Ra for computer A and Rb can access this for computer B).

Because only one computer can have access to the one printer at a time, the system is designed so that if computer A has access to the printer then it has priority and even if computer B wants to access the printer it must wait until computer A has completed its access to the printer before it can gain access itself.

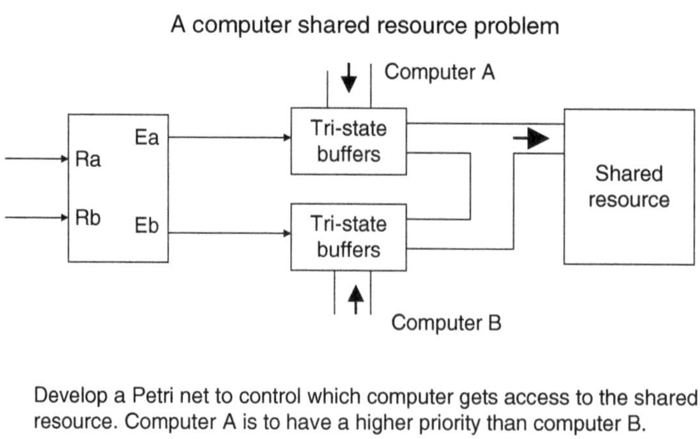


**Figure 7.40** The shared printer between two computers.

Also, if computer B has access to the printer and computer A wants to use the printer then computer B must stop accessing the printer to give computer A printer access. Only when computer A has completed its access to the printer can computer B access the printer.

Thus, computer A has priority over computer B. This allows the system to work correctly.

This sort of priority access is common in microprocessor systems when using interrupts and can lead to a priority scheme.

Most computer systems have a priority interrupt controller to take care of the priority scheme.

However, when using Petri nets, you can make use of the enabling or disabling arc available to the Petri net.

In this case the disabling arc is used to perform this task.

In Figure 7.41 there are two Petri nets consisting of P1, P2, T1, and T2 and of P3, T3, P4, and T4.

Then if computer A wants access to the printer it raises its Ra signal and T1 fires to move the token to P2.

In a similar way, when computer B wants access to the printer it raises its Rb signal to fire T3. However, the additional disabling arc from P2 can disable T3 to prevent T3 from firing so as to block the P3, P4 transition.

Note that if computer A already has access to the printer and computer B is accessing the printer, T4 will cause the computer B access to be cancelled, returning the token in P4 back to P3 via T4, and so allowing computer A to gain access to the printer.

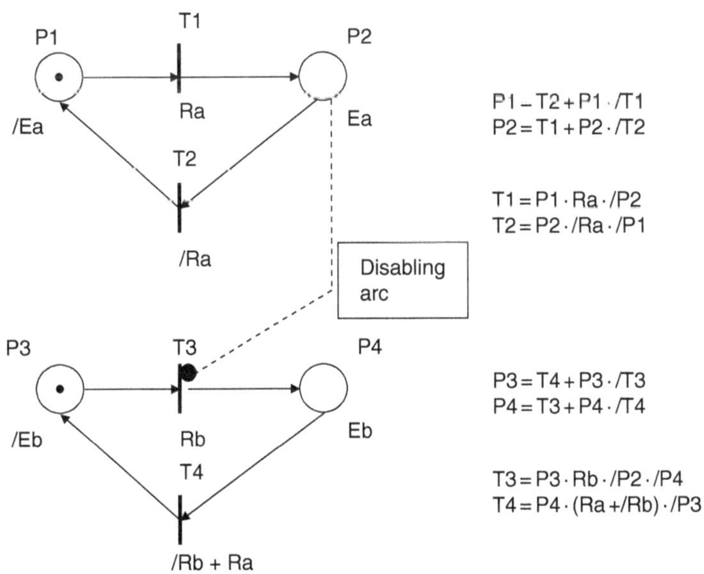


**Figure 7.41** The shared printer solution.

This connection arrangement allows computer A to have priority over computer B. **You could use this arrangement to build up an interrupt priority scheme in hardware systems.** Note the connection arrangement for:

$$T3 = P3 \cdot Rb \cdot / P2 \cdot / P4.$$

Also:

$$T4 = P4 \cdot \left(/Rb + Ra\right) \cdot / P3.$$

This makes use of the disabling arc connection from P2 to T3 in equation T3 (/P2).

This is a very subtle idea used in Petri nets and makes them a very powerful method for control in hardware systems.

You might like to go back and re-read Frames 7.35 and 7.36. to refresh yourself on how to use Petri nets with disabling arcs before moving on.

Now we can look at the SynaptiCAD simulation in Figure 7.42.

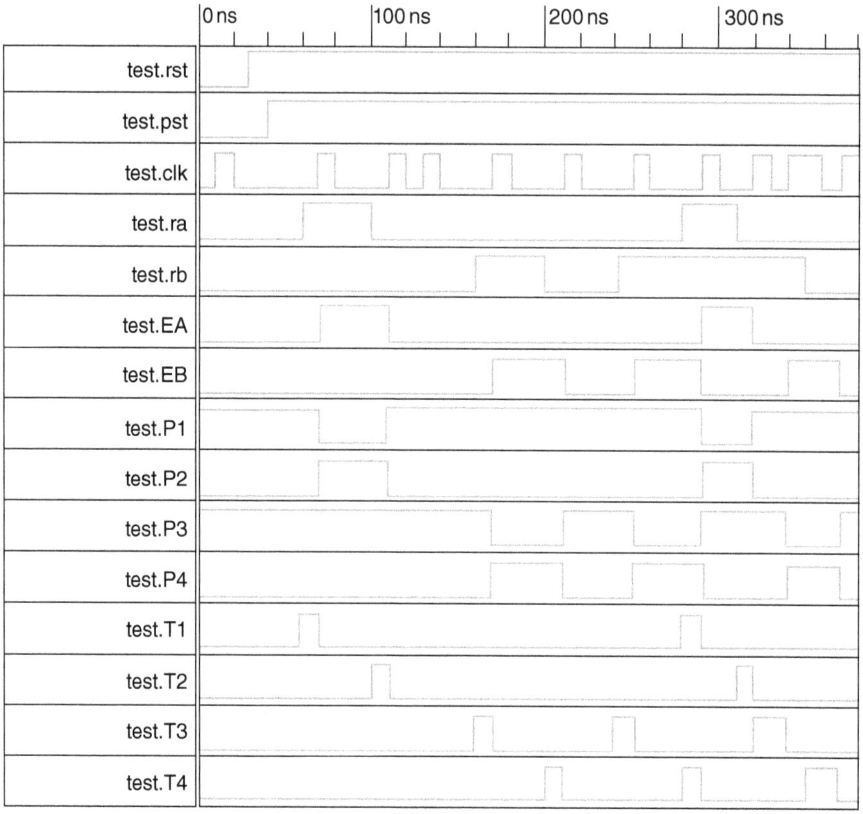


**Figure 7.42** The simulation output.

You can try this out yourself by using SynaptiCAD Verilog HDL simulation. The Verilog HDL code is shown in Listing 7.1.

**File can be found under the name sharedresource.v.**

```
/////////////////////////////////////////////////////////////
// Petri Net 1 Verilog HDL file of Shared Resource Controller.
//
// The Petri Net diagram in this section should be consulted
// to see the operation of the two Petri Nets.
// Author: PDM.
/////////////////////////////////////////////////////////////
// In the following flip-flop module use has been made of the
// Verilog case statement to
// define the performance of the flip-flop in terms of its //
// overall operation.
// This allows it to define asynchronous reset and preset as
// well as its normal DFF operation.

// Note in this version 'preset' and
// 'clear' both active low makes the flip-flop preset & clear.

// Flip-Flop module with preset & clear.
module D_FF(q,d,clk,rst,pst);
 input d,clk,rst,pst;
 output q;
 reg q;
 always @ (posedge clk or negedge rst or negedge pst)
  case ({rst,pst})
    2'b00: q<=0;   //If both rst and pst are zero so clear
the flip-flop.
    2'b01: q<=0;   //If rst is zero reset.
    2'b10: q<=1;   //If pst is zero preset.
    2'b11: q<=d;   //If neither are zero perform normal DFF
action.
 endcase
endmodule

module fsm(rst,pst,clk,Ra,Rb,Ea,Eb,P1,P2,P3,P4,T1,T2,T3,T4);
  input rst,pst,clk,Ra,Rb;
  output P1,P2,P3,P4,T1,T2,T3,T4,Ea,Eb;
  wire p1d,p2d,p3d,p4d;

// see how the DFF devices are set up below so that ffp1 and
// ffp3 can be preset
```

```
 // while ffp2 and ffp4 can be cleared.
  D_FF ffp1(P1,p1d,clk,1'b1,pst);
  D_FF ffp2(P2,p2d,clk,rst,1'b1);
  D_FF ffp3(P3,p3d,clk,1'b1,pst);
  D_FF ffp4(P4,p4d,clk,rst,1'b1);

// Note in this example some of the variables have changed as
// in Ra, Rb, when they were used.
// This is OK and will not alter the simulation.


 Assign
     p1d=T2|P1&~T1,
     p2d=T1|P2&~T2,
     T1=P1&Ra&~P2,
     T2=P2&~Ra&~P1,

     p3d=T4|P3&~T3,
     p4d=T3|P4&~T4,
     T3=P3&Rb&~P2&~P4,
     T4=P4&(~Rb|Ra)&~P3,
     Ea=P2,
     Eb=P4,
     ~Ea=P1,
     ~Eb=P3;
 end            // assignment block.
endmodule

// The test module defines the tests to be carried out in
// this simulation.

module test;
  reg rst,pst,clk,Ra,Rb;

  fsm uut(rst,pst,clk,Ra,Rb,Ea,Eb,P1,P2,P3,P4,T1,T2,T3,T4);
  initial
    begin
    $dumpfile("fsm.vcd");
    $dumpvars;
    clk=0;
    rst=0;
    pst=0;
    Ra=0;
    Rb=0;
    #10 clk=~clk;        // pulse the clock
    #10 clk=~clk;        // pulse the clock again
    #10 rst=1;
```

```
   #10 pst=1;
   #10
  //    #10 clk=~clk;   // not used.
  //    #10 clk=~clk;   // not used.

   #10 Ra=1;
   #10 clk=~clk;
   #10 clk=~clk;

   #10
   #10 Ra=0;
   #10 clk=~clk;
   #10 clk=~clk;
   #10 clk=~clk;
   #10 clk=~clk;
   #10
   #10 Rb=1;
   #10 clk=~clk;
   #10 clk=~clk;

   #10
   #10 Rb=0;
   #10 clk=~clk;
   #10 clk=~clk;

   #10
   #10 Rb=1;
   #10 clk=~clk;
   #10 clk=~clk;

   #10
   #10 Ra=1;
   #10 clk=~clk;
   #10 clk=~clk;

   #10 Ra=0;
   #10 clk=~clk;
   #10 clk=~clk;
   #10 clk=~clk;
   #10 Rb=0;
   #10 clk=~clk;
   #10 clk=~clk;          // pulse the clock.
   #10 $finish;           // finish the module.
  end
endmodule
```

**Listing 7.1** Petri net shared resource controller.

The second example illustrates the effect of using the Petri net with enabling arcs. This example is for a serial receiver using 8-data bits, a start bit and a stop bit.

### 7.2.2 Serial clock-driven Petri net FSM

In this example the eight-bit clock-driven receiver system is looked at using a Petri net based FSM. This is made up of the FSM, a shift register, a divide by 11 counter and an eight-bit buffer to hold the received data bits. It is done because it is rather different.

Next we need to look at the serial signal protocol in Figure 7.43 before looking at the block diagram in Figure 7.44 and the Petri net diagram in Figure 7.45.

Figure 7.43 illustrates the protocol structure for the Peri net controller. This is very similar to the one shown in Chapter 3 in Figure 3.33, except that it is for a Petri net state diagram not a DFF-based state diagram.

The block diagram in Figure 7.44 is very similar to Figure 3.34 in Chapter 3, except that lower case outputs are used. Note all outputs are in lower case.

The state diagram is quite a lot different to Chapter 3's Figure 3.35.

This makes use of three enabling arcs to help control the Petri net.

In Figure 7.45 the placeholders P1–P4 control the generation of the rxck pulse, while placeholders Pm1 to Pm6 deal with the working of the serial receiver.

The three enabling arcs control the operation of P1–P4 and Pm1–Pm6 Petri nets that are seen to be separate parallel Petri nets. For example, to move P1 to P2, T1 cannot fire until the enabling arc from Pm1 to Pm2 happens. Note that the enabling arc cannot fire Tm2 until a token reaches P4 (after providing a pulse to rxck). This synchronizes the rxck pulse point. T4 cannot fire until Pm3 has a token to prevent the rxck from happening in the P1–P4 loop.

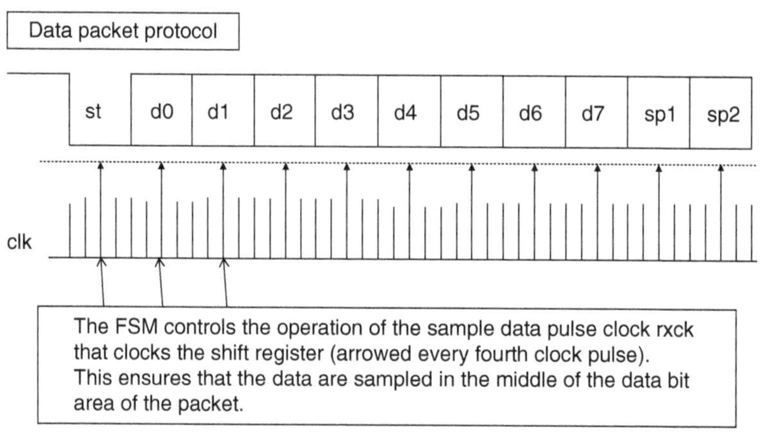


**Figure 7.43** The protocol structure for the Petri net controller.

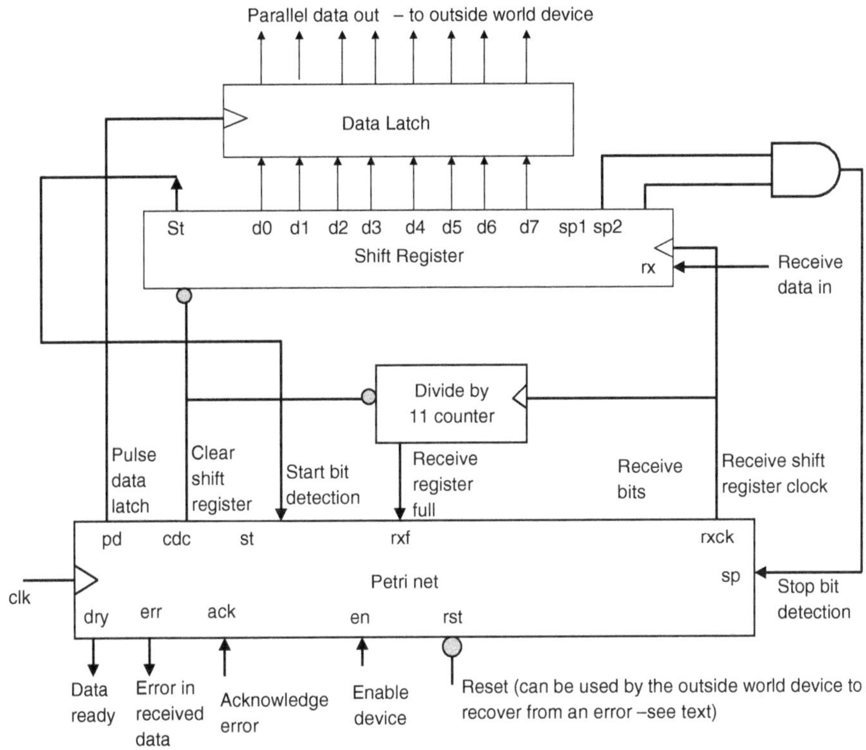


**Figure 7.44** Block diagram for the Petri net serial clock driven controller.

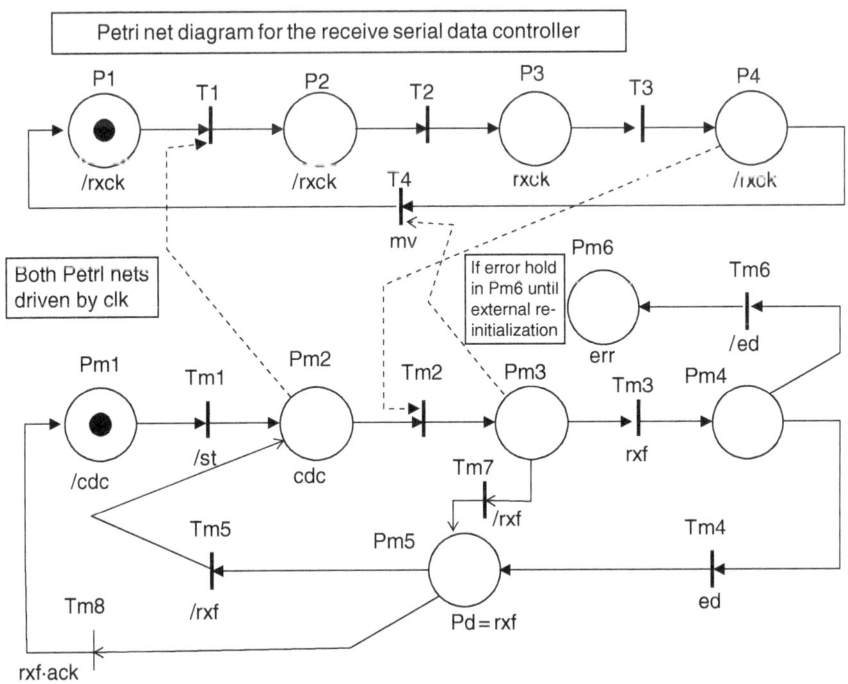


**Figure 7.45** The Petri net FSM – note the use of enabling arcs to synchronize both Petri nets.

T4 cannot fire until the token reaches Pm3, another synchronizing point. Then in the Pm1–Pm6 sequence, errors are detected (or not) before the received packet is sent to the output data latch. The system uses the three enabling arcs to keep both Petri nets in synchronism.

This example illustrates how these two Petri nets are locked together. Note that if an error is detected the system locks in placeholder Pm6. The system must be reset before it can be used again.

## The Petri net equations

There are two Petri net sets of equations:

### *Petri net 1*

P1 = T4 + P1 · /T1  T1 = P1 · Pm2 · /P2
P2 = T1 + P2 · /T2  T2 = P2 · /P3
P3 = T2 + P3 · /T3  T3 = P3 · /P4
P4 = T3 + P4 · /T4  T4 = P4 · /P1 · Pm3

For Petri net 2 we have:

### *Petri net 2*

Pm1 = Tm8 + Pm1 · /Tm1
Pm2 = Tm1 + Tm5 + Pm2 · /Tm2
Pm3 = Tm2 + Pm3·/Tm3 · /Tm7
Pm4 = Tm3 + Pm4·/Tm4 · /Tm6
Pm5 = Tm4 + Tm7 + Pm5 · /Tm5·/Tm8
Pm6 = Tm6 + Pm6  Reset from here

Tm1 = Pm1 · /st · /Pm2
Tm2 = Pm2 · P4 · /Pm3
Tm3 = Pm3 · rxf · /Pm4
Tm4 = Pm4 · ed · /Pm5
Tm5 = Pm5 · /rxf · /Pm2
Tm6 = Pm4 · /ed · /Pm6
Tm7 = Pm3 · /rxf · /Pm5
Tm8 = Pm5 · (rxf·ack) · /Pm1

Outputs for Petri net 2 are:

cdc = /Pm2 active low to reset shift reg and divide by counter
Pd = Pm5·rxf Mealy active high to latch data latch
err = Pm6 active high

This makes up the complete Petri net system.

This can be simulated with the SynaptiCAD software using the file **petri_rx.v**.

The Listing 7.2 now follows:

```
/////////////////////////////////////////////////////////////
// Petri Net Verilog HDL file for the Asynchronous Serial
// Receiver
//
//
// The asynchronous receiver FSM
// Author: PDM.
/////////////////////////////////////////////////////////////
// Uses the Verilog HDL case construct to condition the D_FF
// module.
// Flip Flop module with preset & clear.
module D_FF(q,d,clk,rst,pst);
 input d,clk,rst,pst;
 output q;
 reg q;
 always @ (posedge clk or negedge rst or negedge pst)
  case ({rst,pst})
    2'b00: q<=0 //will not operate if both rst and pst are
active.
    2'b01: q<=0;
    2'b10: q<=1;
    2'b11: q<=d;
 endcase
endmodule

// Note Pm1 to Pm6 are to be the same as Pm1 to Pm6 in Petri
// Net Diagram.
// It is just that this Verilog HDL program uses the former.
// Same for Tm1 to Tm8.

module fsm(rst,pst,clk,P1,P2,P3,P4,P5,T1,T2,T3,T4,T5,Pm1,Pm2,
Pm3,Pm4,Pm5,Pm6,Tm1,Tm2,Tm3,Tm4,Tm5,Tm6,Tm7,Tm8,st,ed,ack,rxf,
rxck,cdc,pd,ack,err);
  input rst,pst,clk,st,rxf,ed,ack;
  output
P1,P2,P3,P4,P5,T1,T2,T3,T4,T5,Pm1,Pm2,Pm3,Pm4,Pm5,Pm6,
               Tm1,Tm2,Tm3,Tm4,Tm5,Tm6,
               Tm7,Tm8,rxck,cdc,pd,err;

wire p1d,p2d,p3d,p4d,p5d,pm1d,pm2d,pm3d,pm4d,pm5d,pm6d;

  // receive clock petri net
  D_FF ffp1(P1,p1d,clk,1'b1,pst);
  D_FF ffp2(P2,p2d,clk,rst,1'b1);
```

```
D_FF ffp3(P3,p3d,clk,rst,1'b1);
D_FF ffp4(P4,p4d,clk,rst,1'b1);
D_FF ffp5(P5,p5d,clk,rst,1'b1);

// main petri net
D_FF ffpm1(Pm1,pm1d,clk,1'b1,pst);
D_FF ffpm2(Pm2,pm2d,clk,rst,1'b1);
D_FF ffpm3(Pm3,pm3d,clk,rst,1'b1);
D_FF ffpm4(Pm4,pm4d,clk,rst,1'b1);
D_FF ffpm5(Pm5,pm5d,clk,rst,1'b1);
D_FF ffpm6(Pm6,pm6d,clk,rst,1'b1);

assign
   // rxclk sub-petri net.
   // placeholders.
   p1d=T4|P1&~T1,
   p2d=T1|P2&~T2,
   p3d=T2|P3&~T3,
   p4d=T3|P4&~T4,
   // transitions.
   T1=P1&Pm2&~P2,
   T2=P2&~P3,
   T3=P3&~P4,
   T4=P4&~P1&Pm3,
   // its output.
   rxck=P3,
   //----------------------------------
   // main petri net.
    // placeholders.
   pm1d=Tm8 | Pm1&~Tm1,
   pm2d=Tm5 | Tm1 | Pm2&~Tm2,
   pm3d=Tm2 | Pm3&~Tm3&~Tm7,
   pm4d=Tm3 | Pm4 & ~Tm4&~Tm6,
   pm5d=Tm4 | Tm7 | Pm5&~Tm5&~Tm8,
   pm6d=Tm6 | Pm6,          //reset from here.
   // transitions.
   Tm1=Pm1&~st&~Pm2,
   Tm2=Pm2&P4&~Pm3,
   Tm3=Pm3&rxf&~Pm4,
   Tm4=Pm4&ed&~Pm5,
   Tm5=Pm5&~rxf&~Pm2,
   Tm6=Pm4&~ed&~Pm6,
   Tm7=Pm3&~rxf&~Pm5,
   Tm8=Pm5&rxf&ack&~Pm1,
```

```
    // outputs.
    cdc=~Pm2,                 // active low.
    pd=Pm5&rxf,
    err=Pm6;
  //--------------------------------
endmodule



module test;
    reg rst,pst,clk,st,ed,rxf,ack;
    fsm uut(rst,pst,clk,P1,P2,P3,P4,P5,T1,T2,T3,T4,T5,Pm1,
Pm2,Pm3,Pm4,Pm5,Pm6,
    Tm1,Tm2,Tm3,Tm4,Tm5,Tm6,Tm7,Tm8,st,ed,ack,rxf,rxck,cdc,
pd,ack,err);

    initial
    begin
    clk=0;
    repeat(2500)
    begin
        #10 clk=~clk;
        #10 clk=~clk;
        end

    end

    initial
    begin
    $dumpfile("fsm.vcd");
    $dumpvars;
    clk=0;
    rst=0;
    pst=0;
    st=1;
    ed=0;
    ack=0;
    rxf=0;
    #10 rst=1;
    #10 pst=1;
    #10
    #20
    #10 st=0;
    #10 rxf=1;
```

```
    #10 ed=1;
    #10 ack=1;
    #40 ack=0;
    #20 ed=0;
    // #20 ed=0;
    #100
    #100 $finish;
    end
endmodule
```

**Listing 7.2** Petri net asynchronous serial receiver.

The simulation of the Petri net system is illustrated in Figure 7.46. You can see the operation of the system as it follows through the Petri net.

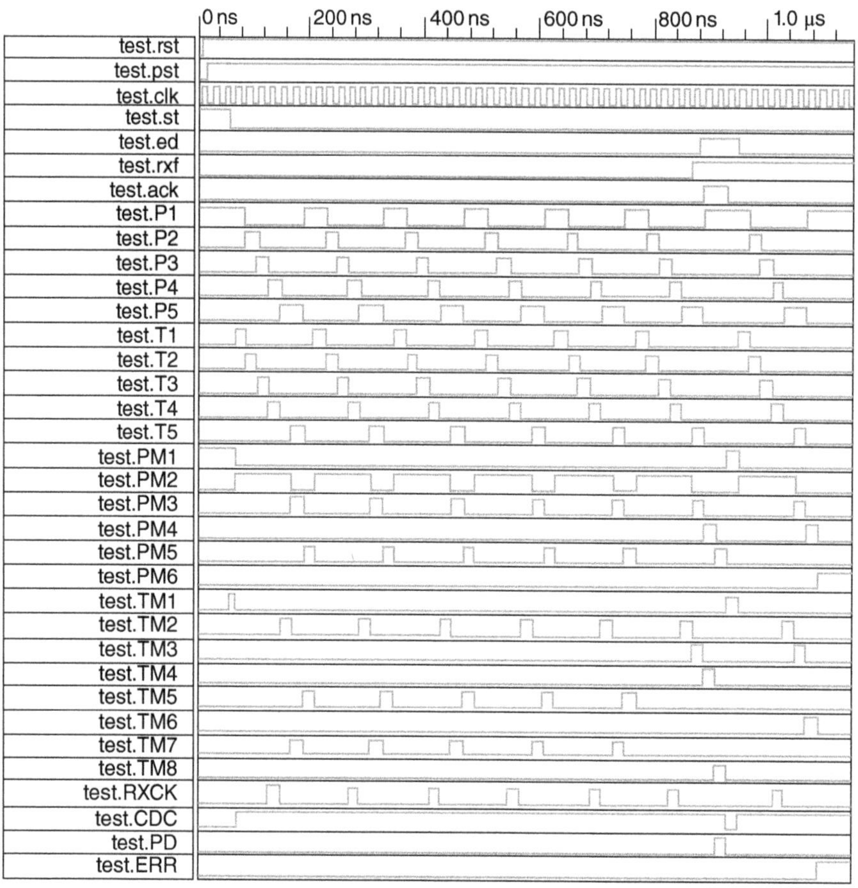


**Figure 7.46** Simulation of the system.

### 7.2.3 Using asynchronous (event-driven) design with Petri nets

In Yakovlev and Koelmans (1998) the ideas behind the need for asynchronous design is introduced and the working of the C gate and Request/Acknowledge handshakes is introduced.

The paper considers the need for unclocked designs.

In clock-driven systems the clock is used to allow the design to settle down after asynchronous inputs have changed. However, this wastes considerable time, also the clock itself consumes up to half the system power and adds little to the working of the system.

Yakovlev and Koelmans (1998) introduce the working of the Muller C gate, with Request/Acknowledge handshakes to eliminate the need for a clock. This improves the operation of the system as well as reducing the overall power requirements.

Figure 7.47 illustrates the action of a C gate and shows how a simple example with four inputs *X* and *Y* and *W*, and *Z* can arrive from asynchronous sources and be controlled with Request/Acknowledge handshakes with a C gate so that the arrival on the signals *X*, *Y*, W, and *Z* can be accommodated without the need for a clock.

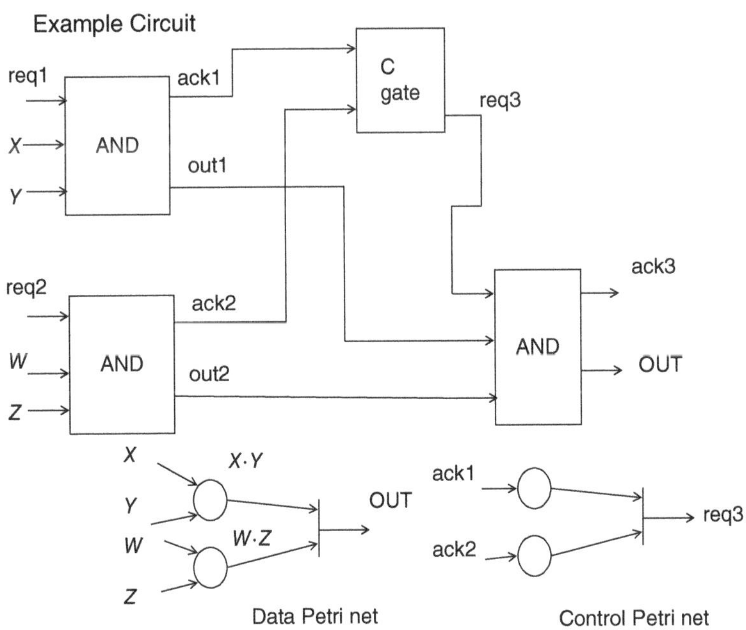


**Figure 7.47** The use of a C gate to control signals.

The two Petri net examples clearly show that in the data Petri net *X* and *Y* as well as *W* and *Z* need to be completed before the output OUT can occur.

The control Petri net shows the processing of the ack1 and ack2 signals in order to generate the request signal at the output.

This simple example illustrates the activities required for this mechanism to operate and avoid glitches in the operation of the system by using the C gate. The C gate generates the req3 output only when out1 and out2 are completed. When ack3 is active, the output OUT can be available.

Clearly, the Petri net is a very useful way to implement such systems.

Yakovlev and Koelmans (1998) provides further information on how and why this is needed to make reliable asynchronous Petri net systems. The idea is to work with larger systems, rather than the small systems looked at in this chapter.

The reference paper considers abstract levels and design stages with design examples.

There are now logic synthesization tools available, such as Pertify, that can be downloaded using Linux, OS X, or Windows to help with the design of these systems.

The paper contains many references (up to 81) that the reader can access in order to acquaint themselves with the techniques.

Finally, Figure 7.48 shows the Workflow Petri Net Designer program (WoPeD) that can be used to help design and simulate Petri nets. The figure shows a traffic lights controller being simulated. Just follow the 'placeholder tokens' showing the traffic lights sequence.

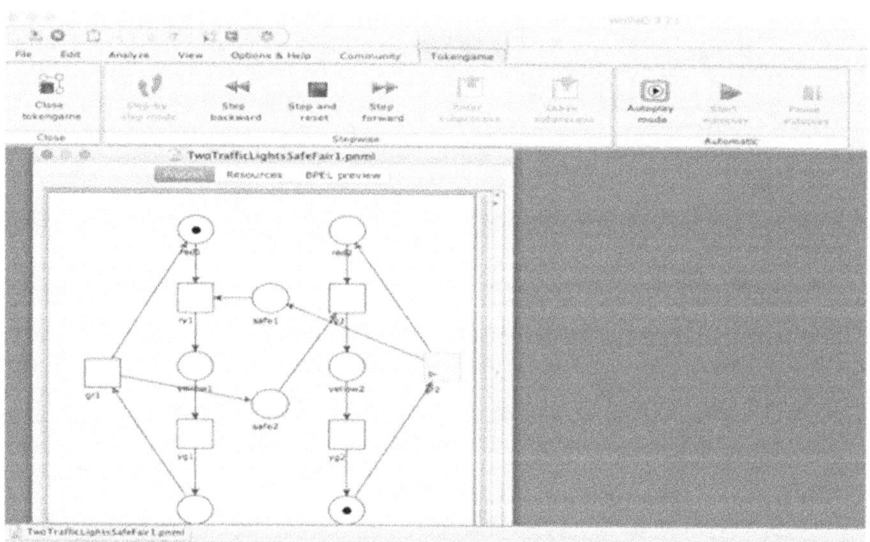

**Figure 7.48** Workflow Petri Net Designer (WoPeD).

The program can be obtained from http://www.woped.org. Note you will need to have Java installed on your computer.

## 7.3 CONCLUSIONS

In this chapter the Petri net has been covered in some detail for the single token Petri net.

The idea that the Petri net can be developed for both clocked and unclocked systems has been looked at, as well as the use of disabling arcs and enabling arcs to help to synchronize parallel systems. Some interesting examples have been considered in the tutorial section.

Finally, the use of Yakovlev and Koelmans (1998) has been looked at to see how to make use of C gates and Request/Acknowledge handshakes in order to produce reliable asynchronous (event-driven) systems.

# Appendix A1: Boolean Algebra

This appendix provides a brief set of notes on Boolean algebra laws and their use.

It is assumed that the reader already has knowledge of these laws. This appendix is provided as a reference only for the Boolean algebra used in this book.

## A1.1 BASIC GATE SYMBOLS

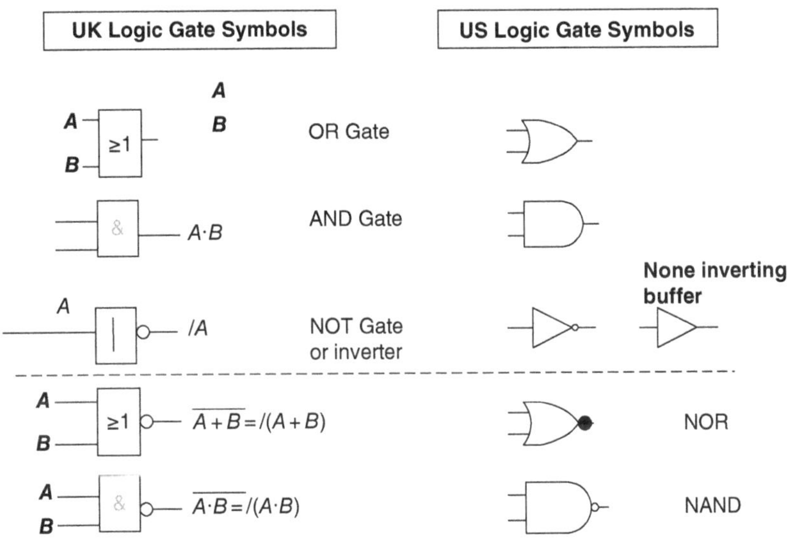


**Figure A1.1** Basic gates using UK and USA symbols.

*Digital System Design using FSMs: A Practical Learning Approach*, First Edition. Peter D. Minns.

Companion website: www.wiley.com/go/minns/digitalsystemdesign

## A1.2 THE EXCLUSIVE OR AND EXCLUSIVE NOR

This gate is made up from AND/OR/NOT gates from the Boolean equation

**F = A·/B + /A·B**

| A | B | F | /F |
|---|---|---|---|
| 0 | 0 | 0 | 1 |
| 1 | 0 | 1 | 0 |
| 0 | 1 | 1 | 0 |
| 1 | 1 | 0 | 1 |

Note that not *F* (/*F*) is the exclusive NOR

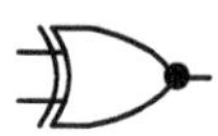

**Figure A1.2** Symbols for UK and USA gates and their characteristics.

## A1.3 LAWS OF BOOLEAN ALGEBRA

These are presented in terms of the Boolean logic equation and gate circuit.

### A1.3.1 Basic OR rules

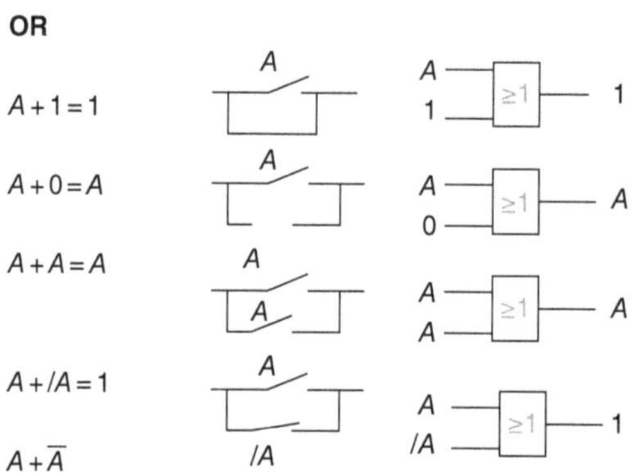


**Figure A1.3** For OR gates using Boolean algebra.

### A1.3.2 Basic AND rules

$A = //A = \overline{\overline{A}}$

**AND**

$A \cdot 1 = A$

$A \cdot 0 = 0$

$A \cdot A = A$

$A \cdot /A = 0$

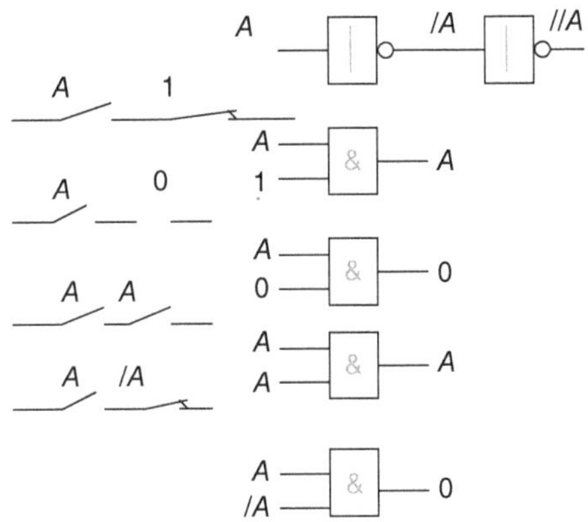


**Figure A1.4** For AND gates using Boolean algebra.

### A1.3.3 Associative and commutative laws

Associative Laws:

$(A+B)+C = A+(B+C) = A+B+C$

$(A \cdot B) \cdot C = A \cdot (B \cdot C) = A \cdot B \cdot C$

Commutative Laws:

$A+B = B+A$

$A \cdot B = B \cdot A$

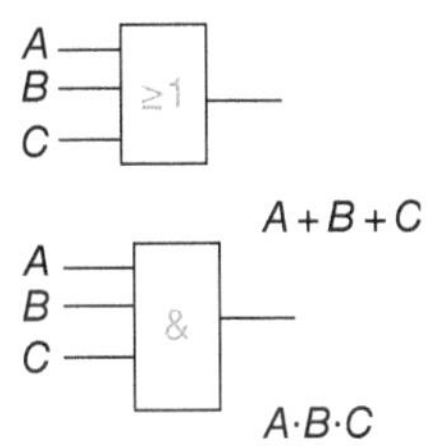


**Figure A1.5** The effect of associate and commutative laws.

### A1.3.4 Distributive laws

Distributive Laws:

$F = A \cdot (B+C) = A \cdot B + A \cdot C$ (as with normal algebra)

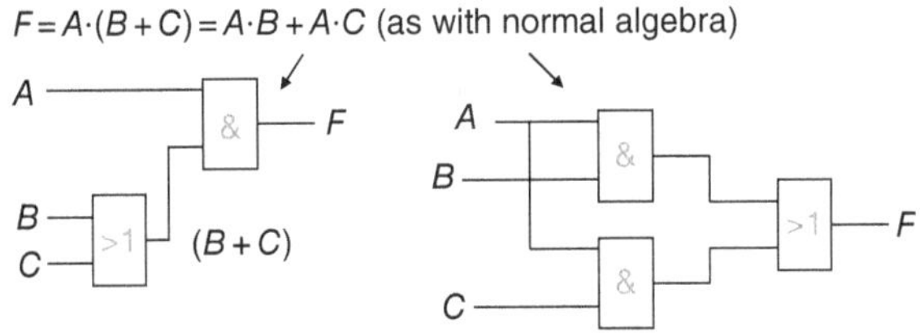


**This rule is also like normal algebra**

**Figure A1.6** Distributive laws and their use.

### A1.3.5 Auxiliary rule for static 1 hazard removal

Auxiliary Law:

$A+(B\cdot C)=(A+B)\cdot(A+C)$ **and** $\mathbf{(A+B)\cdot(A+C)=A+(B\cdot C)}$

**The right-hand side shows a circuit reduction.**

However, a more interesting example is...

$\mathbf{A+(/A\cdot B)=(A+/A)\cdot(A+B)=1\cdot(A+B)=A+B}$

**which also illustrates a reduction.**

**Try: $X+(/X\cdot Y)=?$ and $/P+(Q\cdot P)=?$**

**Figure A1.7** Auxiliary rule in action.

The auxiliary (aux) rule is particularly significant and much used in this book.

#### A1.3.5.1 Proof of the Auxiliary Rule

$$Y = x + /x\cdot z = (x + /x)\cdot(x + z) = x\cdot x + x\cdot z + /x\cdot x + /x\cdot z.$$
$$Y = x + x\cdot z + 0 + /x\cdot z = x\cdot(1 + z) + /x\cdot z = x + /x\cdot z = x + z.$$

The answer to the question $/p + (q\cdot p)$ is of course $/p + q$.

Note that the aux rule is just a special case of the consensus theorem.

### A1.3.6 Consensus theorem

Consider the equation $Y = x + /x\cdot \mathrm{w}\cdot z$. In the equation if $\mathrm{w} = z = 1$ then:

$$Y = x + /x\cdot 1\cdot 1 = x + /x.$$

which is 1.

However, it is possible under some conditions of gate delay that a logic 1 glitch can occur (this is known as a 'static 1 hazard'). To avoid this, a cover term can be added, made up from the literals $w$ and $z$ to make $Y$ stay at logic 1:

$$Y = x + /x\cdot \mathrm{w}\cdot z + \mathrm{w}\cdot z$$

so that when $w$ and $z$ are both 1

$$Y = x + /x + 1.$$

thus covering the potential terms $x + /x$ and preventing the glitch. **Note that the glitch still occurs even though it is covered and this gives rise to a small power increase. In a very large logic system this could be a problem (Rice 2008).**

Now consider the equation:

$$Y = x + /x\mathrm{w}.$$

Now add a cover term thus:

$$Y = x + /x\mathrm{w} + \mathrm{w}.$$
$$Y = x + \mathrm{w}\left(/x + 1\right) = x + \mathrm{w}.$$

So applying the consensus theorem to $Y = x + /x{\cdot}\mathrm{w}$ results in $Y = x + \mathrm{w}$ as obtained using the aux rule.

In effect the aux rule is just a special case of the consensus theorem, that is it is an **auxiliary to the consensus theorem**.

Another example (using the aux rule):

$$Y = /x + x{\cdot}z = /x + z \text{ since } /x + x{\cdot}z = \left(/x + x\right){\cdot}\left(/x + z\right) = 1{\cdot}\left(/x + z\right) = /x + z.$$

In the following two examples the term to be removed is crossed out thus $\not{/}\!\!\not{r}$.

$$P = q{\cdot}r + q{\cdot}/r{\cdot}s = q{\cdot}\left(r + /r{\cdot}s\right) = q{\cdot}\left(r + \not{/}\!\!\not{r}{\cdot}s\right) = q{\cdot}\left(r + s\right) = q{\cdot}r + q{\cdot}s.$$
$$Y = st{\cdot}/x + st{\cdot}x{\cdot}z = st{\cdot}\left(/x + x{\cdot}z\right) = st{\cdot}\left(/x + \not{x}{\cdot}z\right) = st{\cdot}\left(/x + z\right).$$

Therefore:

$$Y = st{\cdot}/x + st{\cdot}z.$$

Note that the reduction here is in the number of inputs. This might be relevant in some cases but when implementing with programmable logic device (PLD) devices it may not be so relevant owing to the large number of inputs available in the AND gate array of the PLD.

**Remember, however, that the application of the aux rule eliminates a static 1 hazard and hence a potential glitch.**

In the case of $P$ the term $r + /r$, in the case of $Y$, $/x + x$.

### *A1.3.7 The effect of signal delay in logic gates*

Figure A1.8 can be obtained by looking at the effect of gate delays due to inverted gate inputs to propagation delays in the gate. Note that the effect of output delay propagation is ignored in Figure A1.8 to simplify the effect but in a real system there would be a time shift in the output of each gate. It is left to the reader to look at the effect in NAND and NOR gates.

In Chapters 3, 4, and 5, signal delay can affect the behaviour of a circuit. The basic effect of signal delay can be clearly seen in Figure A1.8, where the points on the two input signals $x$ and $/x$ where an output change can occur is seen for each gate type.

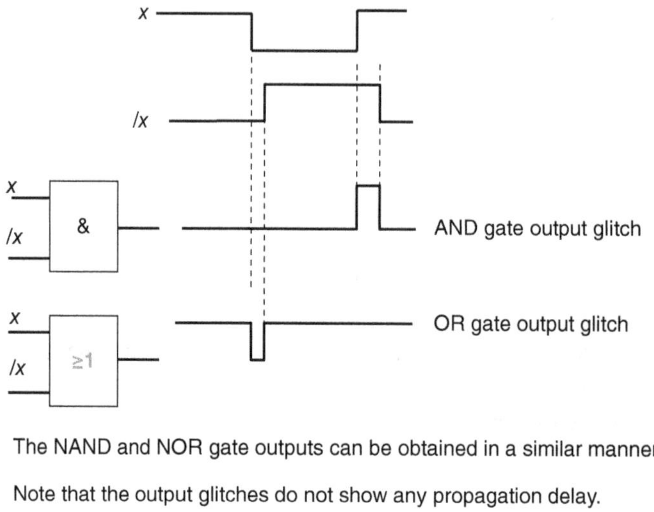


**Figure A1.8** The effect of signal delay in AND/OR gates.

In the case of an AND gate it is the point at which the two overlapping signals are both at logic 1.

In the case of OR (or NOR) gates, it is the point at which the two overlapping signals are both at logic 0.

The effect is to produce an unwanted glitch at the gates output. These so-called glitches can result in maloperation in an FSM. They can occur in a logic system where two signals (they do not need to be the same signals inverted, as shown in Figure A1.8) change state at the same time. This could be a change of two state variables, for example when unit distance coding is not used in a state diagram.

This kind of behaviour can manifest itself within a logic block where two signal changes occur due to the delays through the logic gates. Further work on this type of behaviour can be found in most textbooks on advanced digital design. These are not discussed any further as they are beyond the scope of this book.

## A1.3.8 De-Morgan's theorem

De-Morgan's theorem is used significantly (Figure A1.9), particularly in Chapter 4.

De-Morgan's rules are often used to convert from NAND [**/(A·B)** ] to NOR [**/A + /B**] and NOR [/(**A** + **B**) ] to NAND [**/A·/B**].

De-Morgan's Theorem: very often used.

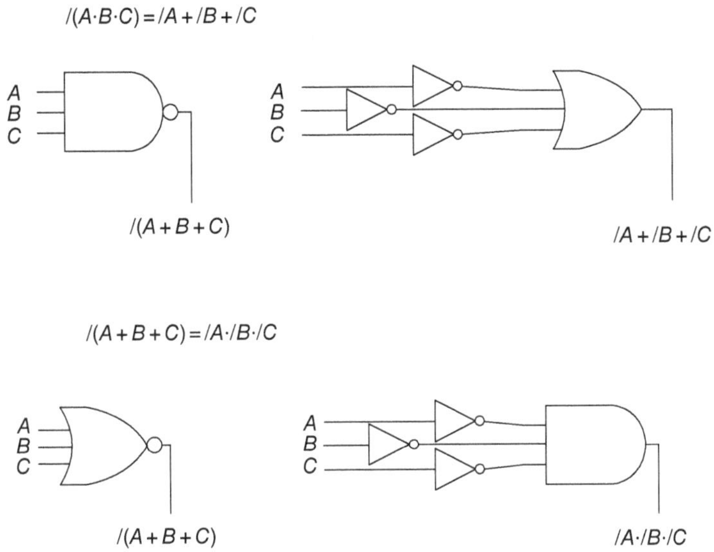


**Figure A1.9** The use of De-Morgan's rules.

## A1.4 EXAMPLES OF APPLYING THE LAWS OF BOOLEAN ALGEBRA

### A1.4.1 Converting AND-OR to NAND

$$Z = x{\cdot}y + /x{\cdot}/y.$$

Use De-Morgan's rule: ***/(A + B) = /A·/B*** where ***A*** and ***B*** can be any product term. So in the following example ***A*** is replaced by ***x·y***, and ***B*** is replaced by ***/x·/y***:

$$Z = //Z = //\left(x{\cdot}y + /x{\cdot}/y\right) = /\left(/\left[x{\cdot}y\right]{\cdot}/\left[/x{\cdot}/y\right]\right).$$

### A1.4.2 Converting AND-OR to NOR

Use the De-Morgan rule: ***/(A·B) = /A + /B***. where ***A*** and ***B*** can be any product term. So in the following example ***A*** is replaced by ***x·y*** and ***B*** by ***/x·/y***.

$$\begin{aligned} Z &= x{\cdot}y + /x{\cdot}/y = //Z = //\left(x{\cdot}y + /x{\cdot}/y\right) = //\left(//\left[x{\cdot}y\right] + //\left[/x{\cdot}/y\right]\right) \\ &= //\left(/\left[/x + /y\right] + /\left[x + y\right]\right). \end{aligned}$$

### A1.4.3 Logical adjacency rule

This rule is used in the Karnaugh map method of Boolean algebra minimization and works on the idea that $b + /b = 1$, so in the equation:

$$F = b{\cdot}c + b{\cdot}/c = b{\cdot}(c + /c) = b{\cdot}1 = b.$$

Thus eliminating the literal $c$ (in this case).

The following example makes use of the logical adjacency rule:

$$X = a{\cdot}/b{\cdot}c + a{\cdot}b{\cdot}c = a{\cdot}c(/b + b) = a{\cdot}c{\cdot}1 = a{\cdot}c.$$

In the next example, the rule is applied twice:

$$\begin{aligned} Q &= /a{\cdot}/b{\cdot}/c + /a{\cdot}/b{\cdot}c + a{\cdot}b{\cdot}/c + a{\cdot}/b{\cdot}/c = /a{\cdot}/b{\cdot}(/c + c) + a{\cdot}/c{\cdot}(b + /b) \\ &= /a{\cdot}/b + a{\cdot}/c. \end{aligned}$$

The logical adjacency rule is used in most of the work on synchronous FSM design, and also in establishing correct operation in asynchronous (event) FSM design in Chapter 4.

Both the aux rule and the logical adjacency rule are often used in the reduction of the flip-flop $D$ and $T$ equations, i.e. typical equation from a state diagram could be:

$$A{\cdot}d = /A{\cdot}B{\cdot}\mathrm{st} + A{\cdot}B + A{\cdot}/B{\cdot}\mathrm{sp} = B{\cdot}\mathrm{st} + A{\cdot}B + A{\cdot}\mathrm{sp}.$$

This is obtained from:

$$\begin{aligned} A{\cdot}d &= B(/\mathrm{A}\mathrm{st} + A) + A(B + /\mathrm{Bsp}) = B(\mathrm{st} + A) + A(B + \mathrm{sp}) \\ &= \mathrm{Bst} + AB + AB + \mathrm{Asp}. \end{aligned}$$

Now:

$$AB + AB = AB.$$

Therefore:

$$A{\cdot}d = \mathrm{Bst} + AB + \mathrm{Asp}.$$

## A1.5 SUMMARY

This appendix has looked briefly at the basic laws of Boolean algebra and discussed some of the Boolean techniques used in this book. It provides a reference source for those readers who have not used Boolean algebra for some time. Further information can be found in most books on digital logic.

The consensus theorem attempts to 'cover up' potential glitches. However, the glitch still occurs and in logic systems of immense size in the future (perhaps using neuromorphic computing) this would result in large energy waste that cannot be ignored due to the effects of reversible engineering (Rice 2008). Note that the aux rule removes the glitch effect.

# Appendix A2: Use of Verilog HDL and Logisim to FSM

This appendix describes FSM models in Verilog code and then simulates them using the SynaptiCAD VeriLogger Extreme system and the Logisim gate level simulator. It also provides an explanation to the workings of the Verilog HDL language used to describe the structure and operation of the FSM to help the reader's understanding.

It then looks at how to use the Logisim logic simulator. This is very handy to use to simulate most FSM systems and can complement the development process a great deal. Let's start with the use of the Verilog HDL system and the use of SynaptiCAD.

The SynaptiCAD software can be downloaded from htpp://www.syncad.com/syn_down.html or Google Synapticad.co.uk, search for SynaptiCAD Tools for the Thinking Mind. This is designed to work with Microsoft Windows. You can, of course, use the MAC computer with Parallels and Window 10.

A more detailed account of Verilog HDL is provided in Chapters 6–8 in Minns and Elliott (2008). These three chapters can be purchased from Wiley at a separate cost (much lower than the total cost of the book). Also, the same book is available in the IET library.

This new book looks at Verilog HDL as it progresses. Following this approach, the reader can learn much.

The book covers a lot of the design methods used at the end of Chapters 3, 4, 5, and 7 as well as Appendix A2, A3, A4, A5, and A6.

## A2.1 THE SINGLE-PULSE GENERATOR WITH MEMORY CLOCK-DRIVEN FSM

The design of a single-pulse generator with memory is outlined and then a Verilog HDL file created. Figure A2.1 shows the Verilog file used in the design.

*Digital System Design using FSMs: A Practical Learning Approach*, First Edition. Peter D. Minns.

Companion website: www.wiley.com/go/minns/digitalsystemdesign

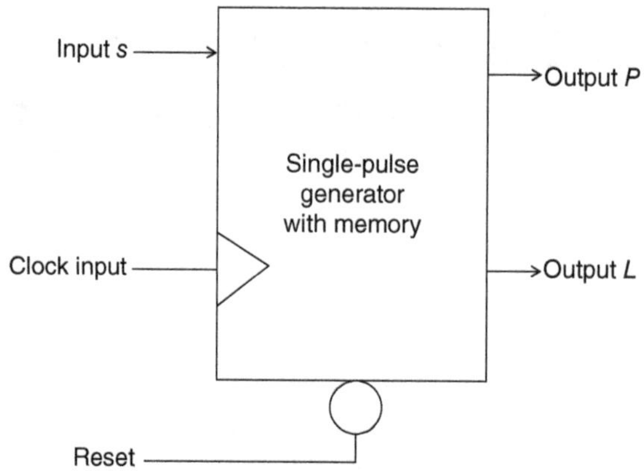


**Figure A2.1** Single pulse with memory block diagram.

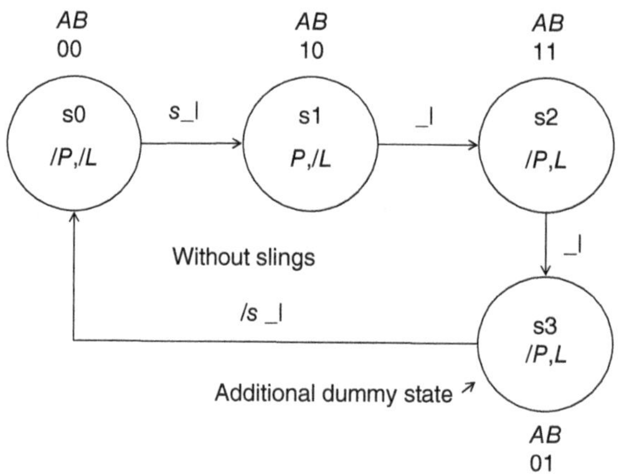


**Figure A2.2** State diagram to help generating the equations.

This consists of an FSM with input *s* and an input clock, and two outputs: *P* and *L*. The *L* output indicates that a *P* pulse has been sent (it could be very fast so you may not be able to see it).

The FSM state diagram for this system is illustrated in Figure A2.2.

In the state diagram you can see the transition from s0 to s1 when the *s* input is pressed. In s1, a *P* pulse is produced, then the FSM moves to s2, switching off the *P* signal and turning on the *L* signal. Thus, in s3 the *L* signal tells the user that a *P* pulse has been sent. The state diagram remains in s3 until the *s* input returns to zero. At this point the FSM moves back to s0 ready for the next operation.

Note how the state diagram shows the sequence of the FSM.

Having done this, each state in the state diagram is given an *AB* secondary state variable (SSV) value that is unique to that state. Also note that the SSV is to follow a unit distance coding so as to avoid glitch problems.

The next stage is to produce the Boolean equations from the state diagram as you learnt to do in Chapters 1–3.

The equations for this example are:

$$A{\cdot}d = \mathrm{s0}{\cdot}s + \mathrm{s1} = /A/B{\cdot}s + A/B = /B{\cdot}s + A/B. \tag{A1}$$

$$B{\cdot}d = \mathrm{s1} + \mathrm{s2} + \mathrm{s3}{\cdot}s = A/B + AB + /AB{\cdot}s = A + B{\cdot}s. \tag{A2}$$

The outputs are:

$$P = s1 = A/B.$$

and

$$L = \mathrm{s2} + \mathrm{s3} = B.$$

You can draw the logic circuit from these Boolean equations if you want to.

**In Verilog HDL, AND (·) is '&', OR (+) is '|', and negate (/) is ' ~'.**

**Also note that when using the assign block all Boolean lines end with (,) apart from the last line ending with a semicolon (;).**

The Boolean algebra equations describe the logic circuit of Figure A2.3 and the Verilog HDL conveys to SynaptiCAD the FSM circuit.

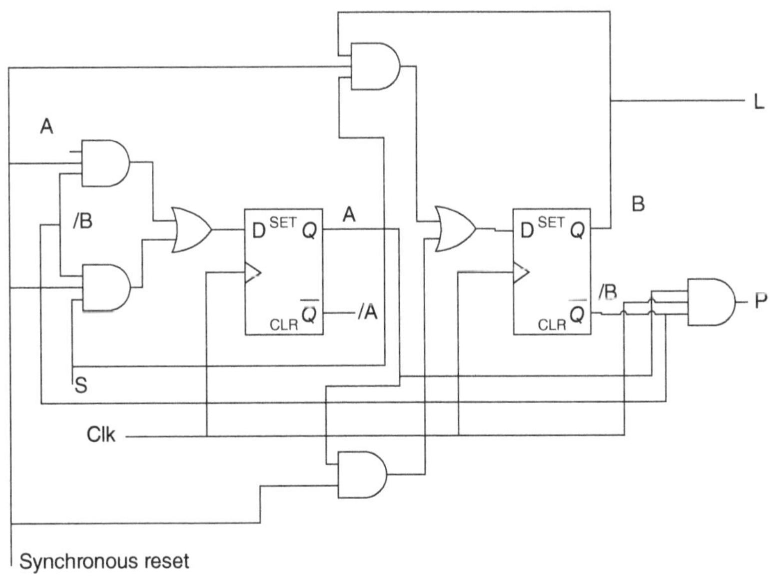


**Figure A2.3** Circuit diagram of the final system.

You could use the diagram of Figure A2.3 to create a Logisim simulation of the design. See end of this appendix for information about using Logisim. You can go to www.logisim.com to download this simulator. Note it has a very good help file on how to use it.

However, you do not need to use the circuit diagram if you are going to use Verilog HDL code. In fact, a lot of modern designers who use Verilog HDL may not look at the circuit at all!

The Verilog code here is shown but you can find it in the Verilog simulation single pulse with memory file as **sngplsmem.v**.

```
// The single pulse with memory Verilog HDL example as used
   in Appendix A2.

// First there is a need to describe the D Type flip-flop to
   be used in the FSM circuit.
// This is illustrated below in what is a Behavioural Verilog
   description:

// The D type Flip-Flop.   // This is the description of the
                              D Type Flip-Flop.
module D_FF(q,d,clk,rst);  // This is defined as a module
                              D_FF with one output
       output q;           // called q,
       input d,clk,rst;    // three inputs d, clk and a reset
                              rst.
       reg q;              // then q defined as register type
                              for storage.
       always @ (posedge clk or negedge rst)  //Always run the
                                                if else code if
                                                clk=1 or
                           // always trigger following at
                              positive edge of the clock, or the
                           // negative edge of rst.
       if (rst==0)         // rst=0
            q <=1'b0;      // if rst=0 reset Flip-Flop Q=0.
       else
            q <=d;         // else make q=whatever D input is.
endmodule                  // defines the end of the D type
                              Flip-Flop.

// Note that the always @ statement id is a parallel
   operation.

// The fsm module. This describes the FSM.
```

```
module fsm(s,clk,rst,P,L,A,B); // defines all inputs and
                                  outputs.

// Now define the FSM module used in the design. This is what
   is called the
// Data structure of Verilog HDL, and uses the Boolean
   Algebra equations A1 and A2.

 input s,clk,rst;          // First define the inputs to the
                              FSM s, clk, and rst.
 output P,L,A,B;           // Now define the FSM outputs P, L,
                              A and B.
 wire ad,bd;

// now we need to define the Flip-Flops used in our design.
 D_FF FFA(A,ad,clk,rst); // Now define an instance of D_FF FFA.
 D_FF FFB(B,bd,clk,rst); // Then an instance of D_FF FFA
// What follows under assign are the Boolean Equations we
   defined A1 and A2.
// Remember, in Veriog HDL, AND (.) is &, OR (+) is |, and
   negate (/) is ~.
// Also note that when using the assign block all Boolean
   lines end with a comma ',' (apart from
// the last line ending with a semicolon ';'.

Assign
     ad=~B&s | A&~B,       // A1
     bd=A | B&s,           // A2
     P=A&~B,               // The P output
     L=B;                  // The L output
endmodule

// Note the ',' after each equation, and the ';' in the last
   L=B statement.
// This saves having to type assign in each line.

// Now need to define the tests we want the simulator to
   follow.

// The Test Bench module.

module test;
      reg s,clk,rst;
      fsm single_pulse(s,clk,rst,P,L,A,B);
```

```
initial
begin
$dumpfile("single_pulse.vcd"); // dump file to pulse.
                                  vcd.
$dumpvars;                     // dump all variables.
                               //initialize the inputs.
s=0;                           // set s to zero.
rst=0;                         // rst=0 so should not
                                  move from s0.
clk=0;
//should stay in s0 since rst still on.
#10 clk=~clk;                  // the #10 is delay
                                  10ns then invert clk
                                  signal.
#10 clk=~clk;                  // do it again so as to
                                  create a single
                                  clock pulse.
//release rst, should stay in s0.
#10 rst=1;                     // after a further 10ns
                                  make signal rst=1.
#10 clk=~clk;                  // now do another clock
                                  pulse.
#10 clk=~clk;                  // on change from s0.
// set s to 1 to move to s1 after 10ns.
#10 s=1;
#10 clk=~clk;
#10 clk=~clk;
// move to s2 on next clock pulse.
#10 clk=~clk;
#10 clk=~clk;
// and on to s3 on next clock pulse.
#10 clk=~clk;
#10 clk=~clk;
// should stay in s3 on next clock pulse
// since s still 1.
#10 clk=~clk;
#10 clk=~clk;
// let s=0 to allow fsm to return to s0
// on the next clock pulse.
s=0;
#10 clk=~clk;
#10 clk=~clk;
// go around the loop again using
// repeat loop with 4 clk pulses.
```

```
        #10 s=1;
        repeat(8)
                #10 clk=~clk;
                //#10 clk=~clk;
                // back to s3
        #10 s=0;
        #10 clk=~clk;
        #10 clk=~clk;
        // back to s0.
        // finish the simulation.
        #10 $finish;                // Finish the simulation
        End                         // End all the tests
        Endmodule                   // End the Verilog HDL file
```

**Listing A2.1** The Verilog HDL code.

**Note that in some Verilog HDL listings you may find inputs in capital letters and outputs in lower case. This is OK as long as you keep to the same letters throughout.**

So you should be able to follow the Verilog HDL code as included comments (//) are added to help you follow the key parts of the file.

**Note that the test bench part of the code was obtained by following the state diagram in Figure A2.2 around the loop and selecting tests to perform.**

## A2.2 TEST BENCH MODULE AND ITS PURPOSE

Figure A2.4 illustrates the arrangement of the test bench module in relation to the FSM module, from which it can be seen that the test bench module provides test signals (outputs that are registered) to the FSM module. These outputs from the test bench module are used to test the FSM by applying the signals *s*, rst, and clk in such a way as to test the operation of the state diagram and hence the FSM.

The test bench sequence is created by observing the requirements of the state diagram and applying the signals *s*, rst (reset), and clk so that it can test for all conditions. It is the test bench module that will define the sequence of signals that will be applied to the FSM in order to verify that the state diagram structure is followed correctly. Test bench modules can be defined in a much more concise form than the one shown here.

If you want to learn more about these, see Chapters 6–8 of Minns and Elliott (2008) or read an account of Verilog HDL. You could always look at other examples used in this book.

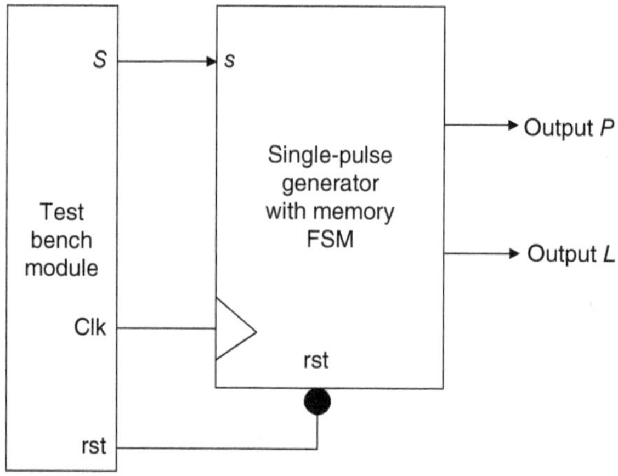


**Figure A2.4** The addition of the test bench to the FSM.

## A2.3 USING SYNAPTICAD SOFTWARE

The SynaptiCAD software is easy to use and is described here. There is a different approach shown later.

1. Choose Start>SynaptiCAD>Simulation Debug>VeriLogger Extreme + BugHunter menu to launch the graphical debugger.
2. Notice that the HELP>BugHunter VeriLogger Manual menu launches a help program with full simulator instructions.
3. Also note that the Help>Tutorials>Basic Verilog Simulation is a tutorial on how to use the graphical interface and test bench generator features of the VeriLogger Extreme.
4. Create a project to store the list of files to be simulated.
5. Choose the Project>New Project menu to open the **new project wizard** dialog (see Figure A2.5).

It is best to use this carefully if you have not much experience using Verilog HDL.

You need to provide the project directory path and its location. Also, make sure that you select the project language (Verilog) and the simulator (VeriLogger Extreme).

You then need to select the correct Verilog HDL **.v file.** In this case navigate to the file **sngplsmem.v** using the Editor.

Now you need to do the following if you are using the Verilog HDL code.

In Project Hierarchy, select **User Source file .hpj** copy to HDL files to source folder, as shown in Figure A2.6. Or create a file called **.v** by choosing the

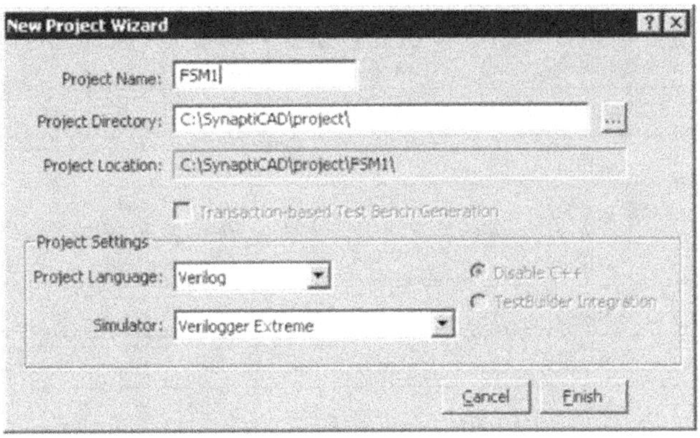


**Figure A2.5** Screenshot of new project wizard using SynaptiCAD.
*Source:* SYNAPTICAD SALES, INC.

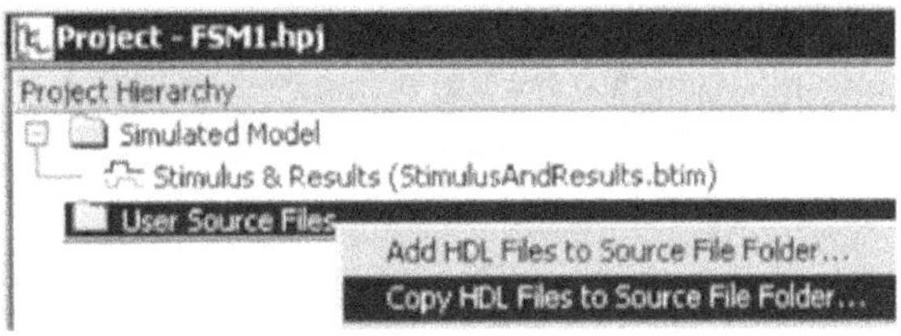


**Figure A2.6** Screenshot of project window to show some of the options available.
*Source:* SYNAPTICAD SALES, INC.

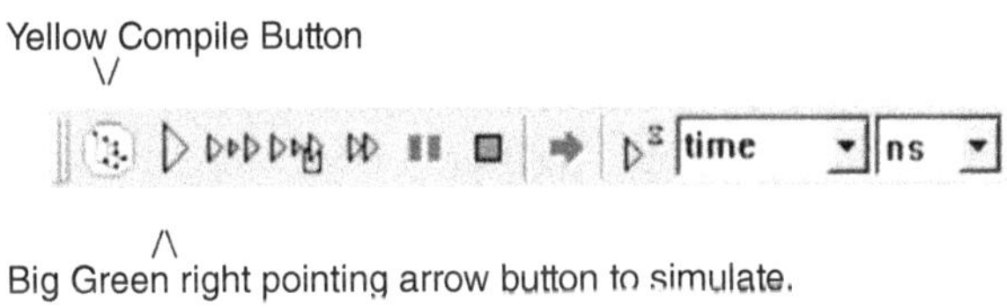


**Figure A2.7** The yellow compile and green simulate buttons in SynaptiCAD.
*Source:* SYNAPTICAD SALES, INC.

Editor>New HDL file option menu and open the file to copy and paste (CTRL C and CTRL V). Now add the file to the project by right-clicking the user source file folder and choosing 'Add HDL File' to your project.

You should see the project name, so now use Figure A2.7 to compile and simulate the file under SynaptiCAD. In Figure A2.7 you need to select the Yellow Compile button, then the big right pointing arrow to run the simulator.

Simulate with the 'Big Green' right pointing arrow button. You should now be able to see the waveforms from the simulator. These are shown in the Figure A2.8.

In Figure A2.8 you can see the expected simulation of the code in Listing A2.1.

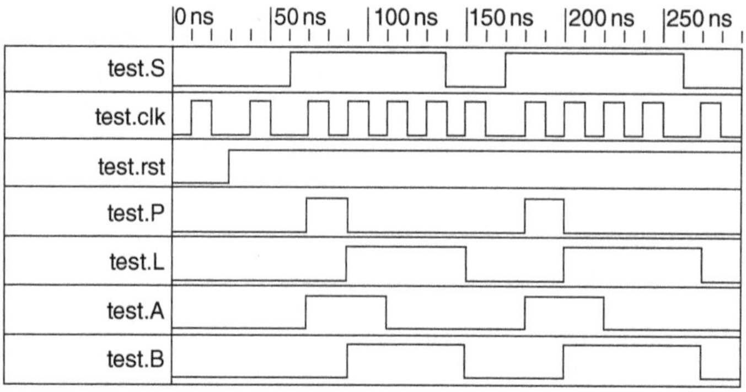


**Figure A2.8** Simulation of the FSM in SynaptiCAD.

The *s* button (shown as 'S' in Figure A2.8) is pressed to cause the FSM to move into s1 (*A/B*) then in the next state s1 (*A·/B*) the output *P* = 1. The FSM then moves into s2 (*AB*) where *P* = 0, then on into s3 (*/AB*). There it waits for *s* = 0 to move into s0 (*/A/B*) again to wait for another *s* = 1 to occur.

Note this is a static simulation showing the timing relationship. That is, it is not **dynamic**, as you would see if you were using Logisim (see later).

## A2.4 MORE DIRECT METHOD

The following method is one that you could also use as it is a direct method which should work on any version of SynaptiCAD that you download from the Web using www.syncad.com, then select SynaptiCAD and choose 'downloads'. You can also browse through the literature to find out more about the program.

So now let us look at how it could be used.

**Assume that your examples are all kept in folder examples, or it might be examples/FSM_based_Digital_Design, whatever.**

These are usually of type **.hpj** (project files) or **.v** (Verilog files). If you have just made up a .v file in Verilog then the .v file format is what you will have.

Here is the sequence:

1. When you first enter the SynaptiCAD system, select 'Projects' and then 'New Project Wizard'.
2. Select the project you want to use by navigating to the example files that you have on your computer. They are under the Projects menu and are in .hpj format.
3. When you have found the project you are looking for, place it into the Projects window.
4. Now click on the 'Compile' then 'Simulate' buttons (see Figure A2.7) and you will see the timing waveform diagram in the waveform box of Figure A2.8.

**Note: make sure you close down SynaptiCAD after each file you simulate so as to empty the project file window box. Otherwise, this will still be selected.**

You could instead do the following if you are making a new example not in the Examples folder.

1. Use an Editor (could be part of Windows or the MAC) to create a new .v Verilog HDL file and store it into the Example folder (or somewhere else if you like).
2. Now you can store this using the Project Wizard to create a new file in the Project window. Assume that here you are making use of the .v Verilog HDL file that you may have written yourself (you could also select the .v file in Edit from the files already there).
3. Choose the .v file you want to work with.
4. You can click on this file if you want to view it (optional).
5. Move this file by clicking on the (-) tab to move it out of view.
6. Select 'Projects' then 'AddFile(s)'.
7. Click on the .v file (for example you might have selected the Motor Fault FSM so when you open this folder you see motfl.v). Now click on it and it will move into the Projects window.
8. Now click on the 'Compile' and then 'Build', as shown in Figure A2.7.

You should see the Diagrams window now has a waveform displayed in it showing the Verilog HDL simulation like in Figure A2.8.

**Note: remember to make sure you close down SynaptiCAD after each file you simulate so as to empty the project file window box.**

Now that you have done this, try performing it a few times to get to know how to use the system. After a while, you will find that it is quite easy.

The test bench code at the bottom of Verilog Listing A2.1 is important to understand since it is used to test out the design.

It contains Verilog HDL code to select and operate the test bench input values in sequence according to the required state diagram sequence of Figure A2.2.

## A2.5 A VERY SIMPLE GUIDE TO USING THE LOGISIM SIMULATOR

### A2.5.1 The Logisim top level menu items

Run the Logisim Simulator program. You should see the top labels running along the Logisim window, as shown in Figure A2.9.

These run along the top of the program and are labelled as:

**Logisim File Edit Project Simulate Window Help**

Click on any of these to open up the appropriate menu. For example, open **File** to reveal other options like **New**, **Open, Close**, **Save**, and so on.

The **Simulate** option, when opened, contains **Simulate Enabled**, **Reset Simulator**, **Tick Once**, **Tick Enabled**, **Tick Frequency**, and **Logging**. More on this one later.

The **Help** option is very useful and contains items such as **Search**, **Tutorial**, and **Library**.

Figure A2.9 illustrates the Logisim Simulator program with a view of the one-pulse FSM using two *D* type flip-flops. Later you could try to simulate this example. The wording shown in the circuit helps you to follow how to use it.

1. The 'Hand' icon is for using the inputs and looking at the logic level on the wires.
2. The 'Arrow' (left-pointing arrow) icon is pressed if you want to make changes to the circuit and is usually used for editing text in the circuit.
3. The left-pointing arrow icon is for editing the circuit diagram (i.e. gates, wires, etc).
4. The square icon is to select 'Input Pins' that you want to add to the circuit.
5. The circle icon is for adding 'Output Pins' to the circuit.
6. Then follows selection of gates, such as 'Inverter', 'AND', or 'OR' gates. However, in the left-hand side of Figure A2.9 (in blue) you can see a number of menus for selecting different components for the circuit. The item marked 'Gates' can be pressed to open and view all the gate types. The item marked 'Memory' contains all the flip-flop types.

Figure A2.9 illustrates what the top window might look like with a circuit (right-hand side) showing in the main circuit window.

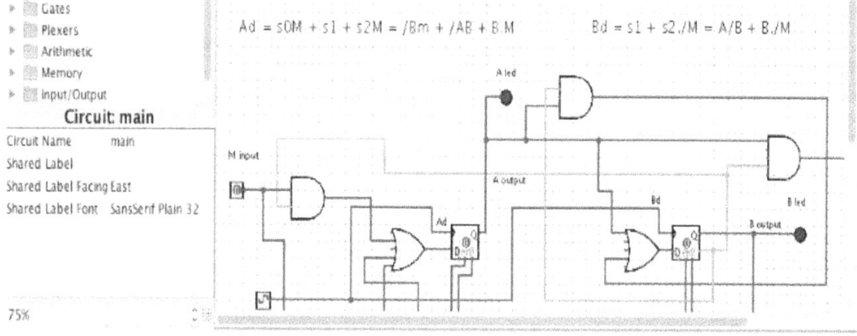


**Figure A2.9** Screenshot of Logisim window with menu and part of circuit. *Source:* Carl Burch.

In the Top Menu window you can see some of the options. These are:
**Logisim File Edit Projects Simulate Windows Help**
In the top left-hand window you can see the following options:

- **Main** – This is the set of menu items listed below.
- **Wiring** – For adding wires and other things to do with wiring.
- **Gates** – All the different type of gates you can use in your circuit.
- **Plexers** – These are multiplexers you can use.
- **Arithmetic** – Adders subtractors.
- **Memory** – For flip-flops.
- **Input/Outputs** – For the circuit.
- **Bases** – Is used for editing any text you might want to add to the circuit (not seen in Figure A2.9). This is what was used to make the text in Figure A2.10.

The idea is to create a circuit of the system you want to simulate and use the **Hand Tool** to simulate the circuit by pressing inputs in the circuit diagram to make things happen.

Incidentally, the circuit partially shown in Figure A2.9 is the circuit called: figA2.14b **onepulsedff.circ** that you can open later and try to run (see Section A2.7).

The idea is to create a circuit of the system you want to simulate and use the **Hand Tool** to simulate the circuit (top left-hand side hand tool in Figure A2.9), and by pressing inputs (the next pointer icon to make things happen). The A icon is for editing text in your circuit diagram. Beyond this are the circuit input and circuit output tools. Play around with these later to become familiar with them, as you will use them a lot.

## A2.6 USING FLIP-FLOPS IN A CIRCUIT

In the Memory Item (left-hand window in Figure A2.9), you will find the different types of flip-flops, such as *D* type (DTT), *T* type (TFF) and *JK* type (JKFF). In this book we tend to use *D* type and *T* type flip-flops.

Figure A2.10 illustrates what a DFF looks like in Logisim. This can be difficult to see in other Logisim diagrams, unless you make them bigger.

You should look carefully at the circuit of Figure A2.10 to find out how to use the Logisim *D* type flip-flops. It is easy once you get familiar with it.

**You can load up Logisim and look in the Help item for more detailed instructions on how to use Logisim. For example, when you select a gate from the Gates menu you select the logic gate with five inputs (Figure A2.11).**

You can select the gate in the menu item (in our case it is an AND gate), then in the main circuit area (with the gate still selected) select number of inputs (see Figure A2.12) and click on this to change it to provide, say, two inputs (or whatever

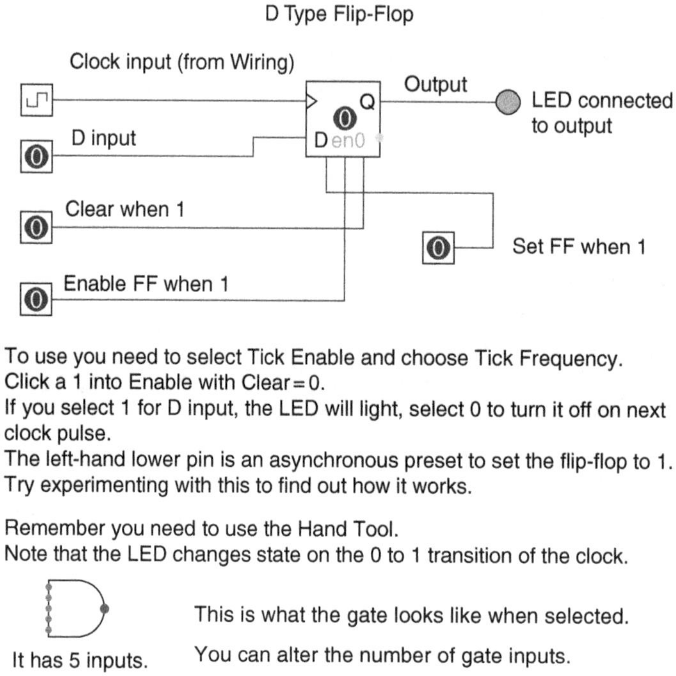


**Figure A2.10** The operation of a *D* type flip-flop.

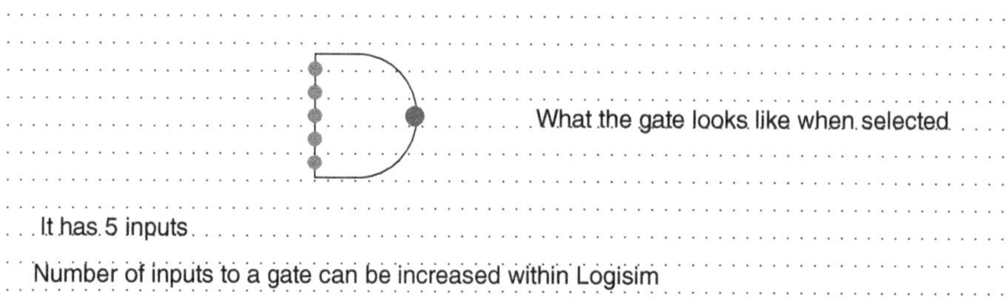


**Figure A2.11** What a logic gate will look like on the screen.

the number of inputs you want). The number of inputs can be over five, to accommodate larger designs if required.

This can be used to change the number of inputs to only two, or any number (greater than five).

**Also note that the input in Figure A2.12 'Output Value' can be used to change the 'Active' output of the gate to active high or active low.**

As you can see, there are other things you can do with the gate symbol.

**Figure A2.12** The gate menu and some of the items that can be changed. *Source:* Carl Burch.

Remember, when you are making up a circuit you need to select the left-pointing **Arrow** symbol (next to the Hand symbol) to make up the circuit, then when you want to simulate it you need to use the **Hand** symbol.

Note you can also make use of the Help menu to find out more about Logisim.

This is a very useful tutorial showing you how to use the Logisim system (see Figure A2.13).

It is worth reading though the beginner's tutorial to find out how to use Logisim (or if you really get stuck!).

There are lots of other areas you can look at, but it is best to read though the tutorial section and try a few examples it order to learn how to use the simulator before you start. You could try their examples first.

## *A2.7 EXAMPLE SINGLE-PULSE FSM*

This uses *D* type flip-flops.

Consider Figure A2.14a.

In Figure A2.14a there is an input *s* and an output *P*. The state diagram is similar to Figure A2.2 but with only a single output. In this case a press of $s = 1$ causes the FSM to move to s1 where output $P = 1$, then on the next clock pulse the FSM moves

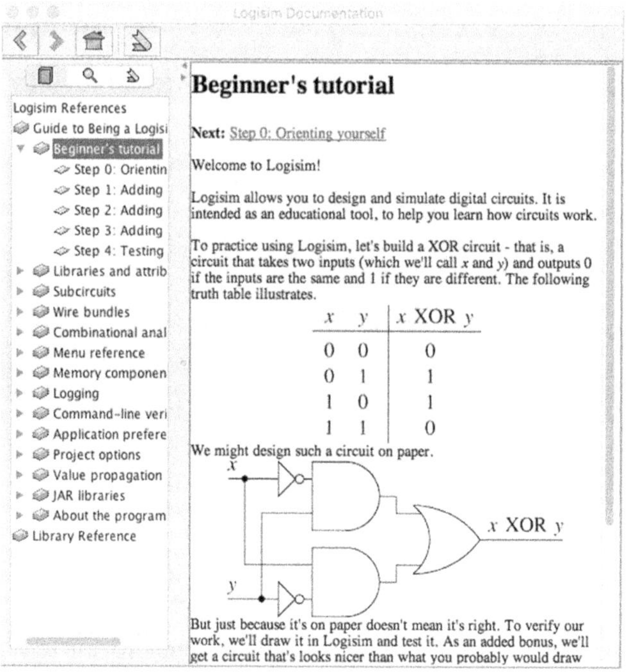


**Figure A2.13** Part of the beginner's tutorial.
*Source:* Carl Burch.

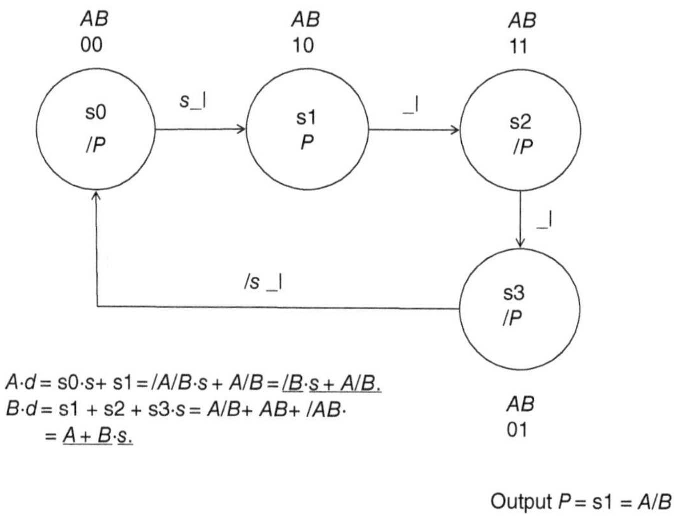


**Figure A2.14a** The state diagram with design equations.

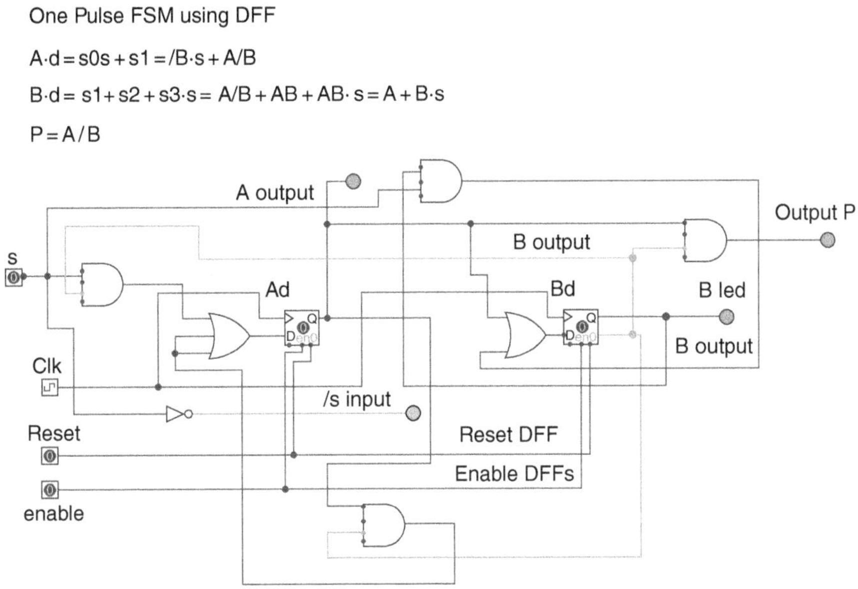


**Figure A2.14b** Complete circuit of the FSM drawn in Logisim.

through s2 to s3 where it waits for input *s* to go to zero before moving back to s0 to wait for *s* = 1 again.

The equations for *A·d* and *B·d* are shown and minimized to those in Figure A2.14b. Figure A2.14b illustrates the Logisim circuit details.

Find and select the **A2.14b onepulsedff.circ** file for this and click on it to open.

Assuming that you have installed Logisim onto your computer (Windows, MAC, or Linux), the system will start up Logisim and you should see in the Logisim window a figure the same as Figure A2.14b: a single-pulse FSM together with the equations.

The size can be altered by clicking on the item in the left-hand of the screen with the up and down arrows (shown just before the bottom of the circuit window). See help for further guidance.

Click on the HAND icon and now click on the **Simulate** then click on the **Tick Enable**, you will notice that the clock is changing between dark green (0) and light green (1) as the clock goes between 0 and 1 (assuming the clock frequency is slow enough to see this).

Now click on **Simulate** then **Tick Frequency** and select the **1 Hz frequency** to make the clock run very slowly (so you can see the clock changing state), you can change this later.

With the HAND icon still selected click on the **Enable DFF** to allow it to work (you need to do this with Logisim before anything can simulate).

## A2.8 HOW TO USE THE SIMULATOR TO SIMULATE THE SINGLE-PULSE FSM

- With the HAND icon still selected click on the *s* input.
- The *A* LED will turn on, as will the *P* output LED. The FSM is now in state s1.
- After a single clock pulse the *P* LED will turn off as the FSM moves to s2 and the *B* LED will be on.
- After another clock pulse the FSM moves to s3 where *A* LED turns off.
- Now click on the *s* input again to turn it to 0, the next clock pulse will move the FSM to s0 and turn the *B* LED off. The FSM is now in state s0.
- In this state only LED */s* will light up.

**You should try the example again a few times to become familiar with it. Try comparing it with the state diagram illustrated in Figure A2.14a.**

To end the simulation, click on **File** then **Exit** to terminate the Logisim Simulator.

### A2.8.1 Using Logisim with the truth table approach

It is possible to use Logisim with the truth table approach.

Look under the window menu item and open the Combinational Analysis item. Figure A2.15 shows what you will get.

In this you will find an 'Inputs', 'Outputs' and 'Table' tab to use to access input/output items of your design. These can be added to your design, as shown in Figure A2.15. You can enter Boolean equations using the 'Expression' tab in Figure A2.15.

When entering Boolean expression you can use:

<table>
<tr><th>Verilog Symbols</th><th>Logic</th><th>Precedence</th></tr>
<tr><td>~ ! NOT</td><td>NOT</td><td>Highest precedence</td></tr>
<tr><td>&, &&, (none)</td><td>AND</td><td></td></tr>
<tr><td>^ XOR</td><td>EXOR</td><td></td></tr>
<tr><td>+ ! | ||</td><td>OR</td><td>Lowest precedence</td></tr>
</table>

So *A* & *B* is the same as *AB*, and */AB* is the same as *~AB*.

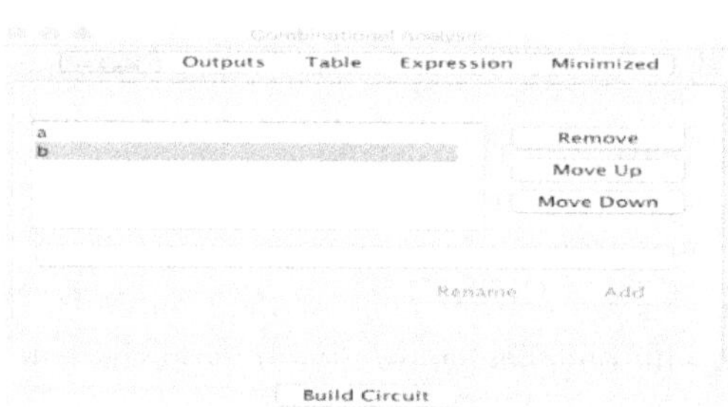


**Figure A2.15** What a Logisim screen might look like using the truth table approach
*Source:* Carl Burch.

## A2.9 USING LOGISIM WITH THE TRUTH TABLE APPROACH

Truth table:

| | *A* | *B* | *S* | *P* |
|---|---|---|---|---|
| s0 | 0 | 0 | 0 | 0 |
| s1 | 1 | 0 | 1 | 1 |
| s3 | 0 | 1 | 1 | 0 |
| s2 | 1 | 1 | 0 | 0 |

The state diagram sequence flow is s0 → s1 → s3 → s2.

The truth table is in binary form as s0 → s1 → s2 → s3 (i.e. *AB* 00 → 10 → 01 → 11).

**Remember: the state diagram should follow a unit distance coding to avoid glitches in the outputs and also avoid potential race issues as it follows the state diagram sequence.**

**A lot of people tend to make use of the binary sequence when using clocked-driven FSMs. But, as already stressed, the unit distance coding is more reliable. Note however, when using event-driven FSMs that the unit distance sequence must ALWAYS be used.**

So you need to establish the state diagram flow.

**Finally, it is suggest that you make use of the state diagram to obtain the equations, then use these to obtain the Verilog data flow for the Verilog HDL (as used in Chapters 1–3). Or translate the equations into logic gates (as used in Chapter 4), or draw the Logisim circuit. Always try to use unit distance coding as this can result in fewer binary transitions (reducing power levels).**

One point: Logisim is constantly being updated so it is worth looking on line for the latest version.

I hope you find this all useful and interesting. I think you may well find you will be using it a lot!

Figure A2.16 shows a circuit produced by **Logisim-Evolution** using TFFs as an up counter connected to the scope. **Note the different flip-flop arrangement used by Logisim-Evolution.**

Figure A2.16 illustrates the use of the oscilloscope in Logisim-Evolution.

In Figure A2.16 you can see that the user can change the number of channel inputs to the oscilloscope.

One of the problems encountered when using Logisim to draw circuits of FSM systems is the possibility of needing many wire loops in a circuit. This can make the FSM look very confusing, particularly in more complex FSM systems.

Logisim has a means of helping by using **Tunnels** (see **Wires → Tunnels** in **Logisim**).

The Tunnel provides a means to connect individual wires between places without having to draw wires between them (try using **Tunnel** in Logisim Help).

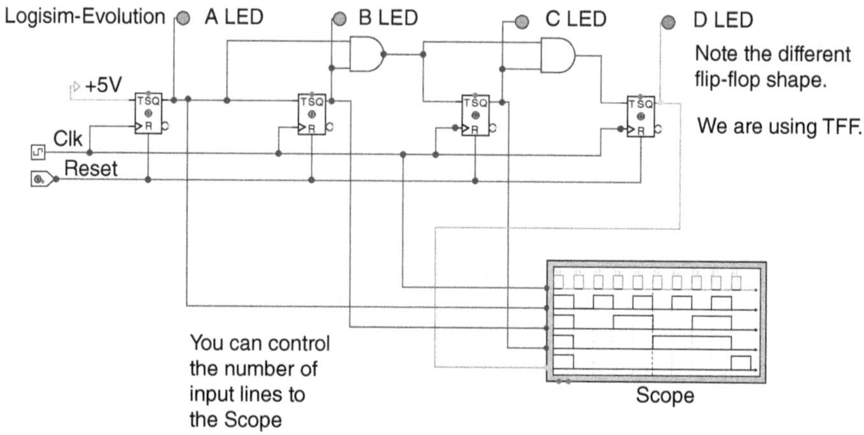


**Figure A2.16** Example screenshot of counter with use of oscilloscope.

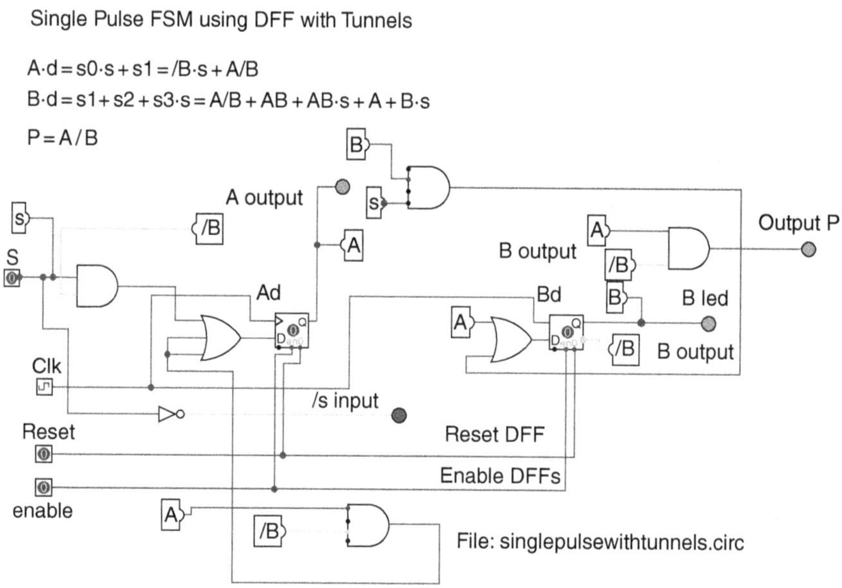


**Figure A2.17** The same FSM as Figure A2.14b but using Tunnels.

Using **Logisim-Evolution** for the one-pulse FSM you need to alter the Figure A2.14b so that it uses the correct flip-flops and Tunnels.

Figure A2.17 uses Burch's Logisim program (older flip-flop version) for the example of Figure A2.14b but using Tunnels. This makes the circuit easier to follow.

As you can see this looks a lot clearer. More on this later.

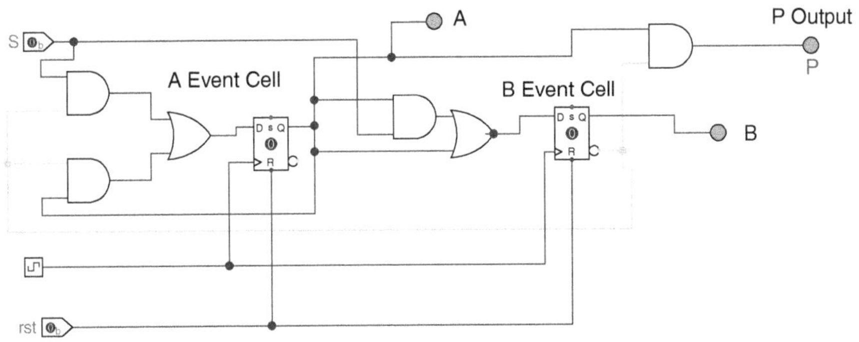


To Operate: rst = 0, s = 1, then s = 0

The Single-Pulse DFF Clocked FSM using Logisim-Evolution

SinglePulsedffevol.circ

**Figure A2.18** This version uses Logisim-Evolution – note different flip-flops.

Figure A2.17 is similar to the circuit in Figure A2.14b. Figure A2.18 shows the same FSM using Logisim-Evolution (without Tunnels). Note the different *D* flip-flop symbols used in this design.

Tunnels can be used in Logisim-Evolution, so you could try this yourself. Also you could try adding **Tunnels** to the *AB* SSVs in examples.

Then add more **Tunnels** as inputs to gates connected elsewhere to connect these other points without the need for additional wire loops. There are examples using Tunnels in the Logisim files that can be entered into Logisim (and indeed some examples use Logisim-Evolution).

Figure A2.19 shows an example of Logisim (not Logisim-Evolution) using a Shift Register with Tunnels.

The best way to see this is to follow the instructions in Figure A2.19. The example in Figure A2.19 is called **shift_reg_with_tunnels.circ** in Logisim examples.

### *A2.9.1 Useful note*

To obtain Logisim-Evolution, visit: **https://github.com/reds-heig/logisim-evolution/releases/tag/v2.150**. Also, visit Google and type in **Logisim-Evolution** and this will also bring you to the same website.

## *A2.10 SUMMARY*

This tutorial looks at only one of many ways in which to develop a Verilog description of an FSM design. Other ways are discussed in Chapters 6–8 of Minns and Elliott (2008).

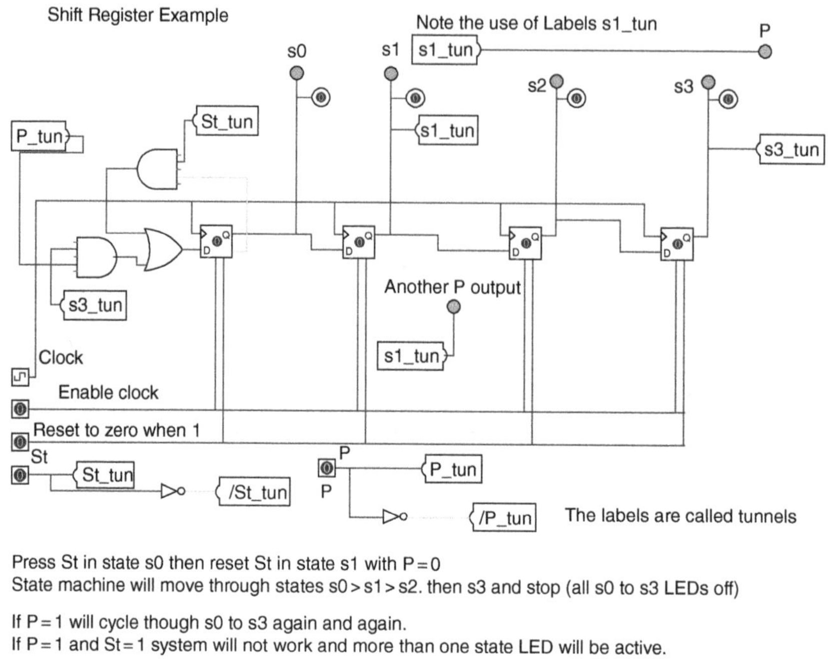


**Figure A2.19** Sample of a shift register using Tunnels.

This book can be obtained from Wiley, or chapters can be bought for a lesser fee if desired. The e-book can be obtained also from Wiley in .pdf format.

The design method is very easy to apply, with all the design information contained within the state diagram. This information can then be extracted via the equations in order to synthesize a given design. Simulation of the circuit can then be used to confirm the design. The latest versions of VeriLogger Extreme may be updated from time to time from SynaptiCAD at http://www.syncad.com.

In addition, the Logisim Logic Simulator has been introduced as an alternative way of testing FSM-based systems. This is a very easy way to test out designs before you commit them to a circuit, a programmable logic device (PLD), complex programmable logic device (CPLD), or a field programmable logic device (FPLD).

You should remember that, when using these programmable devices (most commonly used with Verilog HDL), the propagation delays in such devices is usually a lot smaller than with devices such as TTL and CMOS and easier to control.

The original Logisim developed by Carl Burch is no longer being developed but you can still download it for free.

However, a consortium of people have decided to continue with its development calling it **Logisim-Evolution**. At the time of writing this book the best version of

Logisim-Evolution is version 2.15 and can be downloaded from **https://github.com/reds-heig/logisim-evolution/releases/tag/v2.150**.

One point, however, is that the Logisim-Evolution uses different *D* type flip-flops that will not work with the clocked FSM in this book without modifying the circuits. Once changed, they will work OK.[1] I have tried some of the event FSM (including both event sequential equations and basic one hot event) and they appear to work OK unchanged. This new version includes ITA components such as an oscilloscope that could be very useful.

Another addition is the introduction of VHDL code import, export, and development under HDL-IP. Verilog is still under development. This could develop into an interesting feature.

Yes, you can install Logisim-Evolution with Carl Burch's Logisim version 2.7.1 but best to keep it in its 'Logisim-Evolution.jar' form. This avoids confusion between the two Logisim versions.

**To access the Logisim Simulator files in this book follow the link:** www.wiley.com\go\minns\digitalsystemdesign.

[1] Actually all flip-flops in Logisim-Evolution are different from those used in Carl Burch Logisim and if you try to open with the older version the flip-flops will be missing.

# Appendix A3: Counters, Shift Registers, Input, and Output with an FSM

This appendix contains a number of techniques to help in the development of synchronous binary counters and shift registers. These are used in some of the designs covered in various chapters throughout this book. There is also an example of the memory tester FSM using the Logisim Simulator.

## A3.1 BASIC DOWN SYNCHRONOUS BINARY COUNTER DEVELOPMENT

The development of synchronous pure binary up/down counters can be mechanized to produce a general $n$-stage pure binary counter. This can then be implemented directly using PLD/CPLD/FPGA devices. To illustrate how this is achieved a four-stage down counter is described in Table A3.1.

Table A3.1 represents a down counter with Q0 the least significant bit. This counter is to be designed as a synchronous counter so all flip-flops will be clocked by the same clock edge. Also, the flip-flops will be $T$ flip-flops. Most CPLDs and FPGAs can support the TFF, either directly or using $D$ type flip-flops with an exclusive OR input.

The equation for the $T$ input of each flip-flop can be obtained by inspection of Table A3.1 and entering a product term for every 0 to 1, and 1 to 0 transition required by each flip-flop.

*Digital System Design using FSMs: A Practical Learning Approach*, First Edition. Peter D. Minns.

Companion website: www.wiley.com/go/minns/digitalsystemdesign

**Table A3.1** The binary down counter.

| Q0 | Q1 | Q2 | Q3 | State |
|---|---|---|---|---|
| 1 | 1 | 1 | 1 | s15 |
| 0 | 1 | 1 | 1 | s14 |
| 1 | 0 | 1 | 1 | s13 |
| 0 | 0 | 1 | 1 | s12 |
| 1 | 1 | 0 | 1 | s11 |
| 0 | 1 | 0 | 1 | s10 |
| 1 | 0 | 0 | 1 | s9 |
| 0 | 0 | 0 | 1 | s8 |
| 1 | 1 | 1 | 0 | s7 |
| 0 | 1 | 1 | 0 | s6 |
| 1 | 0 | 1 | 0 | s5 |
| 0 | 0 | 1 | 0 | s4 |
| 1 | 1 | 0 | 0 | s3 |
| 0 | 1 | 0 | 0 | s2 |
| 1 | 0 | 0 | 0 | s1 |
| 0 | 0 | 0 | 0 | s0 |

For example, from Table A3.1 the equation for flip-flop Q0·*t* will be:

$$
\begin{aligned}
&0>1\\
&Q0{\cdot}t = s15+s14+s13+s12+s11+s10+s9+\\
&\qquad s8+s7+s6+s5+s4+s3+s2+s1+s0=1.\\
&1>0
\end{aligned}
$$

Each state where the *T* flip-flop is to change state (0 to 1, or 1 to 0) is entered into the equation.

This can then be written in terms of the Q0Q1Q2Q3 outputs, or simply entered into a Karnaugh map, as illustrated in Figure A3.1.

| Q2Q3 \ Q0Q1 | 00 | 01 | 11 | 10 |
|---|---|---|---|---|
| 00 | s0 | s2 | s3 | s1 |
| 01 | s8 | s10 | s11 | s9 |
| 11 | s12 | s14 | s15 | s13 |
| 10 | s4 | s6 | s7 | s5 |

Karnaugh state map showing all states

**Figure A3.1** A four-bit Karnaugh map and its states.

The Karnaugh state map of Figure A3.1 can then be used to help to minimize the flip-flop equations.

**If you have never used a Karnaugh map before, you can do a Google search to find out how to use them. However, you will be using Karnaugh maps again further on and perhaps this will show you more clearly. Also, Zissos (1979, p. 213) has a very good account.**

**Essentially, owing to the unit distance labelling for Q0Q1 and Q2Q3, combined rows and columns of the Karnaugh map can be used to remove the changing (0 to 1, and 1 to 0) terms (see the examples later). The columns Q0Q1 and rows Q2Q1 follow a unit distance code (Holdsworth 1993, p. 52).**

Since all cells will be filled with 1 for the Q0·*t* equation (every cell whose term appears in the **Q0·*t*** equation) then the *T* input for flip-flop Q0·*t* will be logic 1.

The equation for flip-flop **Q1·*t*** will be:

$$0>1$$

$$\text{Q1}{\cdot}t = \text{s14}+\text{s12}+\text{s10}+\text{s8}+\text{s6}+\text{s4}+\text{s2}+\text{s0} = /\text{Q0} \text{ from the state map.}$$

$$1>0$$

An inspection of the state map of Figure A3.1 shows that Q1·*t* must minimize to /Q0 since cells s14, s12, s10, s8, s6, s4, s2, and s0 all contain a 1.

You can combine all left-hand cells in Figure A3.1 together and loop them all so that all Q2Q3 terms change and Q0Q1Q2 changes, leaving only Q0 that does not change. That is Q0 = 0 or /Q0. See Holdsworth (1993) for details.

Following on in this manner Q2·*t* and Q3·*t* can be obtained thus:

$$\text{Q2}{\cdot}t = \text{s12}+\text{s8}+\text{s4}+\text{s0} = /\text{Q0}{\cdot}/\text{Q1}.$$

$$\text{Q3}{\cdot}t = \text{s8}+\text{s0} = /\text{Q0}{\cdot}/\text{Q1}{\cdot}/\text{Q2}.$$

The patterns of equations follow in a general manner and can be expressed in the form:

$$\text{Qx}{\cdot}t = /\text{Q}_{\text{x}-1}{\cdot}/\text{Q}_{\text{x}-2}{\cdot}/\text{Q}_{\text{x}-3}{\cdot}\ldots{\cdot}/\text{Q}_{\text{x}-\text{x}}. \quad \text{(A3.1)}$$

Equation (A3.1) describes the product *p* terms for a down counter implemented with TFFs. These equations can be directly entered into a Verilog HDL file for each flip-flop.

An up counter can be realized by replacing all the /*Q* terms in Equation (A3.1) with *Q* terms as shown in Equations A3.2a and b and A3.3:

$$\text{Qx}{\cdot}t = \text{Q}_{\text{x}-1}{\cdot}\text{Q}_{\text{x}-2}{\cdot}\text{Q}_{\text{x}-3}{\cdot}\ldots{\cdot}\text{Q}_{\text{x}-\text{x}}. \quad \text{(A3.2)}$$

Or in general:

$$\text{Qn}{\cdot}t = \Pi^{p=n}{}_{\text{p}=1}\left(\text{Q}_{\text{n}-p}\right). \quad \text{(A3.3a)}$$

with:

$$Q0{\cdot}t = 1. \tag{A3.3b}$$

For each flip-flop where Π mathematically is the product (i.e. AND) of each output term. Note that TFF Q0 has its *T* input at logic 1. This is not covered in Equation A3.3a, hence Equation A3.3b.

These equations can be obtained directly from a Karnaugh state map similar to that shown in Figure A3.1 but counting in the opposite direction (up counter).

**You could try to do this yourself in the same manner as the down counter to see if you can get this correct. If you still do not get it correct, try just moving on to see if other examples help you (rather than just giving up).**

## A3.2 EXAMPLE OF A FOUR-BIT SYNCHRONOUS UP COUNTER WITH T TYPE FLIP-FLOPS

Figure A3.2 illustrates a design for a four-bit up counting synchronous counter using the techniques described above.

The equations for each *T* flip-flop for the up counter are:

$$Q0{\cdot}t = 1.$$
$$Q1{\cdot}t = Q0.$$
$$Q2{\cdot}t = Q0{\cdot}Q1.$$
$$Q3{\cdot}t = Q0{\cdot}Q1{\cdot}Q2.$$

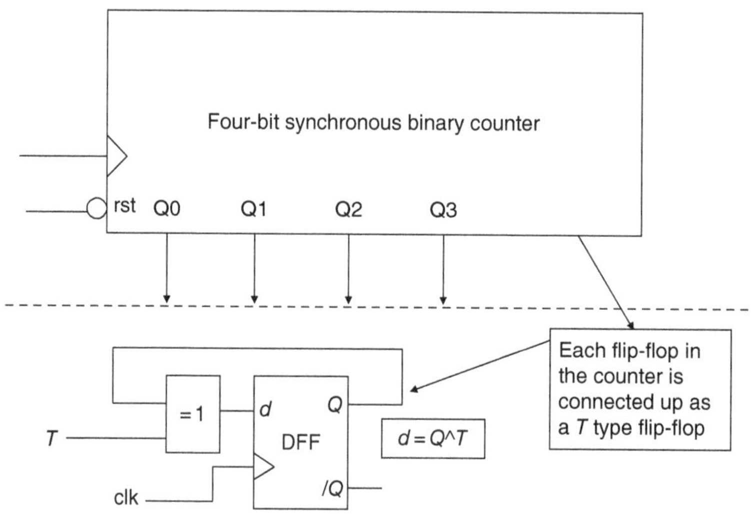


**Figure A3.2** Four-bit up counter of how each *T* type is based around a *D* type flip-flop.

The Verilog HDL source file, with the test bench module for the counter, is shown in Listing A3.1.

```
// Four bit counter design.
// Define the TFF using behavioural method.
module T_FF (q,t,clk,rst);
  output q;
  input t,clk,rst;
  reg q;                    // q output must be registered - to
remember?
  always @ (posedge clk or negedge rst)
     if (rst==0)
          q <=1'b0;
     else
          q <=t^q;          // TFF is made up with EX-OR gate.
endmodule
// Now define the counter.
module counter(Q0,Q1,Q2,Q3,clk,rst);

     input clk, rst;        // clk and rst are inputs.
     output Q0,Q1,Q2,Q3; // all q outputs.
     wire t0,t1,t2,t3;      // all t inputs are interconnecting
wires.

     // need to define instances of each TFF defined earlier.
     T_FF ff0(Q0,t0,clk,rst);
     T_FF ff1(Q1,t1,clk,rst);
     T_FF ff2(Q2,t2,clk,rst);
     T_FF ff3(Q3,t3,clk,rst);

     // now define the logic connected to each t input.
     // we use an assign for this.  What follows in the
assign block
     // is the TFF Equations given above.
     assign
          t0=1'b1,          // this is just following the
technique
          t1=Q0,            // for binary counter design.
          t2=Q0&Q1,         // will generate AND gates.
          t3=Q0&Q1&Q2;      // note the ; here.
     endmodule              // end of the module counter.

     // Test Bench design to test the circuit under
simulation.
```

```
    module test;
        reg clk, rst;  // has two inputs which must be
registers.
        // no wires in this part of the design
        // since counter is not connected to anything.
        counter count(Q0,Q1,Q2,Q3,clk,rst);
            initial
            begin
            $dumpfile("counter4.vcd");    // file
waveforms.
            $dumpvars;          // dump all values to the
file.
            rst=0;              // initialise circuit with
rst cleared.
            clk=0;              // set clk to normally low.
You could try
                                // clocking clk?
            #10 rst=1;          // after 10 time units raise
rst to remove     // reset.
            repeat(17)
            #10 clk=~clk;       // change clk 17 times every
10 time                         // units.
            #20 $finish;        // finish the simulation
after 20 time                   / units.
            end                 // end of test block.
            endmodule           // end of test module.
```

## Listing A3.1 Four-bit synchronous counter (TFF).

This contains the *T* type flip-flop definition (defined using the behavioural method).

This is followed by the counter definition, which makes use of four instances of the *T* flip-flops and also uses an assign block to define the logic connections between the flip-flop outputs and the *T* inputs of each flip-flop. Note: the old-style input and output is used outside the module header.

Following on from this is the test bench module. This contains an instance of the four-bit counter followed by the stimulus to test the counter.

Note, that there are two '$' commands to 'save' the timing diagram of Figure A3.3. This is so that it can be saved to a word document (for printout).

The command `$dumpfile`('counter4.vcd'); names the file to be created with the information.

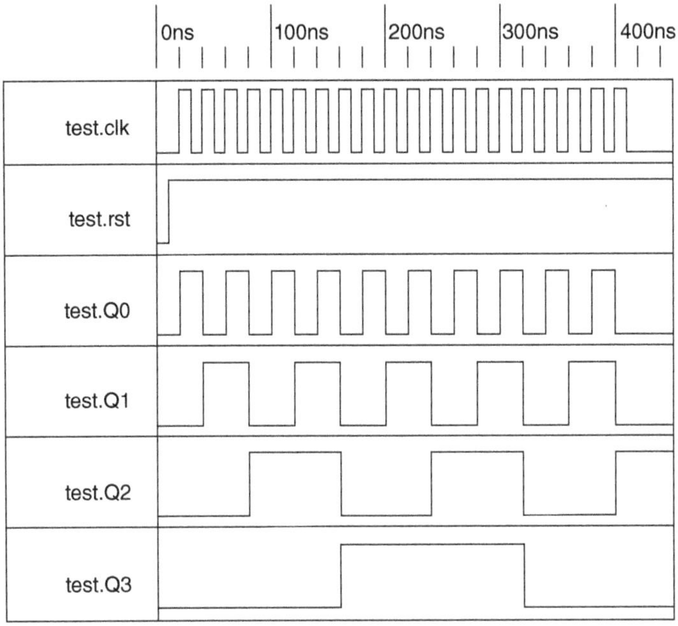


**Figure A3.3** Simulation of Listing A3.1.

The command `$dumpvars;` simply dumps all variables to the file.
The file is saved as a metafile and is illustrated in Figure A3.3.
The waveforms of Figure A3.3 clearly show the binary counter sequence.

## A3.3 PARALLEL LOADING COUNTERS – USING T FLIP-FLOPS

A synchronous parallel input may be required if there is not an asynchronous preset and clear input to the flip-flops.

See the example for this idea in Chapter 3, Frame 3.16 and Figure 3.14. Using additional product terms in the Qx·*t* equations can do this.

A general 'bit slice form' with the additional inputs is shown in Equation A3.4 for a TFFx:

$$Qx\cdot t = \text{ptermx}\cdot/\text{load} + \text{px}\cdot/\text{Qx}\cdot\text{load} + /\text{px}\cdot\text{Qx}\cdot\text{load}. \qquad (A3.4)$$

The load input is used to load the parallel data synchronously into the flip-flop. In this case the load input is active high.

In Equation (A3.4) the product term **ptermx·/load** is the normal product term needed for the counter and is true while the load input is not active. The term **px·/Qx·load** is the parallel input term to set the flip-flop, while the term **/px·Qx·load** is the term to clear the flip-flop.

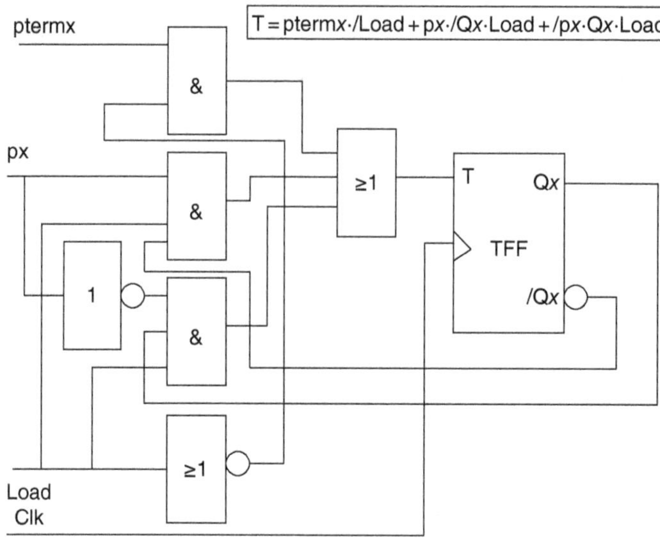


**Figure A3.4** A single flip-flop of the parallel loading counter (bit slice).

Figure A3.4 shows a general structure of a single flip-flop with parallel loading. All other flip-flops follow the same general structure. It is assumed here that the active state for the load input is high. Therefore, during counting mode, load would be low (logic 0).

Note that the TFF shown in Figure A3.4 is produced by the DFF with exclusive OR gate, as shown in Figure 3.2.

Equations (A3.1), (A3.2), and (A3.4) may be used to produce parallel loading up/down counters for many applications, including the address counters for FSMs that control memory. Thus it is possible to create not only sequential control of the access of memory but also random control by way of the parallel inputs.

## A3.4 USING D FLIP-FLOPS TO BUILD PARALLEL LOADING COUNTERS

The DFF can be used in place of the TFF to implement parallel loading synchronous counters that do not have preset or clear inputs. There are lots of cheaper PLD devices that use only DFFs and do not have asynchronous preset and clear, so the idea seems attractive.

Consider the circuit in Figure A3.5.

The bit slice equation for this general model is:[1]

$$\mathrm{Qx} \cdot d = \mathrm{px} \cdot /\mathrm{l} + \mathrm{pterm} \cdot \mathrm{l}. \tag{A3.5}$$

[1] 'Bit slice' refers to the fact that each bit is replicated to form an $n$-bit version.

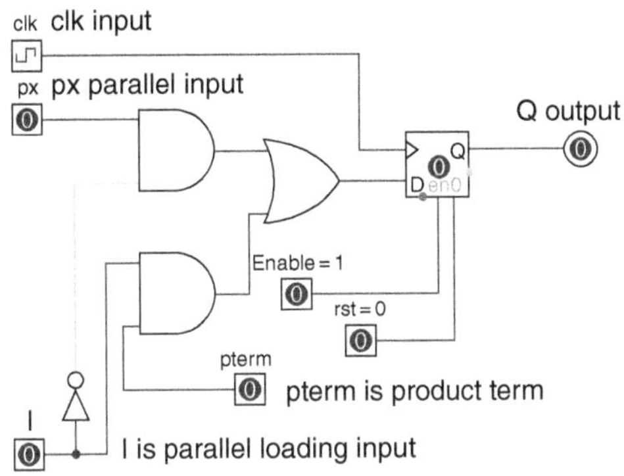


**Figure A3.5** Single bit slice of parallel loading shift register using *D* type flip-flop.

This bit slice defines the single bit of the counter, 'l' is the parallel loading input and '/l' the inverted parallel loading input.

This defines the general form for the equations for each flip-flop in the counter chain.

The individual product pterms here will depend upon the sequence table and there is no simple way to do this; therefore, the method is not as easy to implement as that using TFFs.

As an example, consider a simple three-stage synchronous binary up counter.

## A3.5 SIMPLE BINARY UP COUNTER WITH PARALLEL INPUTS

A physical circuit with a simple three-state parallel loading pure binary counter is illustrated in Figure A3.6.

Looking at Figure A3.6, a three-stage pure binary synchronous counter, the state sequence illustrates the binary sequence. The state map is used to help simplify the pterms (shown here in their simplified form), and finally the full equations for the *D* inputs of each flip-flop.

In Figure A3.6 the three full equations read:

$$\mathbf{Q0}\cdot d=\mathbf{p0}\cdot/\mathbf{l}+(/\mathbf{Q0})\cdot\mathbf{l}.$$

$$\mathbf{Q1}\cdot d=\mathbf{p1}\cdot/\mathbf{l}+(\mathbf{Q0}\cdot/\mathbf{Q1}+/\mathbf{Q0}\cdot\mathbf{Q1})\cdot\mathbf{l}.$$

$$\mathbf{Q2}\cdot d=\mathbf{p2}\cdot/\mathbf{l}+(\mathbf{Q2}\cdot/\mathbf{Q1}+\mathbf{Q2}\cdot/\mathbf{Q0}+/\mathbf{Q2}\cdot\mathbf{Q1}\cdot\mathbf{Q0})\cdot\mathbf{l}.$$

Note that, compared with the method for designing synchronous parallel loading up/down counters using *T* flip-flops, this arrangement requires the development of each flip-flop pterm. In general, there is no systematic way to do this other than work out the logic for each flip-flop.

| Q0 | Q1 | Q2 | state | |
|---|---|---|---|---|
| 0 | 0 | 0 | s0 | |
| 1 | 0 | 0 | s1 | |
| 0 | 1 | 0 | s2 | |
| 1 | 1 | 0 | s3 | State Sequence |
| 0 | 0 | 1 | s4 | |
| 1 | 0 | 1 | s5 | |
| 0 | 1 | 1 | s6 | |
| 1 | 1 | 1 | s7 | |

State Map

| Q2 \ Q0 Q1 | 00 | 01 | 11 | 10 |
|---|---|---|---|---|
| 0 | s0 | s2 | s3 | s1 |
| 1 | s4 | s6 | s7 | s5 |

$$Q0 \cdot d = /Q0.$$

pterms $$Q1 \cdot d = Q0 \cdot /Q1 + /Q0 \cdot Q1.$$

$$Q2 \cdot d = Q2 \cdot /Q1 + Q2 \cdot /Q0 + /Q2 \cdot Q1 \cdot Q0$$

For a parallel loading counter $Qx \cdot d = px \cdot /I + /Qx \cdot I$ where $x$ is 0,1,2.

**Figure A3.6** Counter sequence using *D* type flip-flops.

However, one advantage of using *D* flip-flops is that the count sequence is not restricted to pure binary count sequences (i.e. could develop unit distance code sequences, for example).

Of course, the counter could be developed directly from the Verilog HDL behavioural description direct, and this would be the more usual way of doing it. The above method, however, gives an insight into the Boolean equations involved in such counters.

## A3.6 CLOCK CIRCUIT TO DRIVE THE COUNTER (AND FSM)

There are many circuit arrangements for crystal oscillators, but the one shown in Figure A3.7 is a common one, which is often used. It is included for completeness.

The circuit in Figure A3.7 provides overtone suppression via the two capacitors C1 and C2 with values to keep the capacitive reactance small.

That is Xc1 = 1/(2·Pi·f·C), the capacitor reactance for C1. Note in Figure A3.7 there are no preset or clear inputs.

**In general, flip-flops usually have either active high or active low preset or clear inputs. If they are active low, they are usually shown with a negating circle.**

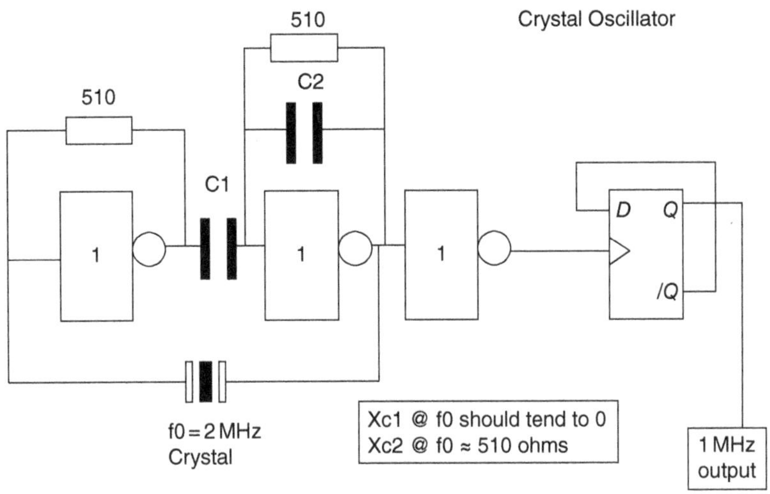


**Figure A3.7** Typical circuit of a gate crystal oscillator with overtone suppression.

## A3.7 COUNTER DESIGN USING DON'T CARE STATES

In some designs use can be made of states that do not appear in the count sequence. This can lead to a reduction in the number of gates used in the logic of the counter.

Consider the twisted ring counter, so called because it has each flip-flop connected in the form of a ring, but with a twist in the connection between the last flip-flop and the first. Figure A3.8 illustrates the state sequence and a design method using a state map to highlight **don't care states**.

The state sequence table in Figure A3.8 shows the required sequence for the twisted ring counter. From this it is apparent that states s2, s4, s5, s6, s9, s10, s11, and s13 are not part of the sequence so these are made don't care terms (marked as X) in the state map.

In Figure A3.8 the equations for the Q0·*d* to Q3·*d* outputs followed as listed here:

$$\mathbf{Q0} \cdot d = \mathbf{s0+s1+s3+s7} + \left(\textbf{don't carc terms}\right) = /\,\mathbf{Q3}.$$

$$\mathbf{Q1} \cdot d = \mathbf{s1+s3+s7+s15} + \left(\textbf{don't care terms}\right) = \mathbf{Q0}.$$

$$\mathbf{Q2} \cdot d = \mathbf{s3+s7+s15+s14} + \left(\textbf{don't care terms}\right) = \mathbf{Q1}.$$

$$\mathbf{Q3} \cdot d = \mathbf{s7+s15+s14+s12} + \left(\textbf{don't care terms}\right) = \mathbf{Q2}.$$

From the state sequence table, and state map of Figure A3.8, the equations for each flip-flop *d* input can be obtained, looking for 0 to 1, and 1 to 1 transitions in each column of the sequence table. The don't care terms are then added to the end of each equation. Finally, the state map is used to obtain the minimized equations (using, as in this case, the Karnaugh map).

| Q0 | Q1 | Q2 | Q3 | State | |
|---|---|---|---|---|---|
| 0 | 0 | 0 | 0 | s0 | Twisted Ring Counter |
| 1 | 0 | 0 | 0 | s1 | |
| 1 | 1 | 0 | 0 | s3 | |
| 1 | 1 | 1 | 0 | s7 | State sequence |
| 1 | 1 | 1 | 1 | s15 | |
| 0 | 1 | 1 | 1 | s14 | |
| 0 | 0 | 1 | 1 | s12 | |
| 0 | 0 | 0 | 1 | s8 | |

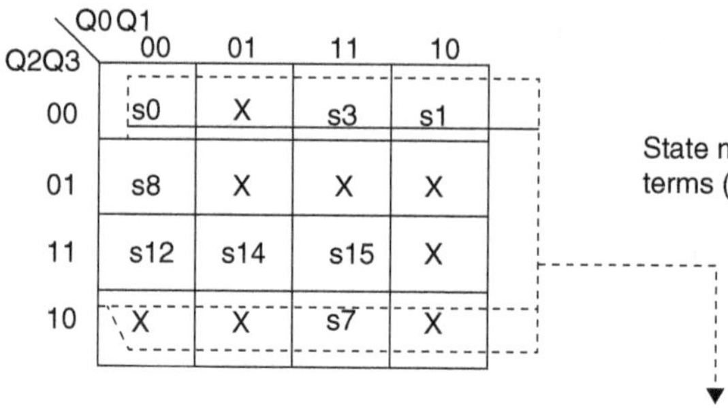


State map with don't care terms (X) included.

$Q0 \cdot d = s0 + s1 + s3 + s7 + (\text{don't care terms}) = /Q3$

$Q1 \cdot d = s1 + s3 + s7 + s15 + (\text{don't care terms}) = Q0$

**Figure A3.8** Details for the design of twisted ring counter using don't care states.

They could, of course be obtained using Boolean algebra instead.

For example, in equation Q0·*d* states s0, s1, s3, and s7 are combined with don't care terms s2, s4, s5, and s6 to obtain /Q3. The Karnaugh map makes this easy to see and minimize.

Note when using Boolean algebra you can enclose don't care terms in parentheses '(' and ')' to identify them.

For Q0·*d* entering s0 and s1 and s3 and s7 as a 1 in the Karnaugh map for Figure A3.8 will result in the reduced result /Q3.

The other equations are dealt with in a similar manner.

## A3.8 SHIFT REGISTERS

A special form of synchronous counter is the shift register. Quite often, a parallel loading shift register is required. The bit slice form for each stage of the parallel loading shift register is obtained from Equation (A3.6a) and (A3.6b) as:

$$\mathbf{Q0} \cdot d = \mathbf{din} \cdot ld + \mathbf{p0} \cdot /ld. \quad \textbf{(A3.6a)}$$

$$\mathbf{Qx} \cdot d = \mathbf{Q}_{x-1} \cdot ld + \mathbf{px} \cdot /ld. \quad \textbf{(A3.6b)}$$

Where in this case the active state for the load input *ld* is low and din is data input.

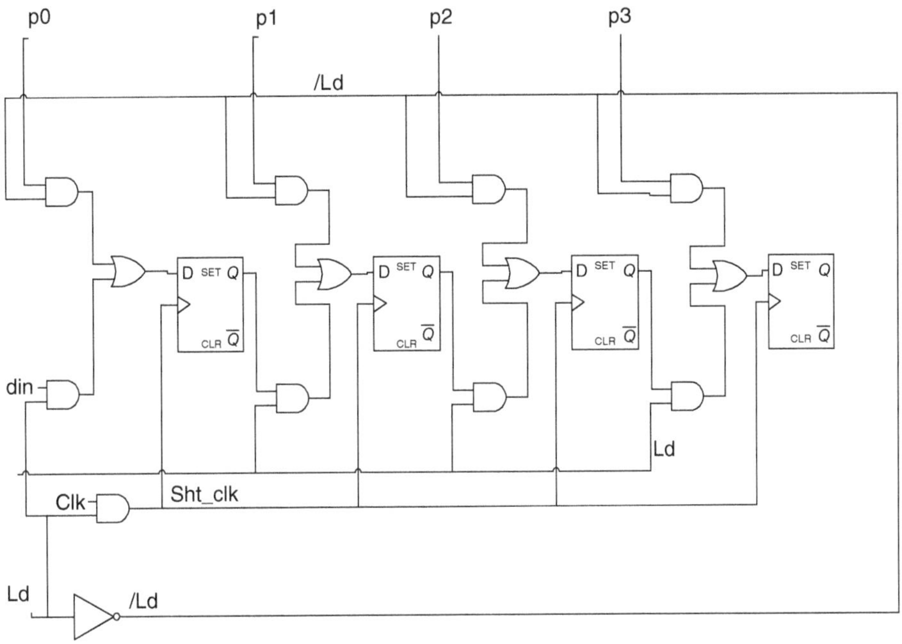


**Figure A3.9** Typical four-bit parallel loading shift register.

Note that if serial input is to be zero, make din = 0. The shift register design is using *D* flip-flops.

These equations could be used to create a four-bit parallel loading counter thus:

$$\mathbf{Q0} \cdot d = \mathbf{din} \cdot \boldsymbol{ld} \quad + \mathbf{p0} \cdot /\boldsymbol{ld}. \tag{A3.7}$$

$$\mathbf{Q1} \cdot d = \mathbf{Q0} \cdot \boldsymbol{ld} \quad + \mathbf{p1} \cdot /\boldsymbol{ld}. \tag{A3.8}$$

$$\mathbf{Q2} \cdot d = \mathbf{Q1} \cdot \boldsymbol{ld} \quad + \mathbf{p2} \cdot /\boldsymbol{ld}. \tag{A3.9}$$

$$\mathbf{Q3} \cdot d = \mathbf{Q2} \cdot \boldsymbol{ld} \quad + \mathbf{p3} \cdot /\boldsymbol{ld}. \tag{A3.10}$$

$$\mathbf{Sft\_clk} = \mathbf{clk} \cdot \boldsymbol{ld}. \tag{A3.11}$$

In Equation (A3.7) the first term is the serial data input. In Equations (A3.8)–(A3.10) the first term denotes that the output of each flip-flop will connect into the input of the next flip-flop (i.e. a standard shift register connection).

In addition Equation A3.11 defines the shift clock. This is disabled during parallel loads.

Figure A3.9 shows the four-state shift register developed from Equations (A3.7)–(A3.11). Note that in practice the equations would be converted into Verilog HDL code direct for synthesization. The equations converted into Verilog HDL are:

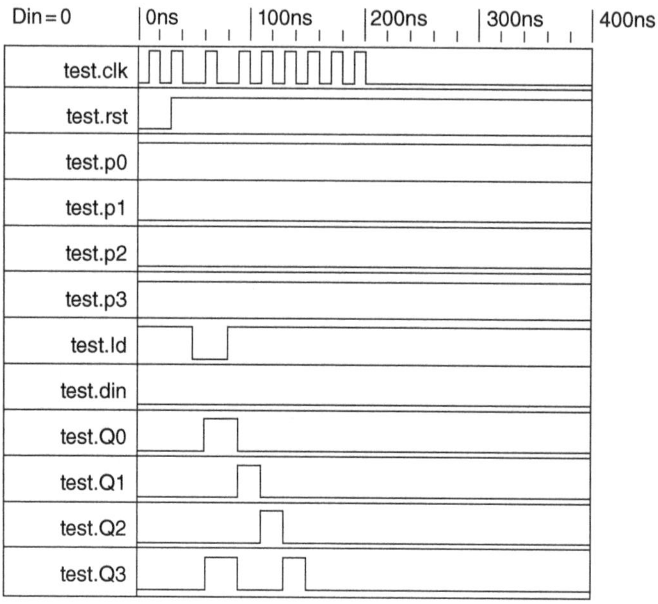


**Figure A3.10** Simulation of a parallel loading shift register.

$$\mathbf{Q0}\cdot d = \mathbf{din}\ \&\ ld \mid \mathbf{p0}\& \sim ld.$$
$$\mathbf{Q1}\cdot d = \mathbf{Q0}\ \&\ ld \mid \mathbf{p1}\& \sim ld.$$
$$\mathbf{Q2}\cdot d = \mathbf{Q1}\ \&\ ld \mid \mathbf{p2}\& \sim ld.$$
$$\mathbf{Q3}\cdot d = \mathbf{Q2}\ \&\ ld \mid \mathbf{p3}\& \sim ld.$$
$$\mathbf{Sft\_clk} = \mathbf{clk}\ \&\ ld.$$

The above shift register, once converted into Verilog HDL code, can then be simulated for correct operation. Figure A3.10 shows such a simulation.

## A3.9 DEALING WITH INPUT AND OUTPUT SIGNALS USING FSM

Often FSM-based systems need to be able to access individual input and output bits.

They may also be arranged in the form of a register of more than one bit if needed. This section illustrates how this can be achieved. We will look at how an FSM could be used to allow an input st to be arranged, as well as outputs.

Figures A3.11–A3.13 could be used to perform these actions.

In Figure A3.11 the st input is addressed via the address decoding logic at address 380h. This latches a logic 1 into the DFF, which is presented to the FSM as a single input.

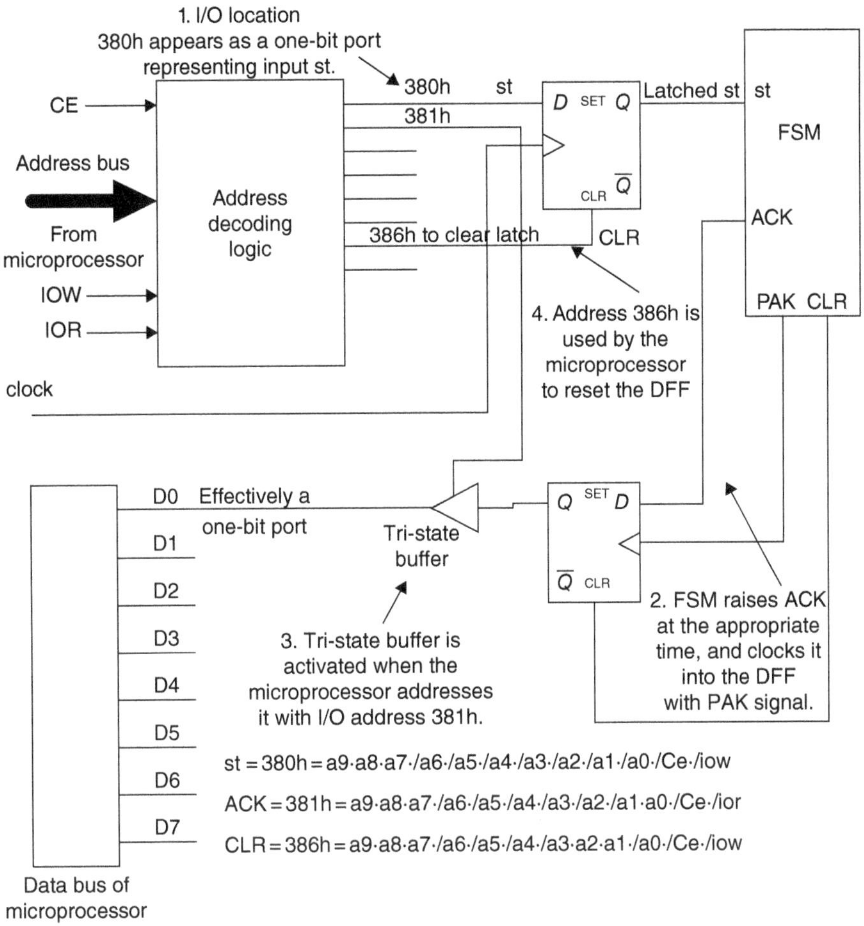


**Figure A3.11** How to deal with input st and output ACK using an FSM.

To clear the st input when address 386h is reached. This is connected to the CLR input of the DFF so as to clear it to logic 0.

To read the ACK signal from the FSM, the ACK output from the FSM is connected to the *D* line of the second DFF so it is latched into the DFF when the input PAK is raised to logic 1 (a 0 to 1 pulse).

To read the state of the ACK signal the address 381h is used to activate the tri-state buffer so that its output connects to the data bus (which could be that of a microprocessor, or perhaps the data input of a microcontroller) so as to be available at its input line D0, as in this case.

Figure A3.12 shows the timing relationship for the access to the input st in terms of the T1 to T4 waveforms. This arrangement is used in Intel microprocessors with correct address decoding logic.

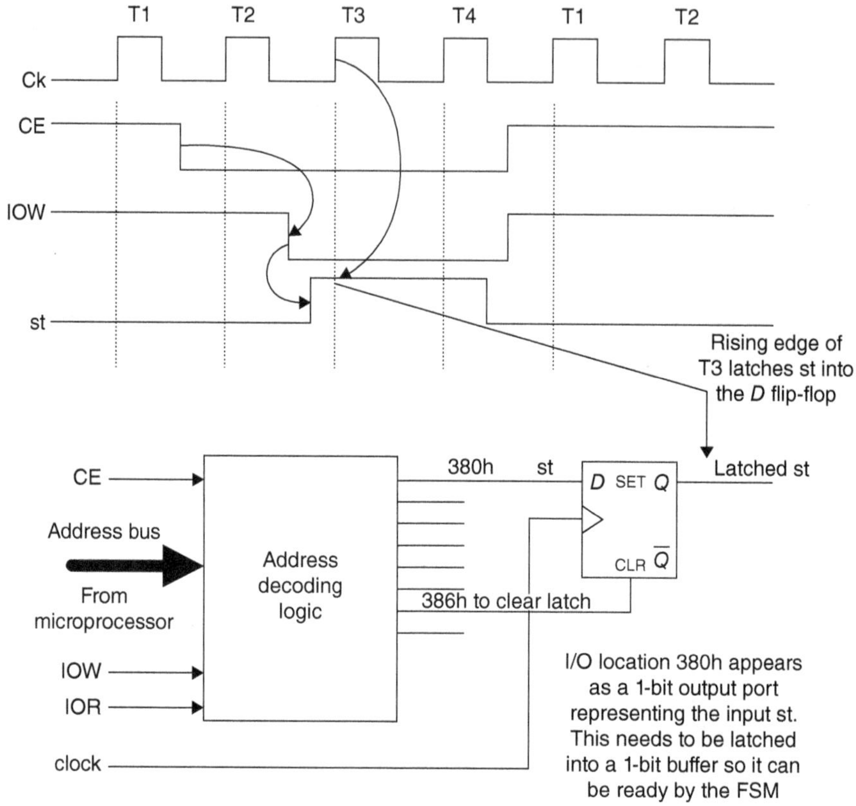


**Figure A3.12** How the timing T1 to T4 relates to the Ck, CE, IOW, and st signals.

Figure A3.13 also illustrates timing waveforms showing how this could work with an Intel processor (such as the 80X86 type).

Note that the Ck signal has a four-state sequence T1–T4, as shown in Figures A3.12 and A3.13.

- T1 is the time to set up the address bus.
- T2 is the time to allow this signal to settle.
- T3 is the time for the data bus to come out of its tri-state condition.

T3 is sometimes referred to as a 'wait state'. T4 is the period for reading or writing the selected input or output signal (st and ACK in our case).

Note that in Figure A3.13 the write operation is being performed so the signal IOW is active low, and the read operation is being performed so it is the IOR signal that is active low.

Note that the write and read actions are performed when the IOW and IOR signals are going from logic 0 to logic 1 (rising edge of the IOW or IOR signal). This is often carried out with the chip select line CS going from logic 0 to logic 1 at the same time.

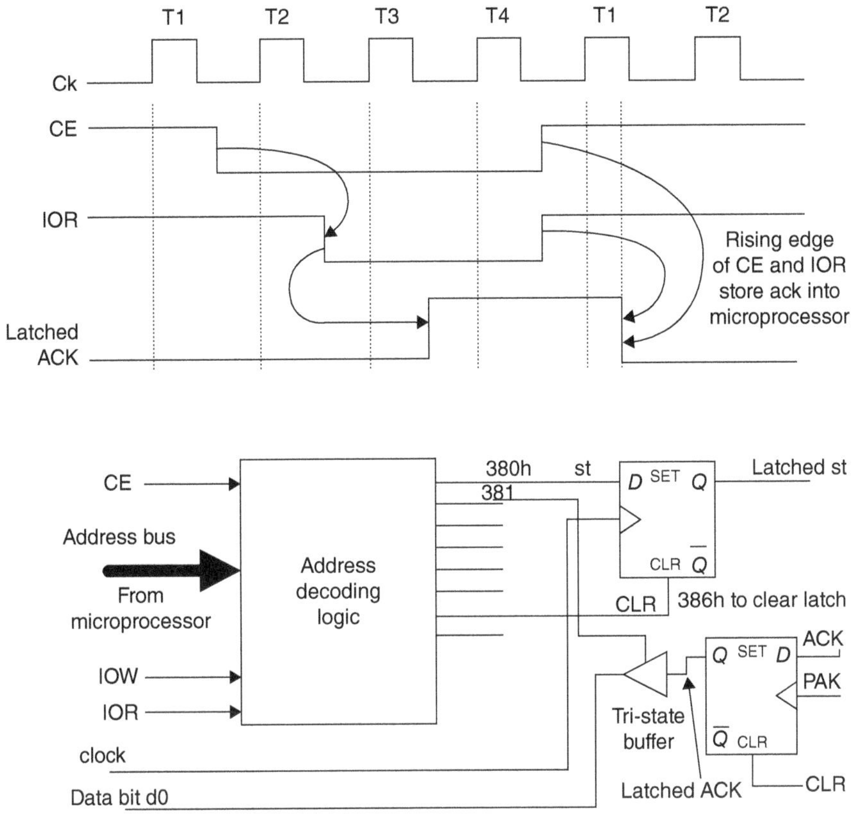


**Figure A3.13** Shows T1 to T4 and the ACK input signal.

Note that the address 381 is a hexadecimal one of 381h, i.e. 0011 1000 0001 binary.

If you look at Chapter 5 of this book on memory operations it will be apparent to you that this is the case.

## A3.10 USING LOGISIM TO WORK WITH LARGER FSM SYSTEMS

In Appendix A2 we looked at how to use Logisim to simulate FSM-based systems. We made use of Tunnels to help connect such systems together to avoid using lots of looping wires. This made it easier to draw and also easier to follow.

Now we look at larger FSM systems in an attempt to create these systems.

You may remember looking at the clocked data acquisition FSM used to store digitized data samples into a memory device in Chapter 5. This used the one hot method.

Now let's look at Figure A3.14 in this example using clocked DFF and SSVs rather than the one hot method.

In this you can see the analogue input Vin is first sampled using a sample and hold device so as to fix its value, then fed into a Flash ADC (very fast). When the

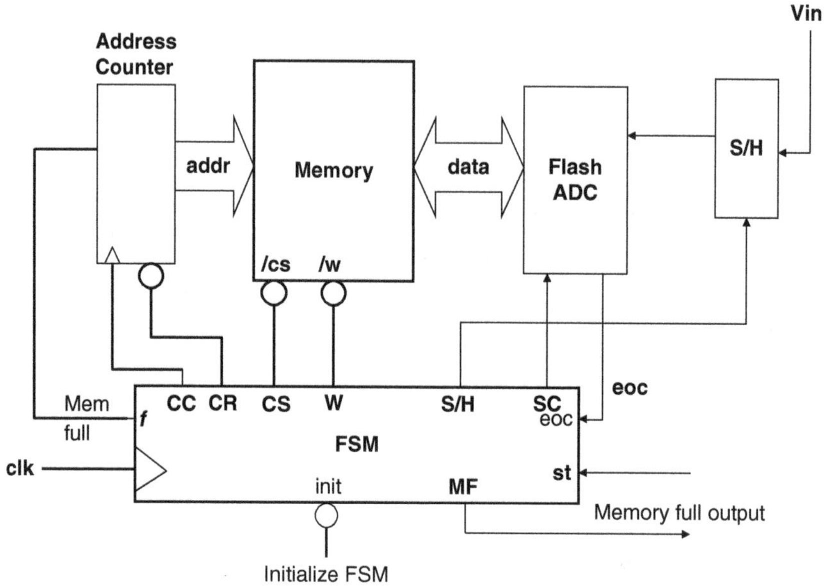


**Figure A3.14** The FSM-based data acquisition system with sample and hold input.

eoc signal from the ADC goes high, the FSM takes the converted output data and places it into the memory device. This is continued until the memory device is full (f = 1) then the memory full (MF) output signal is made high so that the remote end device can know that the memory is full.

Clearly, this system follows a sequential pace that can be described in a state diagram. Figure A3.15 illustrates this state diagram for the FSM.

This state diagram consists of 10 states following the sequence for the FSM.

You will note that each state is assigned a unique four-digit SSV that follows a unit distance code. Note that in Figure A3.15 the input signal eoc is monitored going high in s2 and low in s3.

This needs to be converted into a set of equations using the method described in Chapter 3 for *D* type flip-flops.

## A3.10.1 The equations

The equations for the *D* inputs of the flip-flops are:

$$D \cdot a = \mathrm{s0} \cdot \mathrm{st} + \mathrm{s1} + \mathrm{s2} + \mathrm{s3} + \mathrm{s7} \cdot /\mathrm{f} + \mathrm{s9}.$$

$$D \cdot b = \mathrm{s1} + \mathrm{s2} + \mathrm{s3} + \mathrm{s4} + \mathrm{s7} \cdot \mathrm{f} + \mathrm{s8}.$$

$$D \cdot c = \mathrm{s2} \cdot \mathrm{eoc} + \mathrm{s3} + \mathrm{s4} + \mathrm{s5}.$$

$$D \cdot d = \mathrm{s3} \cdot /\mathrm{eoc} + \mathrm{s4} + \mathrm{s5} + \mathrm{s6} + \mathrm{s7} + \mathrm{s8}.$$

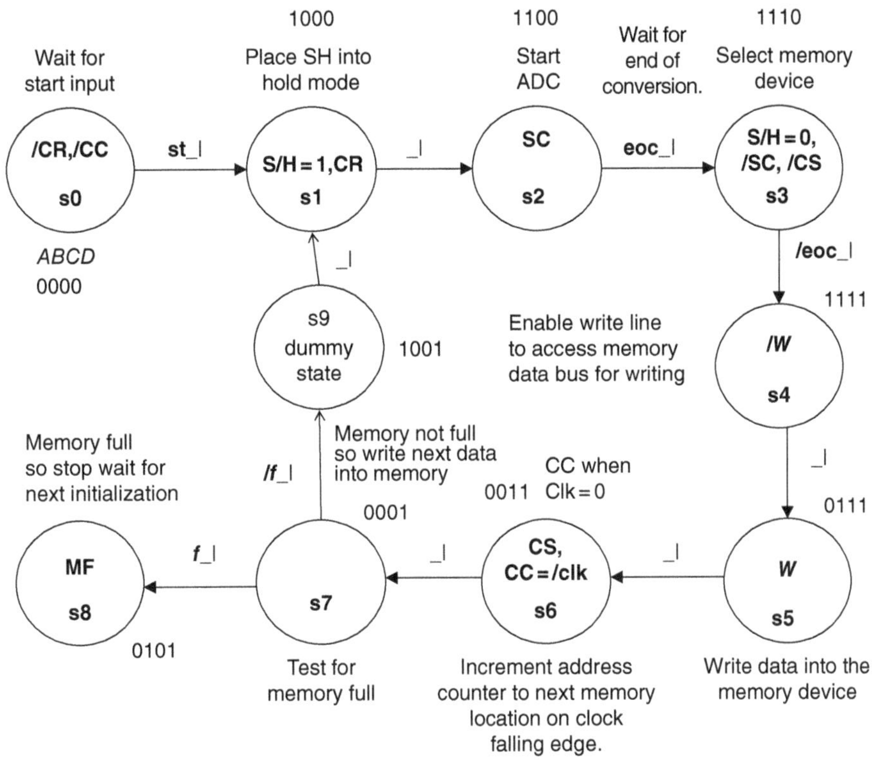


**Figure A3.15** State diagram for the data acquisition system.

The output equations are:

The reset for the address counter /CR = /s0 = /(/*A*/*B*/*C*/*D*).

Sample and hold output is S/H = s1 + s2 = *A*/*C*/*D*.

The start ADC conversion SC = s2 = *AB*/*C*/*D*.

The chip select for the memory is /CS = /(s3 + s4 + s5) = /(*ABC* + *BCD*).

The write signal to the memory is /*W* = /s4 = /(*ABCD*).

The CC signal used to produce the increment for the address counter is CC = s6·/clk.

This becomes CC = /*A*/*B*CD·/clk.

The memory full signal MF = s8 = /*AB*/CD.

You can write down the *ABCD* values for the DFF inputs yourself. For now you might like to keep them in their canonical form.

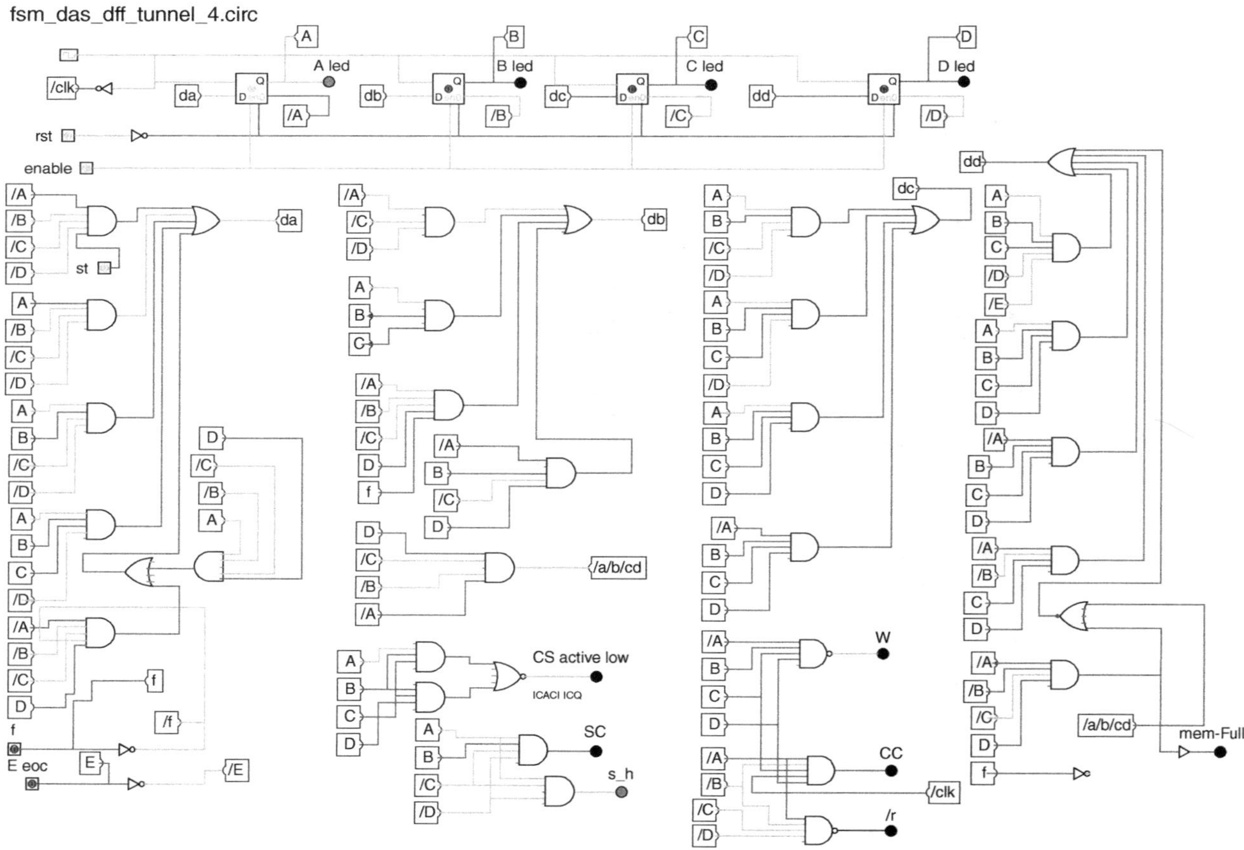


Seems you need to raise E for s2 then lower it for s3 then raise it again befor it returns to s1

**Figure A3.16** Circuit for the Logisim simulation of the FSM.

Now to implement the Logisim circuit you can make use of Tunnels when drawing the circuit. This makes it much easier to follow the circuit.

Figure A3.16 shows the final circuit. Note that the term for CR is missing from Figure A3.16.

One of the nice points of this arrangement is that you can easily alter any point in the circuit without having to redraw the circuit too much (if you have not missed terms out!).

For larger circuits you can make use of sub-circuits (see Appendix A6) and hence build up quite complex systems. Note that in Figure A3.16 use is made of canonical form (no reduction in minterms). You could, of course, reduce the logic.

Note that you will need to add the logic gate for /CR = /(*/A/B/C/D*) to Figure A3.16 to complete.

You could add to the state diagram so as to send the memory content to a PC system for further analysis of the data. In this case the added parts of the FSM could be contained in a sub-circuit so it can be added to the same logic circuit.

## A3.11 SUMMARY

This appendix has introduced simple ways to develop synchronous up and down pure binary counters, with or without parallel loading inputs that can be used in a PLD or FPGA device. It has also described how parallel loading shift registers can be developed and used.

It has also looked at the design of counters with *D* type flip-flops using don't care terms to help in the minimizing of such counters.

These techniques may be used to develop Verilog HDL modules for use in some of the designs covered in this book. Bit slice equations have been developed to allow counters and shift register circuits to be constructed directly in equation form in Verilog HDL.

In Section A3.9 the operation of IO (input/output) are shown for use with an FSM.

Note that all these circuits could be simulated using the Logisim Digital Simulator if you want to play around with these circuits.

Finally, in Section A3.10 a method for implementing larger FSM is explored.

One of the nice things about simulating with Logisim is that you get to see the dynamics of the FSM, rather than just looking at the static behaviour in a logic-timing diagram.

# Appendix A4: Finite State Machines Using Verilog Behavioural Mode

## A4.1 INTRODUCTION

In this book finite state machines (FSMs) have been implemented using the equations obtained from the state diagram/Petri net. This approach ensures that the logic for the state machine is under complete control of the designer.

However, if the state machine is implemented using behavioural mode, the Verilog compiler will optimize the design. Remember that the behavioural method describes the behaviour of the designed system.

There is a very close relationship between the state diagram and the behavioural Verilog description that allows a direct translation from the state diagram to the Verilog code.

## A4.2 THE SINGLE-PULSE/MULTIPLE-PULSE GENERATOR WITH MEMORY FSM

The example of Figure A4.1 illustrates a single-pulse with memory FSM.

In this system there are two inputs: *s* to start the system and *x* to choose either single-pulse or multiple-pulse mode. In single-pulse mode the *L* output is used to indicate to the user that a single-pulse has been generated. In multiple-pulse modes *L* is suppressed (hence the *L=/x* in states s1, s2, and s3).

This example has been used in other chapters, but in this case the behavioural mode will be used, rather than using DFFs with SSV or the clocked one hot method, both of which require Boolean equations.

*Digital System Design using FSMs: A Practical Learning Approach*, First Edition. Peter D. Minns.

Companion website: www.wiley.com/go/minns/digitalsystemdesign

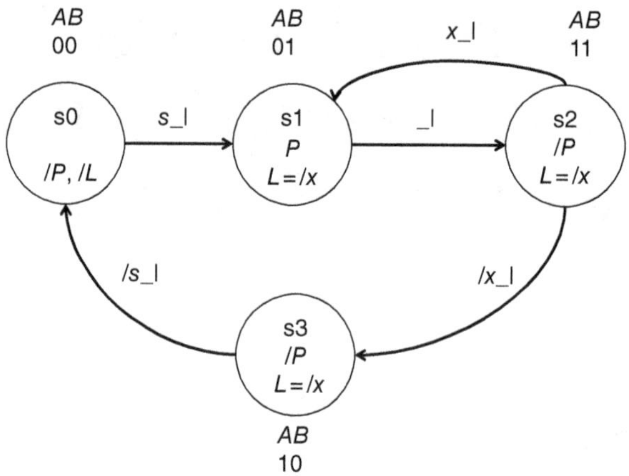


**Figure A4.1** State diagram for single/multiple-pulse generator with memory FSM.

Rather than derive the equations directly from the state diagram, a Verilog description can be obtained from the state diagram of Figure A4.1 from which Listing A4.1 of the Verilog can be made.

```
// Behavioural State Machine

module pulsar(s,clk,rst,P,L,ab);
  input s,clk,rst;
  output [1:0]ab, P, L;
  reg [1:0]state, P, L;

  parameter s0=2'b00, s1=2'b01, s2=2'b11, s3=2'b10;
     // Now define state sequence for FSM (from State Diagram).
     // This can be taken directly from the State Diagram.
  always @ (posedge clk or negedge rst)
     if (~rst)
     state <=s0;
     else
     case(state)
          s0:  if (s) state <=s1; else state <=s0;
          s1:  state <=s2;
          s2:  if (~x) state <=s3; else state <=s1;
          s3:  if (~s) state <=s0; else state <=s3;
     endcase
// Now define the outputs @ each state.
     always @ (state)
```

```
        case(state)
            s0: begin P=1'b0; L=1'b0; end
            s1: begin
            P=1'b1; L=(state==s1) & ~x;   //Mealy output.
            end
            s2: begin L=(state==s2) & ~x; P=1'b0; end
            s3: begin P=1'b0; L=(state==s3) & ~x; end
        endcase
    assign ab=state;
    endmodule
```

**Listing A4.1 Single-pulse multi-pulse generator.**

In Listing A4.1 lines 5–14 (always upto endcase) define the state diagram sequence in terms of states and input signals, while lines 15–23 (always upto endcase) define the outputs in terms of each state. Both of these sets of lines are, of course, happening at the same time (i.e. in parallel).

Note that in this description there are no equations, only a description of the behaviour of the state diagram. So this description is at a higher hierarchical level.

After declaring the module and its signals the inputs and outputs are defined in lines 1 and 2. At line 2, where outputs are defined, note the two-bit vector **[1:0]ab** that will be used to show the state of the FSM (in Figure A4.2 in hexadecimal). Line 3 defines the state vector **[1:0]state** used to keep track of the current state of the FSM, as well as the outputs ***P*** and ***L*** as register type. Line 4 defines the secondary state unit distance variables for each state of the state diagram (Figure A4.1). Note that these follow the binary values defined for each state in Listing A4.1, in line 4 in the parameter list.

In line 5 an **always** statement defines the conditions under which the case statement (lines 9–14) will be used. At line 6 the reset state is defined. If **rst** transition is 1 to 0 **if(~rst)**; otherwise, the **else** will allow the case statement block (lines 9–14) to occur.

The use of the **if else** clause allows this to be defined as a behavioural action.

The **case** statement defines the four possible states that the FSM can reside in and the conditions under which the state transitions occur. The **case** uses the present state to select one of the four possible next states. Initially, after a reset, state will be s0 (2'b00) at line 10 the case statement s0 is actioned and the **if** statement evaluates input *s*.

If input *s* is logic 1, the next state will be s1; otherwise, it will be s0.

On the next **posedge** of the **clock** signal the case statement will be activated again (this is the event marking the action). If current state is s1 then at line 11 the next state will be s2. That is it will always perform this action.

Then, on the next **posedge** of **Clk**, line 12 will ensure that the next state will be either s3 (if $x = 0$) or s1 (if $x = 1$). At line 13 the test of input $s$ will decide whether the FSM moves to s0 ($s$ equal to 0), or remain in s3.

Thus, the state sequence is determined by the case statement and use of the **if** statement to evaluate input conditions, or just define the next state if no input condition is needed.

The output conditions are defined in a separate case block (16–23) using outputs for each state as defined in the state diagram. Note that in line 19 the Mealy output for $L$ in state s1 is defined as:

$L$ = (**state = = s1) & ~x.**

This ensures that $L$ will be determined by the value of input $x$ ($x = 0$), and conditional on being in state s1, i.e. $L$ = **s1·/*x*** as defined in the state diagram.

So in this way output $L$ will only be logic 1 if the input $x$ is zero and the FSM is in s1; otherwise, it is zero. Output $P$ will be assigned logic 1 value.

Other case conditions are defined in a similar manner and assigned the output values shown in the state diagram of Figure A4.1.

In line 24 the value of the vector ab is updated.

Comparing the behavioural statements from lines 5–14 and lines 15–24 with the state diagram of Figure A4.1, it is possible to see a strong relationship between the Verilog description and the state diagram. This, of course, could be automated into a computer program (such as C, C++ or perhaps Python) in order to produce the required Verilog HDL behavioural sequence.

Using this approach allows an FSM design to be created without the need to define any hardware logic. It is in fact a 'terminal behaviour and sequence' for the FSM.

Listing A4.2 defines the test bench module for the system. This follows along the lines of other test bench modules used elsewhere in this book.

```
// Now define the Test Bench fixture

module test;
  reg s,clk,rst;
  wire [1:0]ab;
  wire p, l;
  pulsar uut(s,clk,rst,p,l,ab); //name of the test module.

  initial
     begin
     rst=0;
     clk=0;
```

```
    s=0;
    x=0;
    #10 clk=~clk;
    #10 clk=~clk;
    //stays in s0.
    #10 rst=1;
    #10 s=1;
    #10 clk=~clk;
    #10 clk=~clk;
    // moves to s1.
    #10 clk=~clk;
    #10 clk=~clk;
    // moves to s2.
    #10 clk=~clk;
    #10 clk=~clk;
    // stays in s2.
    #10 s=0;
    #10 clk=~clk;
    #10 clk=~clk;
    // moves back to s0;
    // now make x=1 to produce multiple pulses.
    #10 x=1;
          #10 s=1;
          #10 clk=~clk;
          #10 clk=~clk;
          // move to s1
          #10 clk=~clk;
          #10 clk=~clk;
          // move to s2
          #10 clk=~clk;
          #10 clk=~clk;
          // move to s1 again
          #10 clk=~clk;
          #10 clk=~clk;
          // and back to s1
          #10 clk=~clk;
          #10 clk=~clk;
          // s2 again
          #10 clk=~clk;
          #10 clk=~clk;
          #10 x=0;      //prepare to stop multiple pulse
mode.
          #10 clk=~clk;
          #10 clk=~clk;
          // to s3.
          #10 clk=~clk;
```

```
            #10 clk=~clk;
            #10 s=0;
            #10 clk=~clk;
            #10 clk=~clk;
            // back to s0.
      #10 $stop;                      //stop the simulation.
    end                               //end of test sequence.
endmodule                       //end of test module.
```

**Listing A4.2 The test bench module.**

Figure A4.2 illustrates the simulation waveforms for the design. Note the vector ab defining the state of the FSM at each clock transition. Compare this with the state diagram of Figure A4.1.

The reader should follow the waveform sequence to determine the paths taken through the state diagram. Both modes are tested and are seen to work, as can be seen in Figure A4.2.

- With $x = 0$, the FSM is seen to produce a single output pulse.
- With $x = 1$, the FSM produces a series of output pulses as it moves between s1 and s2 finally returning to state s0 when $x = 0$ (and, of course, $s = 0$).

**Note [1:0]ab uses the left-hand digit as the most significant digit, hence the hexadecimal coding used in Figure A4.2.**

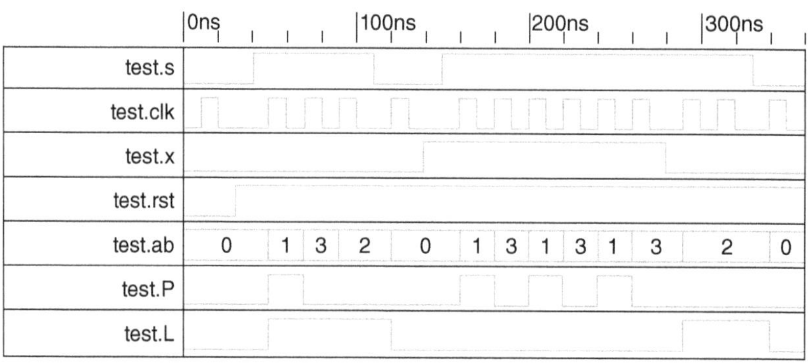


**Figure A4.2** Verilog HDL simulation of its operation as Listing A4.2.

## A4.3 THE MEMORY TESTER FSM REVISITED

This example can be coded directly in Verilog as a behavioural description using the state diagram of Figures 5.16 and 5.17 in Chapter 5. The listing is illustrated in Listing A4.3.

```
module memory_tester_state_machine(clk,rst,st,fab,full,RC,P,C
S,RD,WR,OK,ERROR,abc);
 input clk,rst,st,fab,full;
 output [3:0]abc,RC,P,CS,RD,WR,OK,ERROR;

 reg [3:0]state, RC,P,CS,RD,WR,OK,ERROR;
// assign the secondary state variable values (as used in
chapter 5 figure 5.17).
 parameter   s0=4'b0000, s1=4'b1000, s2=4'b1010, s3=4'b0010,
s4=4'b0110,             s5=4'b1110, s6=4'b1100, s7=4'b1101,
s8=4'b1001, s9=4'b1011,             s10=4'b0100;



// State Machine Sequence

always @ (posedge clk or negedge rst)
  if (~rst)
    state= s0;
  else
    case(state)
     s0: if(st) state <=s1; else state <= s0;
         s1: state <= s2;
         s2: state <= s3;
         s3: state <= s4;
         s4: state <= s5;
         s5: state <=  s6;
         s6: if(fab) state <= s7; else state <= s10;
         s7: state <= s8;
         s8: if(full) state <= s9; else state <= s1;
         s9: state <= s9;
         s10: state <= s10;
    endcase

// the outputs for each state.
always @ (state)
```

```
  begin
  {RC,P,CS,RD,WR,OK,ERROR}= 7'b0011100;
  case(state)
     s0:  begin assign
           RC=1'b0;
           P=0;
           CS=1;
           RD=1;
           WR=1;
           OK=0;
           ERROR=0;
           end
     s1:  begin assign
           RC=1'b1;
                     CS=1'b0;
           P=0; end
     s2:  begin WR=1'b0; end
     s3:  begin CS=1'b0; WR=1'b1; end
     s4:  begin  end
     s5:  begin RD=1'b0; end
     s6:  begin  end
     s7:  begin RD=1'b1; end
     s8:  begin CS=1'b1;
                          P=1'b1;  end //no clock yet could be
added.
     s9:  begin OK=1'b1;P=0; end
     s10:  begin ERROR=1'b1; end
  endcase
  end
  assign abc=state;
endmodule

module test;
     reg st,clk,rst,fab,full;
     wire [3:0]abc;

memory_tester_state_machine uut(clk,rst,st,fab,full,RC,P,CS,R
D,WR,OK,ERROR,abc);
 initial
     begin
          rst=0;
          clk=0;
          st=0;
          fab=0;
```

```
            full=0;
            #10 rst=1;
            #10 st=1;
            #10 repeat(14)
                 #10 clk=~clk;
                 #10 rst=0;
                 #10 rst=1;
            #10 repeat(10)
                 #10 clk=~clk;
                 #10 fab=1;
            #10 repeat(16)
                 #10 clk=~clk;
                 #10 clk=~clk;
                 //#10 rst=0;
                 //#10 rst=1;
                 #10 full=1;
            #10 repeat(20)
                 #10 clk=~clk;
          #10 $finish;
    end
endmodule
```

**Listing A4.3** Memory tester - behavioural mode.

The simulation is shown in Figure A4.3.

The ABC vector illustrates the FSM states in terms of the SSVs allocated in the state diagram of Figure 5.17 in Chapter 5. These were assigned to provide a unit distance code and the numbers shown in the simulation of Figure A4.3 are the hexadecimal values.

The FSM is seen to move around the state diagram from state s0 (0) through s1 (8), s2 (A), s3 (2), s4 (6), s5 (E), s6 (*C*), then s10 (4) the error state. The FSM is then reset back to state s0 and recycles around the state diagram to s8 (9) and on to s9 (B) the OK state.

**Note that vector [3:0]ABC is defined with the left-hand digit being the most significant, hence the hexadecimal codes used in Figure A4.3.**

The write cycle (8,A,2 of test *ABC*) is followed by the read cycle (E,C), and again in (8,A,2,6,E,C,D).

## A4.4 SUMMARY

The use of a behavioural description has the advantage of using the information in the state diagram directly. It can be seen that there is a one-to-one correspondence between the state diagram structure and the Verilog structure, and that the state

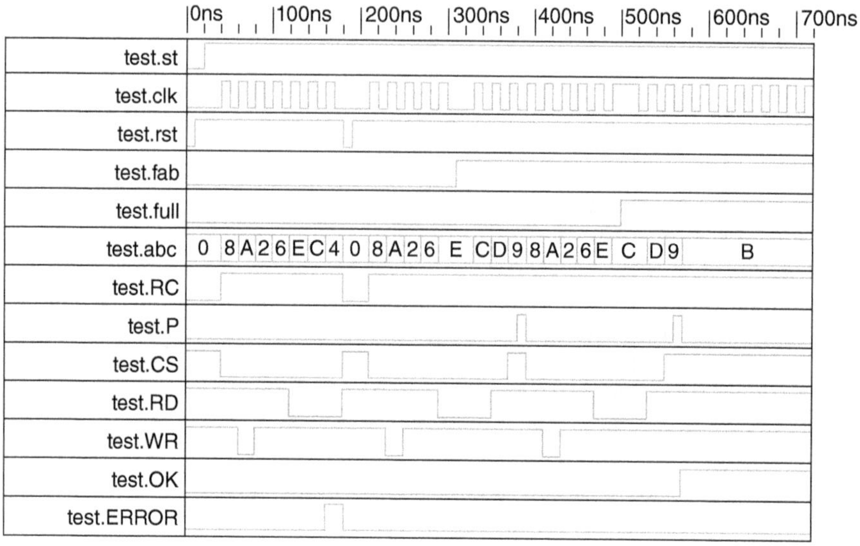


**Figure A4.3** Verilog HDL simulation of Listing A4.3.

diagram method can allow a realization of the design directly (via the behavioural method) or via the Boolean equations.

Both methods, owing to their formal descriptions, are suitable for computer implementation. This has been used in a number of programs developed by postgraduate students.

# Appendix A5: Programming a Finite State Machine

## A5.1 INTRODUCTION

I was once asked by some of my students whether it was possible for software to run without hardware. This was an interesting question, and given that I spend much of my time writing computer programs, I felt well placed to answer it. First, I decided to look into the basic needs of software.

The idea was to look at digital hardware and consider what basic hardware would be needed to produce a programmable system to make a finite state machine (FSM) that could be programmed to produce clock-driven FSMs. It should be possible to use similar ideas for event-driven FSM, but I have not looked into these yet. On reflection, it should be possible to program an FPGA with event cells that could be used with other hardware to make a programmable device.

At first I looked at basic information available in some electronic digital devices, such as parallel loading counters, digital multiplexers, and quad latches, and eventually came up with the arrangement in Figure A5.1 for a programmable FSM.

The system needs to be able to digitally count, so it makes use of a four-stage clocked synchronous up counter.

There is a need to be able to both increment the counter while also allowing the system to branch between states, so the counter needs to have parallel loading inputs so as to branch to these other states.

## A5.2 THE PARALLEL LOADING COUNTER

The counter looked at was a TTL 74163 binary counter with two control inputs: *P* and */Ld*. When $P = 1$ the counter would increment to the next count value, and when $Ld = 0$, (active low) would cause the counter to assume the count value placed onto its four parallel input lines *ABCD* to be loaded into its four output lines QaQbQcQ4. */Ld* has priority over *P*. See Figure A5.2.

*Digital System Design using FSMs: A Practical Learning Approach*, First Edition. Peter D. Minns.

Companion website: www.wiley.com/go/minns/digitalsystemdesign

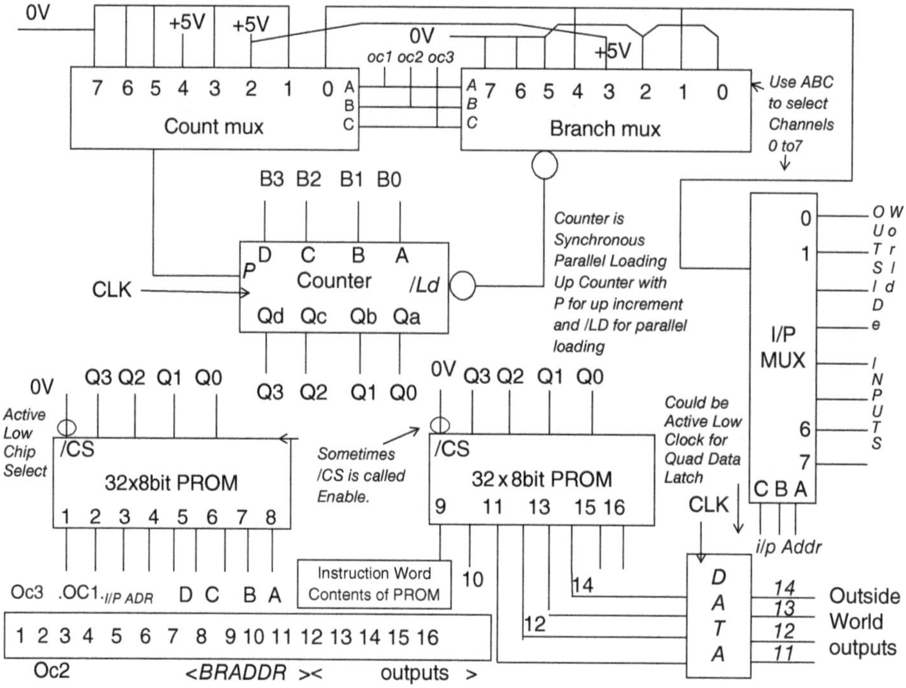


**Figure A5.1** Diagram of the system for the programmable FSM.

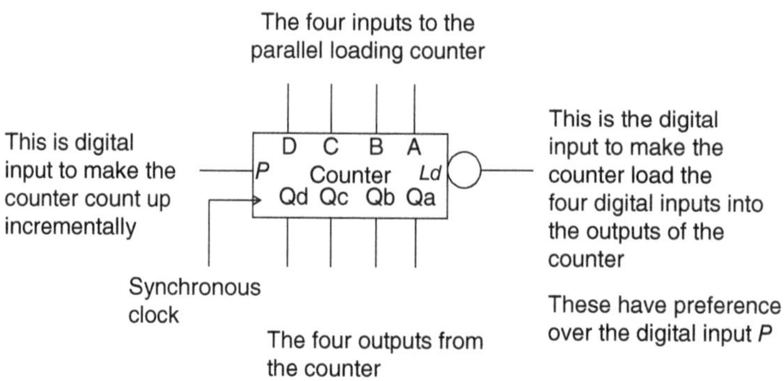


So, the parallel loading counter can be made to count up incrementally by setting $P = 1$, or it can be made to load the four parallel inputs directly into the counter by making $Ld = 0$. Note the *Ld* is an **active low** input.

**Figure A5.2** The parallel loading up counter and its inputs and outputs.

## A5.3 THE MULTIPLEXER

The FSM would need to have inputs from the outside world that could be used to control the operation of the counter (increment or parallel load – branch). The inputs would, of course, be digital and a device that could be used was a multiplexer (in this case it would have eight inputs from eight possible outside world inputs and three digital inputs *ABC* to select any one of the eight possible outside world inputs). The word 'multiplexer' is usually shortened to 'mux'.

In Figure A5.3 the I/P mux performs this action.

You will notice two other eight-channel multiplexers in Figure A5.1, one called the 'count mux', the other the 'branch mux'.

The individual digital inputs to these two muxes are connected to either logic 0, logic 1 (+5V), or to the output of the I/P mux. This is explained later.

Also note that in Figure A5.1 the outputs of the two count and branch muxes are connected to the *P* and */Ld* inputs of the counter. See later for an explanation.

The logic select inputs *ABC* of the two count and branch muxes are connected to oc1, oc2 and oc3. These are the operation code (op code) inputs and are used to provide the system with an instruction select binary code (*ABC*). These are used a little later when we start to look at the instruction format for the controller that will be used to form the software of the system. This is a hardware action to produce some software.

Digital Input

| C | B | A | To select digital input |
|---|---|---|---|
| 0 | 0 | 0 | 0 |
| 0 | 0 | 1 | 1 |
| 0 | 1 | 0 | 2 |
| 0 | 1 | 1 | 3 |
| 1 | 0 | 0 | 4 |
| 1 | 0 | 1 | 5 |
| 1 | 1 | 0 | 6 |
| 1 | 1 | 1 | 7 |

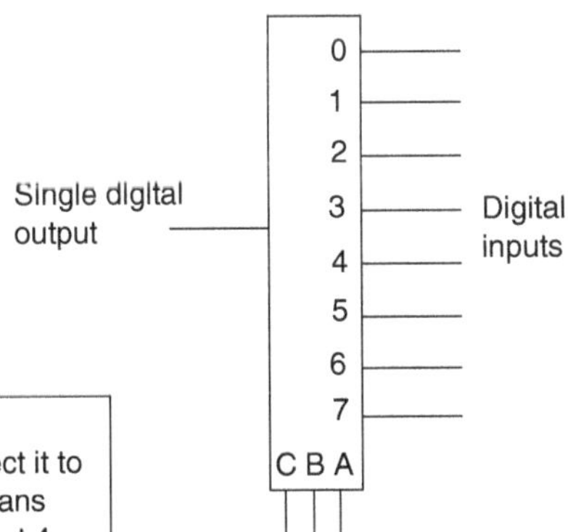


The digital input code 1 0 0 will select digital input 4 and connect it to the digital output line. This means that the selected line digital input 4 will be connected to the digital output.

Select digital input lines C B A to choose which digital input lines 0 to 7 to connect to the single digital output.

In this way any digital input can be selected to be connected to the digital output.

**Figure A5.3** The operation of an eight-input multiplexer.

The mux select inputs *ABC*, the parallel branch select input *ABCD*, and the I/P mux *ABC* form the basis for what is called a 'micro instruction' that is used as the basis for the software to be used by the system so as to create a programmable instruction set for the controller.

You can see the micro instruction (instruction word content of PROM) in Figure A5.1 at the bottom left-hand side of the figure.

## A5.4 THE MICRO INSTRUCTION

The format of the micro instruction is:

| 1 | 2 | 3 | 4 | 5 | 6 | 7 | 8 | 9 | 10 11 12 13 14 15 16 |
|---|---|---|---|---|---|---|---|---|---|
| oc3 | oc2 | oc1 | *C* | *B* | *A* | *D* | *C* | *B* | *A* ←-----------------→ |

For this system to work there is a need to connect the count and branch mux inputs in a certain way in order for the controller to work.

This is considered a little later. However, note the connection arrangement in the above micro instruction.

The micro instruction for the controller is now considered.

| oc3 oc2 oc1 | CBA | DCBA | Outputs |
|---|---|---|---|
| Op-code | I/P | Branch | Outputs |
| | Mux | Mux | from Controller |

If the number of outputs is four then the micro instruction is made up of 14 digital inputs. It could, of course, be longer if we add more outputs or outside world inputs.

## A5.5 THE MEMORY

If we are to design a programmable FSM controller, some memory needs to be added to the controller, and this is shown in Figure A5.1 as the two 32 × eight-bit memory devices. These in the figure are one-time programmable memory devices. However, they could be programmable, or even RAM-based devices (see Figure A5.4 for arrangement).

The purpose of the memory is to hold the micro instruction so that as the memory is advanced it can be used to store the next micro instruction to be executed by the controller. This is illustrated in Figure A5.4 by the memory contents in the right-hand side of the figure with the address and data content.

Then the outside world outputs need to be stored into a four-bit register made up of four *D* type flip-flops as shown by the data block (DATA) in Figure A5.1.

This is the function and arrangement of the controller.

The Memory

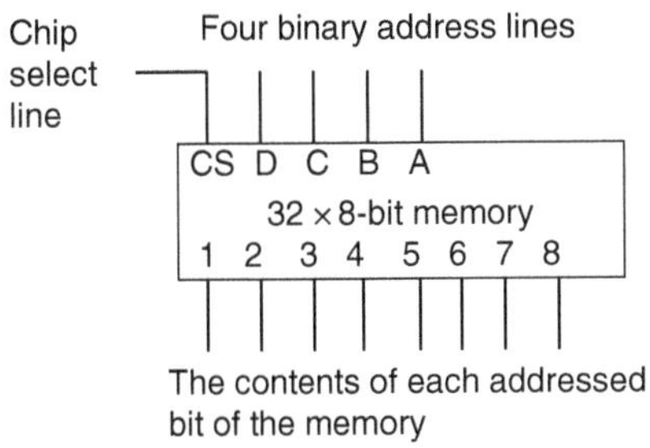


Address 0000 contains 00000000
and address 0010 contains 00000001
likewise address 0111 contains 10100010

The chip select line CS is used to select the memory chip when equal to 0

| Address DCBA | Data in memory 1 2 3 4 5 6 7 8 |
|---|---|
| 0 000 | 0 0 0 0 0 0 0 0 |
| 0 001 | 0 0 0 0 0 0 0 0 |
| 0 010 | 0 0 0 0 0 0 0 1 |
| 0 011 | 0 0 0 0 0 0 0 0 |
| 0 100 | 0 0 0 0 0 0 0 0 |
| 0 101 | 0 0 1 1 0 1 1 0 |
| 0 110 | 0 1 0 1 1 1 1 0 |
| 0 111 | 1 0 1 0 0 0 1 0 |
| 1 000 | 1 1 1 1 0 0 0 0 |
| 1 001 | 0 1 1 1 0 1 1 0 |
| 1 010 | 0 0 0 0 0 0 0 0 |
| 1 011 | 0 0 0 0 0 0 0 0 |
| 1 100 | 0 0 0 0 0 0 0 0 |
| 1 101 | 0 0 0 0 0 0 0 0 |
| 1 110 | 0 0 0 0 0 0 0 0 |
| 1 111 | 0 0 0 0 0 0 0 0 |

In this way the memory can be used to contain whatever data we want it to hold.

**Figure A5.4** The operation of an eight-bit memory device.

The signal lines Qd Qc Qb Qa are the counter outputs that connect to the individual input data lines of the memory and they form the data bus of the controller.

## *A5.6 THE INSTRUCTION SET*

Our programmable controller needs to have an instruction set.

This is 'made up' to suit the requirement for the controller to be able to handle the state diagram requirements. It is shown in Figure A5.5.

HIF 000 **H**old or **I**ncrement on **F**lag

HBF 001 **H**old or **B**ranch on **F**lag

IUC 010 **I**ncrement **U**nconditionally

BUC 011 **B**ranch **U**nconditionally

IBF 100 **Increment** or **B**ranch on **F**lag

HLT 111 **Halt** the controller.

Note that these op-codes are made up as binary numbers $2^2 2^1 2^0$ in that order as the binary coding for each instruction. This becomes the instruction set code for the system.

These instructions can be used to make up a sequence for the controller to follow in order for the FSM to be implemented. The sequence of codes stored in the ROM form the program of the system.

**Figure A5.5** The instruction set for the programmable FSM.

These instructions are used to allow a controller to follow the requirements of an FSM. HIF is the 'hold or increment on flag' instruction. It has an op-code of 000. It is used whenever an instruction needs to wait for an external input to become true, before incrementing the counter to the next instruction. The micro instruction might look like:

**HIF** 000 000 0000 0000

This would signal the **HIF** instruction 000. It would select channel 000 of the I/P mux as the flag bit to look at.

There would be no parallel branch so branch address 0000 would not be used by this instruction. There is no outside world output selected so we default to 0000.

Now consider the instruction:

**HBF** 001 010 0011 0000

This is the **Hold or Branch on Flag**. So in this example the instruction would hold and look at the outside world input on channel 2 of the I/P Mux in Figure A5.1 (010) and if it were true the counter would be loaded with the four-bit value 0011 branch address (this would load the counter with the value 0011). No outside world outputs are used in this instruction so 0000 is at the outputs.

**IUC** 010 000 0000 1010

This is the **Increment UnConditionally** instruction that will cause the controller to increment the counter regardless of any input. It does not use the outside world inputs, or the branch address inputs. In this example the outside world outputs have the value 1010. If you look at Figure A5.1 you can see that this would set output bits 11 and 13, but not bits 12 or 14.

**BUC** 011 000 1100 0001

This is the **Branch UnConditionally** instruction to allow the controller to branch to 'a none sequence' part of the memory. It does not need the outside world input bits and in this case it is branching to 1100 in the memory. Outputs are 0001 used in this example.

**IBF** 100 1010 0000

The **Increment or Branch on Flag** instruction is a special one in that it allows the counter to either increment if the outside world input is not true, or if the outside world input is true, it causes the controller to branch to the address 1010. In this case the outside world outputs are not used.

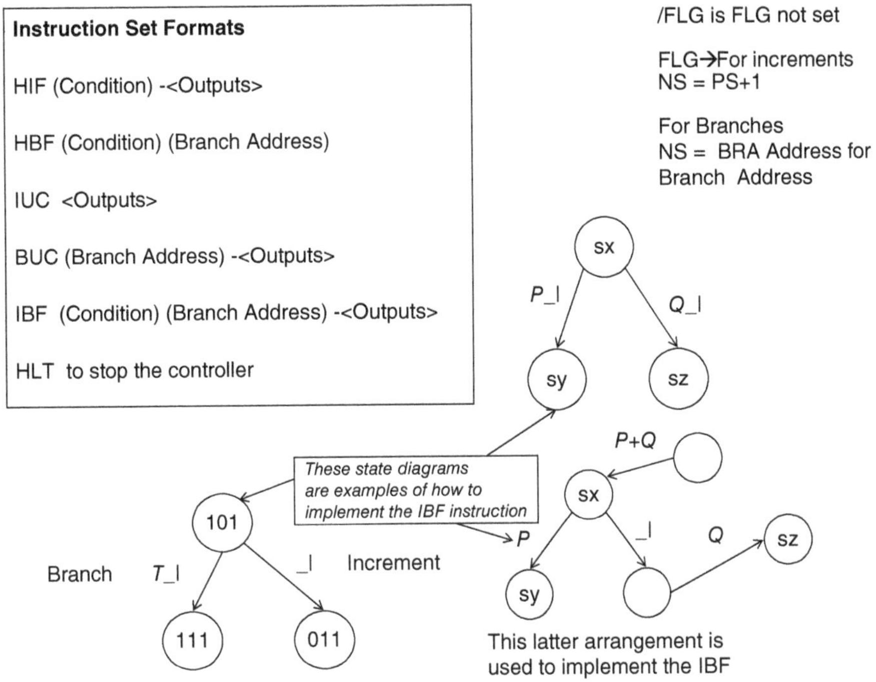


**Figure A5.6** Shows how the IBF instruction can be implemented.

The **IBF** instruction needs to be used with some care. This is illustrated in Figure A5.6 and shows how the controller might use this instruction.

It also shows the instruction format for each instruction.

Note that a basic two-way branch uses this form of instruction as you can see in the three state diagrams in Figure A5.6. NS is next state and PS is present state.

The final instruction is the **Halt** instruction.

**HLT** 111

This instruction simply stops the controller. In a lot of state diagrams there are loops so it may well not stop.

The next thing to do is to show an example of the controller being used to program a simple state diagram.

## A5.7 SIMPLE EXAMPLE: SINGLE-PULSE FSM

This is similar to the single-pulse FSM that has been used from Chapter 1.

In the state diagram in Figure A5.7 the operation of *s* input causes the state diagram to move s0 to s1 to s2 to s3 where it waits for input *s* to go to 0.

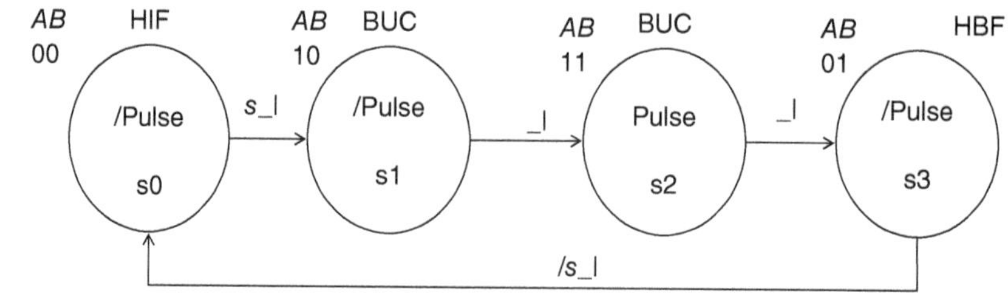


| INS | op code | I/P address CBA | Br DCBA | outputs | |
|---|---|---|---|---|---|
| HIF | 000 | 000 | **0001** | 000000 | Hold or Increment on Flag |
| BUC | 011 | 000 | **0011** | 000000 | Branch Unconditional |
| BUC | 011 | 000 | **0010** | 100000 | Branch Unconditional |
| HBF | 001 | 001 | **0000** | 000000 | Hold or Branch on Flag |

An additional instruction is HLT 111 used to stop the micro instruction microcontroller. Note that the second BUC instruction has outputs 100000; also note the branch (Br) addresses in each instruction and its DCBA label.

Instructions have been added to the figure and these have been used to make up the instruction sequence in the box. Note these are shown above in each state. Note the sequence for outputs and that of Br address.

**Figure A5.7** Example of a single-pulse FSM using the instruction set.

In state s2, the Pulse output goes to 1, then returns to 0 in state s3. When $s = 0$ the state diagram moves back to s0 and then waits for $s = 1$ again. This action produces logic 1 pulse on output Pulse.

The program code in Figure A5.7 is used by the controller to produce the expected code for the controller to follow.

In the state diagram in Figure A5.7 the **HIF** instruction produces a **Hold or Increment on Flag** operation to wait until the **I/P mux** 0 input (in this case *s* input) to go to logic 1. This is connected to the 0 mux counter input and when logic 1 causes the **Pulse** input of the counter to become logic 1 also thus making the **Pulse = 1** and allowing the counter to increment. You can now see how the controller connection mechanism causes the Hold or Increment on Flag operation.

The following **BUC branch unconditional** causes the controller to **branch unconditionally to 0011 or 11** in our state diagram. There is no I/P mux value or any outside world outputs so these are left zero.

The next instruction is **BUC** with an **op-code of 011** and a **branch address** of **0010** or in the state diagram **10**. So getting our controller into state **s3.**

Finally, the last instruction is the **HBF**, which is **Hold or Branch on Flag**.

As you can see, the effect is to move the state diagram to state s0 when the **Flag bit = 1**. This will only occur when the **Flag bit = 1**, which will be when the ***s* input**

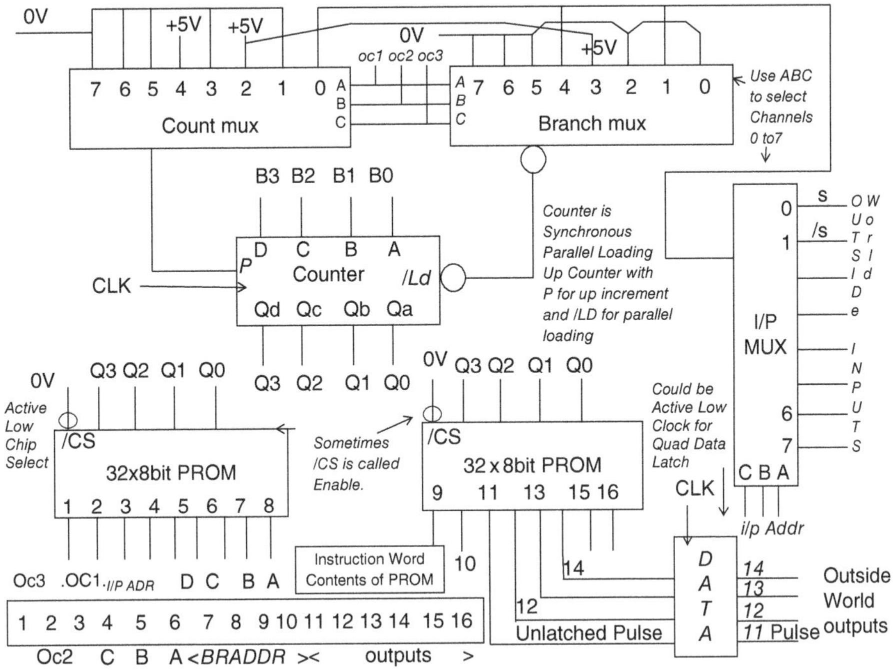


**Figure A5.8** The main circuit diagram for the programmable single-pulse FSM.

**is logic 0**, as can be seen in Figure A5.8. The implication here is that when $s = 0$, $/s = 1$.

Figure A5.8 illustrates the controller diagram with outside world inputs and outputs.

Note the connection of outside world inputs *s* and */s*, and the outside world output Pulse on bit 11 of the data latch.

Remember that the two 32-bit memory devices use an active low chip select, sometimes called an 'enable input'. These are enabled all the time since this is the only memory used in the machine.

It might be useful to read through the working of this program carefully before going on to look at the last example, which is more complicated.

## A5.8 THE FINAL EXAMPLE

This is the motor with fault limit state diagram. The state diagram is modified using the idea illustrated in Figure A5.6 so that the controller can use the **IBF Increment or Branch on Flag** instruction to carry out two-way branch operations.

As you can see in the state diagram of Figure A5.9 the instruction s4 is the IBF instruction and makes use of s6 for the Increment part of this instruction, and the s4 to s5 Branch part of the instruction.

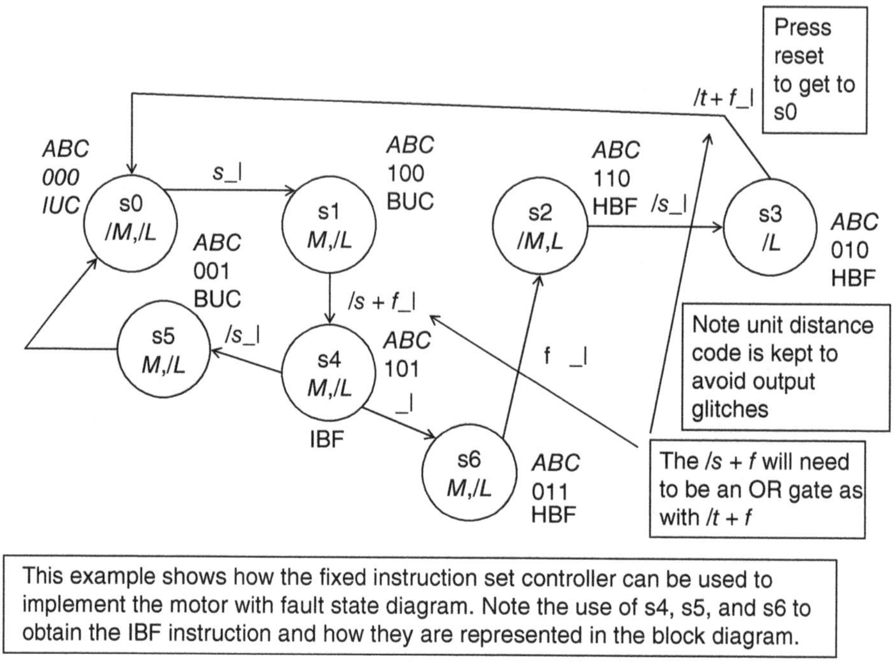


**Figure A5.9** The state diagram for the motor with fault programmable FSM with instructions.

Figure A5.9 also illustrates the controller state diagram for this example. You should be able to follow this state diagram.

Note that the transition between s4 and s6 is not unit distance. This can be tolerated in a clock-driven programmable FSM when the instruction is following a branch (101–001).

Figure A5.10 shows the complete diagram for the controller for this example.

Note the outside world inputs connected to the I/P mux. Input *s* is a single input, but the input to channel 1 is /*s* +*f* which is a logic OR input.

I/P mux channel 2 is also two inputs made up of '*t* +*f*' OR gate. There are two outside world outputs *L* and *M* connected to the data latch outputs 11 and 12. This is a more complex arrangement than the last one.

For this example, Figure A5.10 can be considered.

The program code is shown in Figure A5.11.

You can see how the outside world inputs are combined to form the input logic for this system with /*s* + *f* on input 1 and *f* + *t* on input 2 of Figure A5.10 and Figure A5.11, which follows on from Figure A5.9.

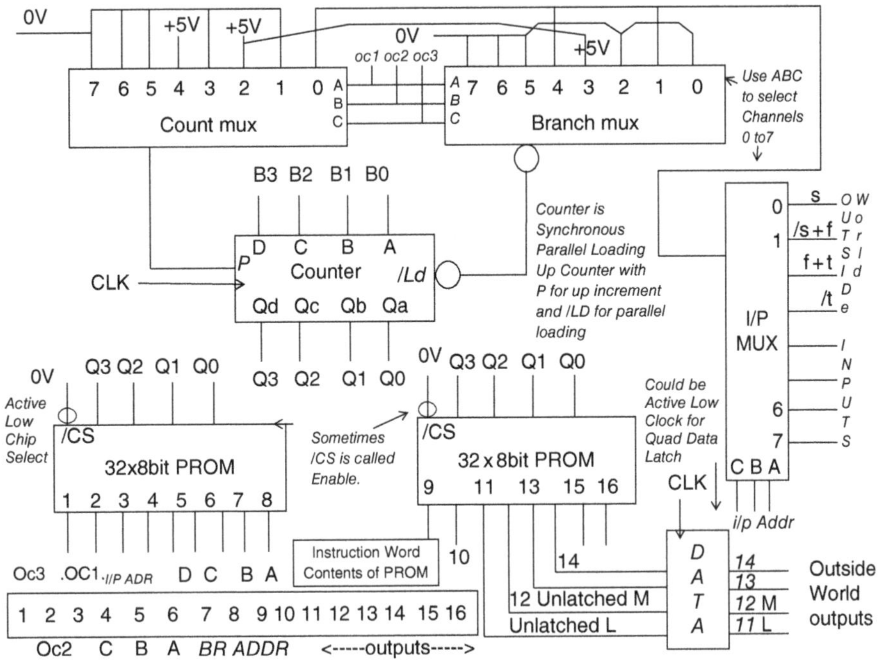


**Figure A5.10** Main circuit diagram for the programmable motor with fault FSM.

The Program Code

| | INS | Op-code | I/P | DCBA | In | | | | Out | |
|---|---|---|---|---|---|---|---|---|---|---|
| | | | | | 0 | 1 | 2 | 3 | 0 | 1 |
| | | | CBA | Branch | *s* | */s+f* | *f+/t* | */t* | *L* | *M* |
| s0 | IUC | 000 | 000 | 0000 | 1 | 0 | 0 | 0 | – | – |
| s1 | BUC | 011 | 001 | 0101 | 0 | 1 | 0 | 0 | – | – |
| s4 | IBS | 010 | 010 | 0100 | 0 | 1 | 0 | 0 | – | *M* |
| s6 | HBS | 001 | 001 | 0011 | 0 | 1 | 0 | 0 | – | *M* |
| s5 | BUC | 011 | 001 | 0000 | 0 | 1 | 0 | 0 | – | *M* |
| s2 | HBF | 001 | 001 | 0010 | 0 | 1 | 0 | 0 | *L* | – |
| s3 | HBF | 001 | 001 | 0000 | 0 | 0 | 1 | 0 | – | – |

Ready to go to s0 again in s3.
Note no HLT in this program.
First column shows state represented by the instructions.
Check this out carefully.

**Figure A5.11** The program sequence for the motor with fault FSM.

## A5.9 THE PROGRAM CODE

Figure A5.11 illustrates the program code for the controller.

Compare the program with the state diagram and the instruction set controller. Study this last example carefully to ensure you understand its action.

## A5.10 RETURNING UNUSED STATES VIA OTHER TRANSITION PATHS

In general, as has already been discussed elsewhere, event-driven FSM **must** follow a unit distance code path, whilst clock-driven FSM could follow any path back to s0.

However, in event-driven systems unused state with two or more SSVs could not be returned to s0 as **each SSV would need to change on its own for each transition back to s0**. Indeed, each state diagram needs to be looked at individually when looking at unused state conditions.

The micro instruction controller will increment in binary and could therefore follow a none unit distance coding. However, it is clock-driven and it is assumed that the clock speed is not too high as to cause a maloperation. The clock speed could, of course, be quite high with today's logic and it is indeed possible that PLD/CPLD/FPGA devices could be used to implement the above micro programmed controller.

Also it may be possible to return an unused state via some other path in a given state diagram, following a unit distance transition. Perhaps you could return unused 101 to s5 (001), as the transition from s5 back to s0 is possible. Clearly, the path and SSV values need to be carefully considered. In general, each path needs to be considered on its own merit.

There is clearly more work to be considered following this approach.

## A5.11 SUMMARY

This appendix has illustrated how a number of common digital devices such as multiplexers, parallel loading counters, memory devices, and data latches can be used to form a programmable FSM system that can use a devised instruction set for the FSM to allow it to be implemented using a devised system of software instructions.

The hardware and software together can be used to implement an FSM.

The hardware is essential for the software to be run in this and indeed any programmable system.

This can be implemented using other more complex hardware such as to form 'bit slice' systems with any number of bits (White 1981). The idea can be taken further so as to form more complex computer systems.

# Appendix A6: The Rotational Detector Using Logisim Simulator with Sub-Circuits

The appendix introduces the rotational detector and its use of event sequential equations and unit distance state variables. This is continued with the development of the rotational detector and how it can be tested using sub-unit clock-driven counters to generate the effect of the optical decoder in order to detect the direction of rotation of the shaft.

Also included is a different state diagram showing how two independent state diagrams can be implemented to perform the same actions. The Logisim Simulator will be used to illustrate its operation.

Figure A6.1 shows the waveforms from the optical detector located on the shaft. There are two patterns: the clockwise pattern and the anti-clockwise pattern. These are used to detect the shaft rotation direction.

Figure A6.1 is similar to that used in figure (a) (Zissos 1979) except using $m$ and $s$ rather than $X$ and $Y$.

Source: *Problems and Solutions in Logic Design* (Zissos 1979), Problem 9 An Incremental Shaft Encoder Figure (a) waveforms, p. 53. Reproduced with permission of the Licensor (Oxford Publishing Limited) through PLSclear.

The event-driven rotational detector FSM is shown in Figure A6.2.

Source: *Problems and Solutions in Logic Design* (Zissos 1979), Problem 9 An Incremental Shaft Encoder Figure (*b*), p. 53. Reproduced with permission of the Licensor (Oxford Publishing Limited) through PLSclear.

The $A$ and $B$ SSVs are also shown in this figure, together with the clockwise (CW) and anti-clockwise (ACW) outputs used to indicate the direction of the shaft.

*Digital System Design using FSMs: A Practical Learning Approach*, First Edition. Peter D. Minns.

Companion website: www.wiley.com/go/minns/digitalsystemdesign

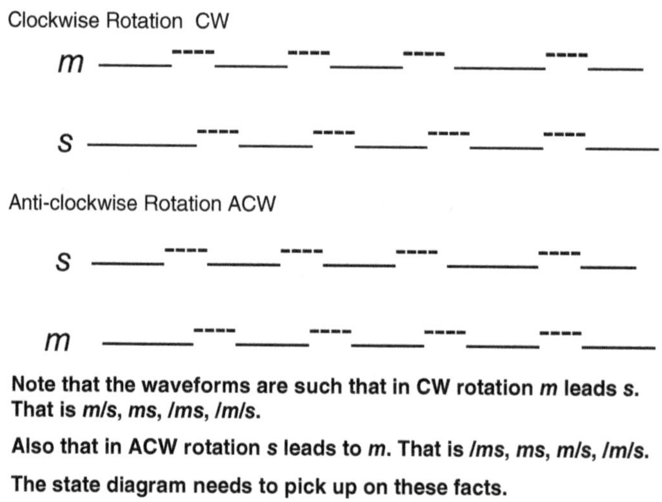


**Figure A6.1** The *m* and *s* waveforms produced by the optical detector.
*Source*: D. Zissos, 1979 / Oxford University Press.

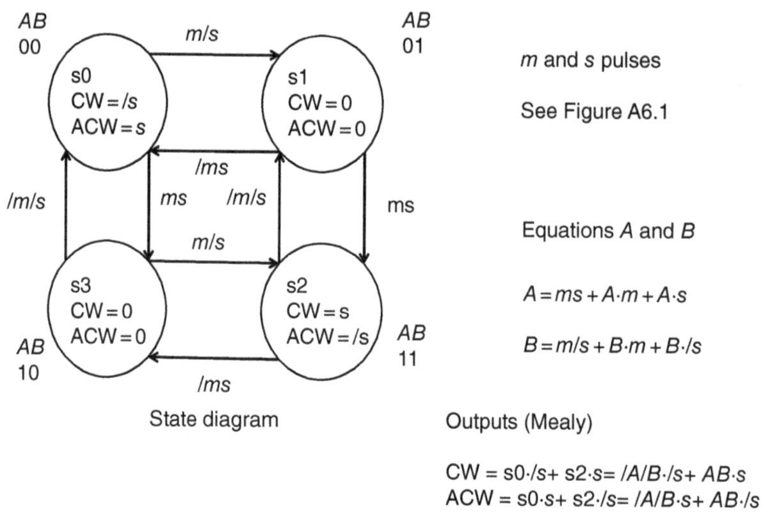


**Figure A6.2** State diagram for the rotational detector FSM.
*Source*: D. Zissos, 1979 / Oxford University Press.

Note: you can see the state diagram shows **two** directions of transitions:

- The clockwise direction CW *m/s*, *ms*, */ms*, and */m/s*.
- The anti-clockwise direction ACW */ms*, *ms*, *m/s*, and */m/s*.

Note that in this state diagram the SSVs *A* and *B* have been included so that it can be designed using the SSVs. The *A* and *B* equations in Figure A6.2 make use of logical adjacency.

The event version using $A$ and $B$ SSVs is detailed here:

$$A = \mathrm{Sum}(\mathrm{Sa}) + A \cdot /\,\mathrm{Sum}(\mathrm{Ra}) = s1 \cdot m{\cdot}s + \mathrm{s}0 \cdot m \cdot \mathrm{s} + A \cdot /\,(\mathrm{s}3 \cdot /\,m\,/\,\mathrm{s} + s2\,/\,m\,/\,s)$$
$$A = /AB \cdot m \cdot s + /A\,/\,B \cdot m \cdot s + A \cdot /\,(A\,/\,B \cdot /\,m{\cdot}\ /\,s + AB \cdot /\,m \cdot /\,s)$$

remove logical adjacency to obtain

$$A = /A \cdot m \cdot s + A \cdot /\,(A \cdot /m \cdot /s) - - - - -$$

now drop the $/A$ and $A$ terms as usual

$$\boldsymbol{A = m\ s + A \cdot m + A \cdot s}$$

the final event equation for $A$.

$$B = \mathrm{Sum}(\mathrm{Sb}) + B \cdot /\,\mathrm{Sum}(\mathrm{Rb}) = \mathrm{s}0 \cdot m \cdot /s + \mathrm{s}3 \cdot m \cdot /\,s + B \cdot /(\mathrm{s}2 \cdot /m \cdot s + \mathrm{s}1 \cdot /m \cdot s)$$
$$B = /A/B \cdot m \cdot /s + A/B \cdot m \cdot /s + B \cdot /(AB \cdot m \cdot s + /AB \cdot m \cdot s)$$

remove logical adjacency to obtain:

$$B = /B \cdot m \cdot /s + B \cdot /(B \cdot /m \cdot s)$$

now drop the $/B$ and $B$ terms as usual:

$$\boldsymbol{B = m \cdot /s + B \cdot m + B \cdot /s}$$

the final event equation for $B$.

From these equations the event-driven FSM of Figure A6.3 can be developed.

Using the event sequential equations a circuit diagram as illustrated in Figure A6.3 can be obtained, which is implemented using Logisim.

This could be tested by entering tests to the $m$ and $s$ inputs but in this example a sub-circuit has been added to figure A6.3 to generate the required sequence of $m$ and $s$ to simulate the effect of the optical detector. This is shown in Figure A6.4.

These are designed in the same way as used in Appendix A3 using clock-driven synchronous counters. Note: remember to enable the clock in Figure A6.4 and set both enable inputs to 1.

The method used is illustrated here.

Sequence for CW and ACW counters:

| CW | $m$ | $s$ | State | ACW | $m$ | $s$ | State |
|---|---|---|---|---|---|---|---|
| | 0 | 0 | s0 | | 0 | 0 | s0 |
| | 1 | 0 | s1 | | 0 | 1 | s2 |
| | 1 | 1 | s3 | | 1 | 1 | s3 |
| | 0 | 1 | s2 | | 1 | 0 | s1 |

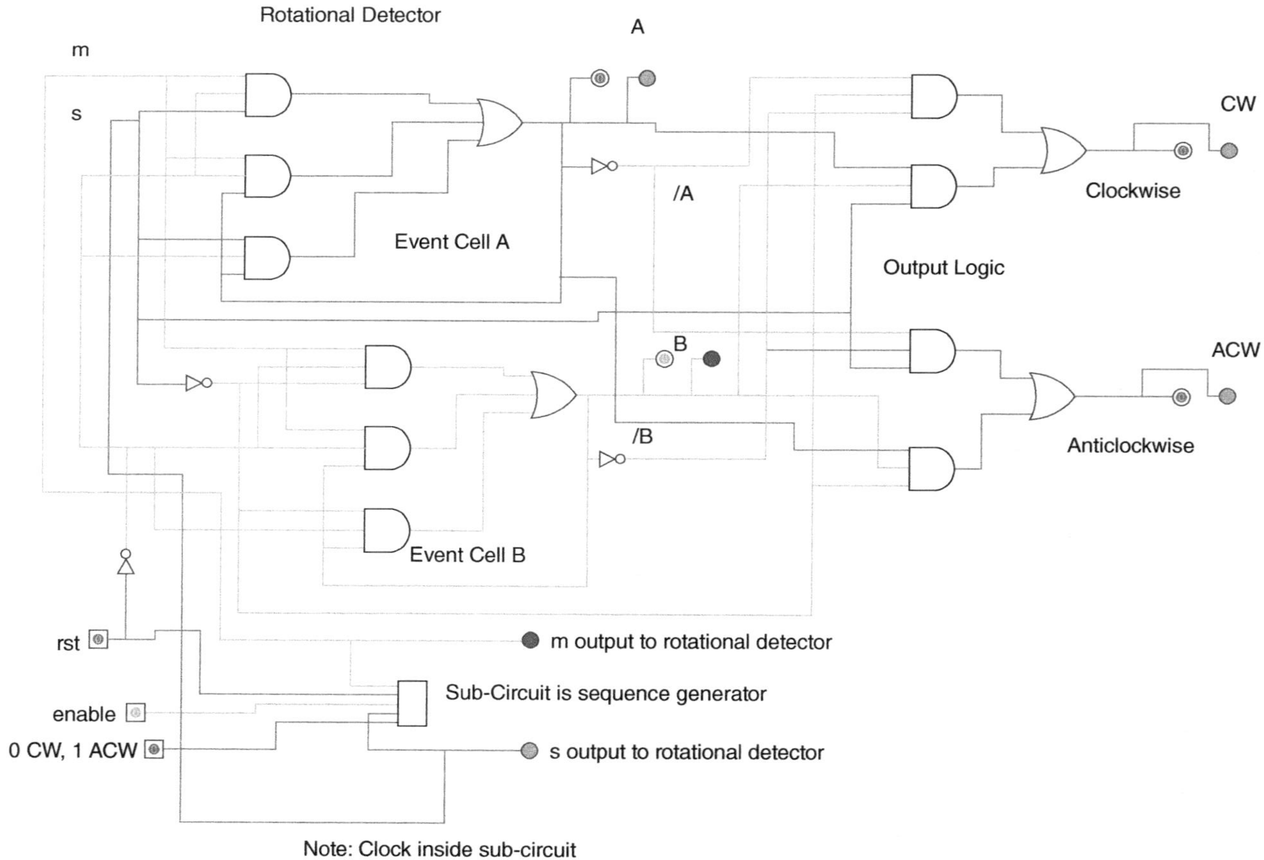


**Figure A6.3** Circuit for the Logisim rotational detector – note sub-circuit for *m* and *s*.

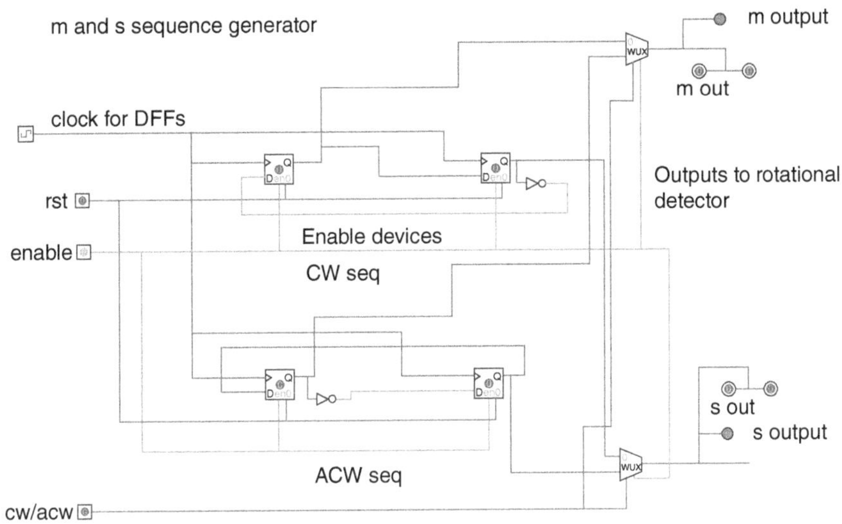


**Figure A6.4** Circuit for Logisim *m* and *s* optical detector

For the CW counter the *D* input for *m* flip-flop is:

$$dm = \mathrm{s0} + \mathrm{s1} = /m \cdot /s + m \cdot /s = /s.$$

For the CW counter the *D* input for *s* flip-flop is:

$$ds = \mathrm{s1} + \mathrm{s3} = m \cdot /s + m \cdot s = m.$$

For the ACW counter the *D* input for *m* flip-flop is:

$$dm = \mathrm{s2} + \mathrm{s3} = /m \cdot s + m \cdot s = s.$$

For the ACW counter the *D* input for *s* flip-flop is:

$$ds = \mathrm{s0} + \mathrm{s2} = /m \cdot /s + /m \cdot s = /m.$$

Looking at the two counter *D* inputs, they agree with these values.

If you want an event-driven one hot system you need to produce the equations for the one hot. However, this could be a challenge without the use of the C gate and handshake signals.

## A6.1 USING THE TWO-STATE DIAGRAM ARRANGEMENT

Consider Figure A6.5.

Following the ACW direction:

| | | | | | | | | | | | | | | |
|---|---|---|---|---|---|---|---|---|---|---|---|---|---|---|
| ACW | *ms,* | *m/s,* | */m/s,* | */ms,* | *ms,* | *m/s,* | */m/s,* | */ms,* | *ms,* | *m/s,* | */m/s,* | */ms,* | *ms..* | |
| ACW | s0 | s1 | s2 | s3 | s0 | s1 | s2 | s3 | s0 | s1 | s2 | s3 | s0.. | State diagram. |
| CW | s4 | s4 | s5 | s4 | s4 | s5 | s4 | s4 | s4 | s5 | s4 | s4 | s4.. | State diagram. |

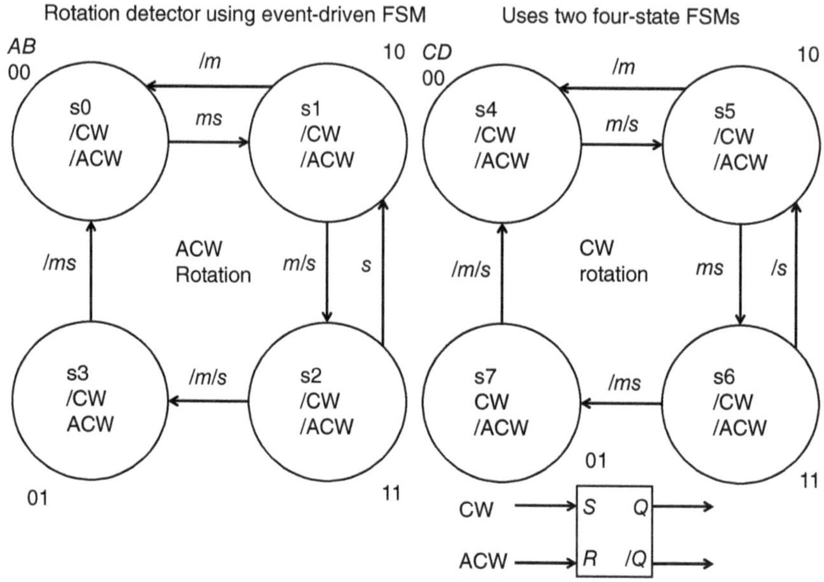


**Figure A6.5** Alternative state diagram for the rotational detector.

Now following the CW direction:

| | | | | | | | | | | | | | |
|---|---|---|---|---|---|---|---|---|---|---|---|---|---|
| CW | *m/s*, | *ms*, | */ms*, | */m/s*, | *m/s*, | *ms*, | */ms*, | */m/s*, | *m/s*, | *ms*, | */ms*, | */m/s*.. | |
| CW | s4 | s5 | s6 | s7 | s4 | s5 | s6 | s7 | s4 | s5 | s6 | s7.. | State diagram |
| ACW | s0 | s0 | s1 | s0 | s0 | s1 | s0 | s0 | s0 | s1 | s0 | s0.. | State diagram |

You can see that the sequence is continually followed for the active state diagram while the other state diagram keeps moving through then gradually resetting to s0.

You could try Logisim-Evolution to perform the same actions.

You could also try to simulate using SynaptiCAD and Verilog HDL for a SynaptiCAD simulation.

# Bibliography

## *REFERENCES*

Burch, C. (2011). Logisim. http://www.cburch.com/logisim/index.html, accessed 10 February 2021.

Carl Burch developed the Logisim Simulator using JAVA. This is now available freely worldwide on Windows, OS X, and Linux. A useful program to test FSM-based systems.

Duncan F.G., Zissos, D. (1971). Gate tolerance in sequential circuits. *Proc. IEE* 118(2):317–320.

This paper details the ideas of gate tolerances in FSM-based digital circuit design.

Fernandes, J.M., Adamski, M., Proença, A.J. (1997). VHDL generation from hierarchical Petri net specifications of parallel controllers. *IEE Proc. Comput. Digital Technol.* 144(2):127–135.

This is a good paper on the use of Petri nets used in Chapter 7.

Holdsworth, B. (1993). *Digital Logic Design*, 3rd edn. Oxford: Butterworth-Heinemann.

A good introduction to Boolean logic design following a similar trend to that used by Zissos.

Minns, P. (1995). Using *D* type flip flops in state machine design: an alternative approach. *International Journal of Electrical Engineering Education* 32(2):124–132.

Note that this is a paper I produced via the *International Journal of Electrical Engineering Education* in April 1995. It is still relevant today and outlines very well the use of DFFs using the clock-driven FSM method discussed in Chapters 1–3 of this book. If you have any difficulties getting a copy of this paper, try onlinesupport@sagepub.co.uk for help. I have found them very good.

*Digital System Design using FSMs: A Practical Learning Approach*, First Edition. Peter D. Minns.

Companion website: www.wiley.com/go/minns/digitalsystemdesign

Minns, P.D., Elliott, I.D. (2008). *FSM-Based Digital Design using Verilog HDL.* Chichester: John Wiley & Sons, Ltd.

This was the original book written mostly in conventional text (Chapters 1, 2, and 3 followed a programmed learning design). Note Ian D. Elliott, the joint author, contributed Chapters 6, 7, and 8 of the book and looked at the more general Verilog HDL language in some detail as used in FSM-based systems. These can be obtained from Wiley, for a fixed fee, as .pdf files. There are other books and sources of information on Verilog HDL that you can peruse.

Myers, C. (2001). *Asynchronous Circuit Design.* Chichester: John Wiley & Sons Ltd.

See pages 226 and beyond for the use of C gates in the design of event-driven FSMs.

Rice, J.E. (2008). *An Introduction to Reversible Logic Fundamentals.* Lethbridge AB, Canada: University of Lethbridge.

Do a Google search on 'reversible logic gates' to find out more about this interesting topic destined to become popular in the future. There is a wide range of papers on this area that you could look at. When people start to use nanosized systems with trillions of gates, reversible logic will be needed to reduce power costs.

White, D.E. (1981). *Bit-Slice Design: Controllers and ALUs.* New York: Garland STPM Press.

This is another quite old book that I used during my lecturing days exploring the use of bit-slice architecture and the 2900 family of microprocessors. It contains some interesting ideas that are well worth looking at.

Yakovlev, A.V., Koelmans, A.M. (1998). Petri nets and digital hardware design: advances in Petri nets. *Lecture Notes in Computer Science* 1492:154–236.

A very good introduction to the idea of asynchronous design using Petri nets.

Zissos, D. (1979). *Problems and Solutions in Logic Design*, 2nd edn. Oxford: Oxford University Press.

A very helpful book on the Zissos method of FSM design. This book was one of the original on FSM systems. It is a very useful way to develop FSM-based digital design.

## FURTHER READING

Bhasker, J. (1992). *A VHDL Primer.* New York: Prentice Hall.

A very good book for those wanting to learn about VHDL as an alternative to Verilog HDL.

Ciletti, M.D. (2003). *Advanced Digital Design with Verilog HDL.* Upper Saddle River, NJ: Person Education.

This is a very good book for those wishing for a deeper understanding of Verilog HDL.

Drexler, E.K. (1991). Molecular machinery and manufacturing with applications to computation. PhD thesis. Massachusetts Institute of Technology.

Klemenchuk, A. (2020). Logisim-Evolution Reference. CSC258H. January 2020.

This is a Tutorial on the use of Logisim-Evolution that you might like to look at. Just type into Google 'Logisim-Evolution Reference' and look for CSC258H to access the paper.

Minns, P. (1995). PLD devices for counting applications. *ElectroTechnology* 6(5):1.

Publication of the Institution of Incorporated Engineers before it became part of the Institute of Engineering Technology which was formerly the Institute of Electrical Engineers (IEE).

# Index




*Digital System Design using FSMs: A Practical Learning Approach*, First Edition. Peter D. Minns.



Companion website: www.wiley.com/go/minns/digitalsystemdesign

www.ingramcontent.com/pod-product-compliance
Lightning Source LLC
LaVergne TN
LVHW062206200926
848484LV00006B/1332